God grant me
the serenity
to accept the
things I cannot
change,
courage to change
the things I can,
& the wisdom
to know
the difference.

Flight Stability and Control

Modern Analytic *and* Computational Methods *in* Science *and* Mathematics

A GROUP OF MONOGRAPHS AND ADVANCED TEXTBOOKS

Richard Bellman, *Editor*
University of Southern California

Published

1. R. E. BELLMAN, R. E. KALABA, AND MARCIA C. PRESTRUD, Invariant Imbedding and Radiative Transfer in Slabs of Finite Thickness, 1963
2. R. E. BELLMAN, HARRIET H. KAGIWADA, R. E. KALABA, AND MARCIA C. PRESTRUD, Invariant Imbedding and Time-Dependent Transport Processes, 1964
3. R. E. BELLMAN AND R. E. KALABA, Quasilinearization and Nonlinear Boundary-Value Problems, 1965
4. R. E. BELLMAN, R. E. KALABA, AND JO ANN LOCKETT, Numerical Inversion of the Laplace Transform: Applications to Biology, Economics, Engineering, and Physics, 1966
5. S. G. MIKHLIN AND K. L. SMOLITSKIY, Approximate Methods for Solution of Differential and Integral Equations, 1967
6. R. N. ADAMS AND E. D. DENMAN, Wave Propagation and Turbulent Media, 1966
7. R. L. STRATONOVICH, Conditional Markov Processes and Their Application to the Theory of Optimal Control, 1968
8. A. G. IVAKHENKO AND V. G. LAPA, Cybernetics and Forecasting Techniques, 1967
9. G. A. CHEBOTAREV, Analytical and Numerical Methods of Celestial Mechanics, 1967
10. S. F. FESHCHENKO, N. I. SHKIL', AND L. D. NIKOLENKO, Asymptotic Methods in the Theory of Linear Differential Equations, 1967
11. A. G. BUTKOVSKIY, Optimal Control Theory for Distributed Parameter Systems, 1969
12. R. E. LARSON, State Increment Dynamic Programming, 1968
13. J. KOWALIK AND M. R. OSBORNE, Methods for Unconstrained Optimization Problems, 1968
14. S. J. YAKOWITZ, Mathematics of Adaptive Control Processes, 1969
15. S. K. SRINIVASAN, Stochastic Theory and Cascade Processes, 1969
16. D. U. VON ROSENBERG, Methods for the Numerical Solution of Partial Differential Equations, 1969
17. R. B. BANERJI, Theory of Problem Solving: An Approach to Artificial Intelligence, 1969
18. R. LATTÈS AND J.-L. LIONS, The Method of Quasi-Reversibility: Applications to Partial Differential Equations. Translated from the French edition and edited by Richard Bellman, 1969
19. D. G. B. EDELEN, Nonlocal Variations and Local Invariance of Fields, 1969
20. J. R. RADBILL AND G. A. MCCUE, Quasilinearization and Nonlinear Problems in Fluid and Orbital Mechanics, 1969
21. W. SQUIRE, Integration for Engineers and Scientists, 1969
23. T. HACKER, Flight Stability and Control
24. D. H. JACOBSON AND D. Q. MAYNE, Differential Dynamic Programming
27. E. D. DENMAN, Coupled Modes in Plasmas, Elastic Media, and Parametric Amplifiers

In Preparation

22. T. PARTHASARATHY AND T. E. S. RAGHAVAN, Some Topics in Two-Person Games
25. H. MINE AND S. OSAKI, Markovian Decision Processes
26. W. SIERPIŃSKI, 250 Problems in Elementary Number Theory
28. F. A. NORTHOVER, Applied Diffraction Theory
29. G. A. PHILLIPSON, Identification of Distributed Systems
30. D. H. MOORE, Heaviside Operational Calculus: An Elementary Foundation
31. S. M. ROBERTS and J. S. SHIPMAN, Two-Point Boundary Value Problems: Shooting Methods

Flight Stability and Control

T. Hacker Institute of Fluid Mechanics
Academy of the Socialist Republic of Rumania, Bucharest

American Elsevier
Publishing Company, Inc.
NEW YORK · 1970

AMERICAN ELSEVIER PUBLISHING COMPANY, INC.
52 Vanderbilt Avenue, New York, N.Y. 10017

ELSEVIER PUBLISHING COMPANY
Barking, Essex, England

ELSEVIER PUBLISHING COMPANY
335 Jan Van Galenstraat, P.O. Box 211
Amsterdam, The Netherlands

Standard Book Number 444–00066–6

Library of Congress Card Number 75–100396

Printed in the United States of America

The art of airplane and missile design is progressing so rapidly, and the configurations and flight régimes of interest are so varied that one is scarcely justified in speaking of typical configurations and typical results. Each new departure brings with it its own special problems. Engineers in this field must always be alert to discover these, and be adequately prepared to tackle them—they must be ready to discard long-accepted methods and assumptions, and to venture in new directions with confidence. The proper background for such ventures is a thorough understanding of the underlying principles and the essential techniques.

From the Preface to
Dynamics of Flight
by B. ETKIN

It should perhaps be added that a certain amount of doubt should accompany confidence, and that the effort for a thorough understanding of principles and techniques should be completed by the desire to further, deepen, and render more accurate the knowledge of the former and to improve the latter.

T. HACKER

Contents

PREFACE xix

CHAPTER 1

Introduction

1.1 General remarks and a few definitions; basic motion and disturbed motion 1
1.2 Flight as a controlled motion 2
1.3 Inherent stability and stability of the control system 3
1.4 State and control variables; the mathematical model of controlled flight 4
1.5 Equations of disturbed motion: Generalities 7
References 9

CHAPTER 2

Mathematical Modeling of Disturbed Flight: Stability Equations

2.1 Longitudinal motion of aircraft 10
2.2 Equations of lateral asymmetrical motion 13
2.3 Nondimensional equations 19
2.3.1 Introductory remarks 19
2.3.2 Longitudinal disturbed motion 21
2.3.3 Dimensionless equations of lateral asymmetrical disturbed motion 25

Appendix: Definition of aircraft orientation by means of the Euler angles; Euler's equations 27
References 30

CHAPTER 3

Generalities and Some Mathematical Aids

3.1 The Liapunov stability: Short-duration disturbances 31
3.2 Stability under persistent disturbances 32
3.3 Some theoretical aspects of the relation between stability and control 34
3.4 How to investigate stability 36
3.4.1 Introductory remarks 36
3.4.2 Stability by the first approximation 37
3.4.3 Method of the Liapunov function 38
3.4.4 Theorems of stability under persistent disturbances . . 40
3.5 The necessity of numerical estimates; the concept of practical stability 41
3.6 Estimating techniques 43
3.6.1 Preliminary remarks 43
3.6.2 Estimate of damping velocity of disturbances 44
3.6.3 Estimate of additional terms: first procedure 45
3.6.4 Estimate of additional terms: second procedure 49
3.6.5 Estimate of allowable velocity of repeated vertical gusts for securing longitudinal stability with respect to the incidence 51
3.7 Stability as a property over an unlimited interval of time . . 59
3.8 Stability with respect to a part of the variables (incomplete stability) 60
References 66

CHAPTER 4

Stability of Partially Controlled Motions

4.1 Introductory considerations 68
4.2 Mathematical statement of the problem of partially controlled motions 69
4.3 Some comments 74

4.4 Constrained stability according to Neumark 79
Appendix: Deduction of the system (4.7) 84
References 86

CHAPTER 5

Linear Systems with Constant Coefficients: Steady Basic Motions

5.1 Introductory remarks 87
5.2 Limits of validity for linear systems of equations with constant coefficients: quasilinear systems 92
5.3 Stability region in space of the parameters 93
5.4 Safe and dangerous zones of the stability boundary 98
5.5 Empirical concept of neutral stability and its mathematical definition in the literature 101
5.6 Mathematical models of neutral dynamic stability; Comments . 104
5.7 Some remarks on the selection of controls: qualitative aspects of the problem of control efficiency 105
5.7.1 Introduction: efficiency of controls 105
5.7.2 Time of response 107
5.7.3 Secondary effects of controls 115
Appendix I: Criteria of Routh and Hurwitz 118
Appendix II: Example to illustrate the notions of safe and dangerous boundary 120
References 122

CHAPTER 6

Longitudinal Stability of Steady Straight Flight

6.1 Stability with locked controls 124
6.2 Stability of partially controlled longitudinal motion 130
6.2.1 Partial control through longitudinal attitude constraint . 130
6.2.2 Case of simultaneous constraint of the angle and angular velocity of pitch 131
6.2.3 Simultaneous control of longitudinal attitude and flight speed 132
6.2.4 Constraint of flight speed 133
6.3 A neutrally stable model for certain longitudinal motions . . . 134
6.4 Effect of perturbation of height on the longitudinal stability characteristics in steady level flight 141
References 154

CHAPTER 7

Lateral Stability of Steady Regime Flight

7.1 Stability with locked controls 155
7.2 Lateral stability in partial control conditions 158
7.2.1 Constraint of the angle of yaw 158
7.2.2 Simultaneous control of both the yaw and the bank angles 162
7.2.3 Yaw and sideslip angles under constraint 165
7.2.4 Constraint of the rotation in yaw. 166
7.2.5 Rapid convergence in pure roll 169
7.2.6 Oscillation in the Oxy plane (yaw and sideslip) 170
7.3 Coupled motions; A particular case: hovering flight stability of VTOL aircraft 172
7.3.1 Coupling of longitudinal and lateral motions 172
7.3.2 Equations of disturbed motion 173
7.3.3 Effect of rotation dampers 175
7.3.4 Simultaneous use of rotation dampers and static stabilizers 177
References 181

CHAPTER 8

Stability of Unsteady Motion: Nonautonomous Systems

8.1 Introduction 183
8.2 Cases reducible to linear systems with constant coefficients . . 186
8.2.1 Small variation of the system parameters 186
8.2.2 Slow variation of the system parameters: the process of the "freezing of coefficients" 187
8.3 Direct investigation of unsteady motion stability 189
References 191

CHAPTER 9

Stability of Unsteady Flight: Varying Undisturbed Velocity

9.1 Rapid incidence adjustment. 192
9.1.1 Assumptions and simplified equations of disturbed motion 192
9.1.2 Sufficient conditions of applicability for the classical model 194
9.1.3 Direct approach via the Liapunov function method . . 200
9.1.4 Comments 205

9.2 Estimates of varying speed effect 210
9.2.1 Damping velocity of incidence deviation 210
9.2.2 Influence of flight speed variation on frequency of phugoid oscillations 214
References 218

CHAPTER 10

Stability of Unsteady Flight: Varying Undisturbed Height

10.1 Introduction: undisturbed height as a time-dependent parameter of the system. 219
10.2 Equations to be considered 220
10.3 Rapid incidence adjustment 223
10.4 Influence of undisturbed flight height variation on phugoid motion 226

CHAPTER 11

Effect of Time Lag on the Stability of Controlled Flight

11.1 The mathematical model. 230
11.1.1 Closed-loop systems with time lag 230
11.1.2 Ideal controls and controls with time lag 231
11.1.3 Equation system with time lag 232
11.1.4 Stability of solution 233
11.1.5 A relation between the stability of an ideally controlled motion and that of a motion controlled with time lag 234
11.2 Artificial stabilization of longitudinal motion allowing for small time-lags; validity limits of the ideal control model; estimates 235
11.2.1 Mathematical aids: lemmata concerning differential inequalities with time lag 235
11.2.2 Removal of phugoid divergence through ideal controls in the case of a steady basic motion 236
11.2.3 Estimation of the validity of condition (11.11) 237
11.2.4 Estimation based on the method of the variation of parameters 238
11.2.5 An estimate based on the method of the Liapunov function 243
11.2.6 The case of unsteady basic motion 244
11.3 Artificial stabilization of longitudinal motion in the case of arbitrary time lag 246

11.3.1 Mathematical aids 246
11.3.2 The case of stationary and quasi-stationary basic motions (autonomous systems) 249
11.3.3 Note concerning nonautonomous systems 250
11.4 Hovering flight of VTOL aircraft. 251
References 253

CHAPTER 12

Optimization Problems

12.1 Introduction 255
12.1.1 Object 255
12.1.2 Problem formulation: the physical model 256
12.2 The mathematical model. 257
12.2.1 State variables and control variables; assumption of piecewise continuity of control variables 257
12.2.2 Control variables 258
12.2.3 Constraints; admissible controls 260
12.2.4 Proposed objective (purpose of the flight mission) . . 261
12.2.5 Performance index. 262
12.2.6 Mathematical formulation of the optimum problem . . 262
12.3 Maximum principle 263
12.3.1 The Pontryagin theorem 263
12.3.2 Comments 264
12.3.3 Switching conditions. 266
12.4 Optimal programming of the basic controls in climb and accelerated flight 267
12.4.1 Introduction 267
12.4.2 Technical problem formulation; assumptions 268
12.5 VTOL aircraft with adjustable orientation of thrust vector . . 270
12.5.1 Equations of motion 270
12.5.2 A simplified case: the constant attitude 272
12.5.3 Variable longitudinal attitude 280
12.6 Orientation of the thrust force fixed against the machine . . 283
12.6.1 Equations of longitudinal motion 283
12.6.2 Optimal climb and acceleration by thrust and longitudinal moment control 284
12.6.3 Minimum time to climb and accelerate by control of the incidence 285
12.6.4 Climb and acceleration in minimum time by path inclination control. 288

12.7 State variable constraint 290

12.8 Optimal stabilization: general 293
12.8.1 Introduction 293
12.8.2 Statement of the Kalman–Letov problem 295

12.9 Hovering flight of VTOL aircraft. 295
12.9.1 Linear models 295
12.9.2 Dynamic programming approach: partially controlled hovering VTOL aircraft modeled as a rigid body with fixed point: A nonlinear model 299

12.10 Optimal stabilization of longitudinal attitude in the phugoid phase 302
12.11 Some conclusions 303

Appendix I. Computational aspects 304

Appendix II. Deduction of certain formulas 310

References 312

Notation 313

Author Index 317

Subject Index 319

PREFACE

The subject of this book belongs to the branch of applied science that is referred to as the theory of flight stability and control. The first part of the Preface is concerned with this theory, while some words on the contents of the book and on the spirit in which it has been conceived are included toward the end.

These few remarks are meant to make clear the point of view from which the theory of flight stability and control is considered in this book, and to outline its presentation, especially with reference to the two aspects of the book that seem least conventional, namely, the neglect of aerodynamics and the technique of occasionally allowing the mathematical theory to control the choice of a physical model.

These two means of approach derive from two quite trivial observations. The first is that the theory of flight stability and control is an autonomous discipline, having its own laws and characteristic features. For some time now it has ceased to be a chapter of aerodynamics and can be studied independently of the latter, as well as of the gas dynamics of the reactive jet, and so forth. (Obviously, though, it cannot be considered independent of the data supplied by these disciplines: stability equations cannot be written without knowing the aerodynamic characteristics of the machine, thrust, or flight medium in the given atmospheric conditions, and so on.) Hence, regarding its specific fundamental aspects there is, for instance, a unique theory of stability and control that is valid for both flight in the atmosphere and extraterrestrial flight.

The second observation relates to the role of mathematics, which is not only a passive auxiliary of an applied science, but also a tool suggesting further development of its content.

Historically the theory of flight stability and control appeared only as a theory of airplane stability, which developed, by means of the small-disturbance method, from rigid-body mechanics (Lanchester moment: F. W. Lanchester, *Aerial Flight*, Part 2: *Aerodonetics* (Constable, London, 1908)) to aerodynamics (Bryan–Glauert moment: G. H. Bryan, *Stability in Aviation* (Macmillan, New York, 1911) and H. Glauert, "A Non-Dimensional Form of the Stability Equations of an Aeroplane," *Aeronaut. Res. Council Rept. & Memo.* 1093 (1927)).

Quite frequently, a new branch of science undergoes a sinuous line of development. It appears as a synthesis of some results obtained in other fields of human knowledge and starts out as a seemingly independent branch. After its first few steps, however, it becomes evident that this new field is not yet able to stand on its own feet; it then becomes an appendix to that discipline, basic or auxiliary, in which the greatest research effort is required to complete the special knowledge intended for use in the new branch. After a further period of development, likewise not necessarily linear, the new field creates its own tools, its own internal logic of development, and acquires its own body of laws. Its independence is thus regained, sometimes in an altogether unexpected form.

The theory of aircraft stability appeared from the outset as a mathematically expressed, and therefore essentially deductive, theory. Relying heavily on rigid-body mechanics, it initially used mainly the small-disturbance method and mathematical theory expounded by E. J. Routh in *A Treatise of the Stability of a Given State of Motion* (1877), on lines suggested by G. H. Bryan and W. E. Williams in their report to the Royal Society in London in 1903 ("The Longitudinal Stability of Aerial Gliders," *Proc. Roy. Soc.* (*London*) **A73**(1904), 100–116). Application of the small-disturbance method implied a command of knowledge from the domain of aerodynamics regarding the dependence of the forces and moments acting on an aircraft on the kinematic parameters. There followed a long period of accumulation of knowledge and data regarding the aerodynamic characteristics that play a part in the definition of the flying qualities, the so-called stability derivatives. Hence, the bulk of research in the theory of stability moved to the domain of aerodynamics, particularly experimental aerodynamics; "static and dynamic stability of aircraft" became, in a way, a chapter of aerodynamics. Nevertheless, this research, as well as flight tests intended to improve stability characteristics, were guided by the existing (Bryan–Glauert) theory; that is, they were undertaken, carried out, and interpreted within its framework.

That this framework would, in time, become too narrow was already anticipated by V. S. Vedrov, as early as 1938, in the introduction to his book on aircraft stability *Dinamicheskaya ustoychivost' samoleta* (Oborongiz, Moscow, 1938). At that time, though, it was still difficult to foresee all the limitations of the classical theory, which was based on Routh's ideas and elaborated for the hypothetical conditions of inactive controls (fixed or free). In the last few years the shortcomings in the initial formulation of the problem and the technique used have become more obvious, notwithstanding the readjustment of aerodynamic data through the introduction of stability derivatives corresponding to speeds in a compressible régime. Indeed, modern aircraft practice has brought to light some new elements that make it necessary to reconsider the classical theory. A few of these are the assurance of flying qualities acceptable for a wide range of speeds (from takeoff speeds of the order of tens of meters per second up to speeds that may exceed a thousand meters per second) and heights (air density decreasing to a tenth or even less in comparison with its value on the ground); implicitly, engine thrust acquires additional importance to the detriment of some aerodynamic characteristics; the necessity increases, at high speeds, for accuracy of evolution; in many cases flight in a variable régime becomes typical; the relationship between the dynamic characteristics of the aircraft and those of the pilot are modified, the adaptation possibilities of the latter being limited. All these elements, and many others, imply new design concepts, a more marked automation of piloting, leading to considerable involvement in the problem of flying qualities.

A first consequence, on the theoretical plane, was to abandon the simplifying schemes of fixed or free controls when treating an increasing number of problems of flight stability.

Neglecting stabilization by controls when considering flight stability has always meant a deliberate concession to the simplification of the theory. Aircraft practice has, from the beginning, rejected the idea of inherently stable aircraft; Draper ("Flight Control," *J. Roy. Aeronaut. Soc.* **59**(**1955**), **451–477**) sees in this renunciation the Wright brothers' fundamental contribution to the development of aviation. Nevertheless, certain agreements being empirically established between the results obtained on the basis of such a simplified theory and actual flying qualities, the hypothesis of inactive controls has long remained acceptable for theory and so far has not been completely replaced. However, the tendency to consider an aircraft as a control system has begun to prevail in the theory of flight stability and control. Concomitantly, the latter is enriched by techniques borrowed from the store of automatic systems theory. This contact of two disciplines that had previously known a parallel development is reflected in modern treatises devoted to flying qualities, among which Etkin's remarkable *Dynamics of Flight; Stability and Control* (Wiley, New York, **1959**) should be mentioned.

However, the basic mathematical model remains the system of linear equations with constant coefficients, and for this reason the new techniques adopted (e.g., those based on operational methods) do not exceed, in principle, the framework offered by Routh's theory. The development of aerospace techniques and the new elements (including the ones mentioned above) make reconsideration of the theory of flight stability by the application of a more comprehensive mathematical theory desirable—perhaps even necessary.

About thirty years ago interest in the theory of the stability of motion on the lines of Liapunov's decisive contribution was renewed in the Soviet

Union by a group of scientists working chiefly at the Kazan Aviation Institute. With all the indisputable merit of the efforts that were then made to foster this powerful new tool for the study of the theory of flight stability, the attempts to apply Liapunov's theory were largely ignored. They were not reflected in the literature, not even in works (such as the books of Vedrov; Ostoslavskiy and Kalachev; Pyshnov; etc.) published in the Soviet Union; nor did they lead to a revision of the mathematical formulation of the problem of flight stability. The explanation must probably be sought in the fact that a theory as simple, universal, and effective as the classical one continued to supply acceptable information about the actual process of flight; moreover, sufficient experience in the application of the Liapunov stability theory had not accrued. Meanwhile, however, the mathematical theory has made some progress, and reports of attempts to apply it in various fields of engineering have been published. Therefore, with the development of aerospace techniques that require more adequate mathematical treatment, the conditions have also been created for the efficient application in this domain, of Liapunov's stability theory.

From the beginning, the theory of flight stability has been influenced by the mathematical method used. Thus, the small-disturbance method played a striking part in shaping the theory, which emphasizes the dependence, expressed in the form of stability derivatives, of stability on the aerodynamic characteristics. It is natural for a new mathematical treatment to influence some aspects of the content of the theory of flying qualities.

The mathematical method may exert an influence on an engineering theory in opposing senses. The more common way is a limiting one. Unfortunately, mathematics is far from omnipotent, and its limitations impose obvious limitations on the content of the theory it serves, thus affecting the entire framework of ideas of this discipline, from the selection of the class of phenomena to be investigated to the posing of the related problems. This relationship is so familiar to any engineering researcher that there is hardly anything new to say about it. Moreover, it constitutes an essential element in making any mathematical model.

The effect of the mathematical method on the contents of a branch of applied science can, however, also contribute to its broadening. Because of their degree of abstraction, mathematical theories have a large sphere of potential application (including some still nonexistent domains of science and practice). Even if they were initially called to life or developed on the basis of certain special requirements of physics or engineering, they soon exceed the sphere of these particular applications. For this reason, examination of some mathematical theories from the standpoint of possible application, combined with an awareness of the physical processes characteristic of a given applied science, may result in improved knowledge about this applied science.

It might be considered more direct, and hence more natural, to proceed only in the opposite manner: to start from the actual (i.e., physical or technical) necessity; model the respective process as faithfully as possible; and then simplify the model until one that can be solved by the available mathematical theory is found. This is in fact the process used whenever possible. We are, however, often hindered in the operations of building a faithful model and deciding on the most readily soluble one by habits acquired in the course of the extended use of a certain type of model. Sometimes these preconceived ideas can mislead us, even in the identification of the most significant actual physical phenomena to be used as a starting point. Because as long as there is no theory, such phenomena do not stand out with sufficient clarity. Bellman, in *Adaptive Control Processes: a Guided Tour* (Princeton Univ. Press, Princeton, New Jersey, 1961) tells the parable of the man who, in the circle of light from a streetlamp, seeks a ring lost elsewhere in a dark part of the street. It is the story of the researcher in an applied field, at whose disposal mathematics places its isolated spots of light in a vast stretch of darkness. Until a more powerful streetlamp is found, and until the place to be investigated is more precisely identified, investigation is carried out within the radius of the available small spots of light, assumed to be the most conveniently located, in the hope of achieving at least partial success. (Sometimes the value of the result of the search exceeds expectations.)

Thus, this book endeavors to consider the theory of flight stability and control as self-supporting at the same time that it stresses the theory's applied mathematical aspect. The problems are presented by applying the Liapunov theory of stability and using some ideas from the theory of control systems. We also attempt to turn to account some suggestions contained in these mathematical theories, with the aim of broadening the framework of the classical theory of flight stability and control and solving some problems of current interest. This effort is carried out within a somewhat subjectively chosen circle of questions, determined by my own preoccupations.

To illustrate the mode in which mathematical theory intervened, reference is made here to some of the main problems dealt with. Thus, starting from the general theory of stability under persistent disturbances, the model of stability of partially controlled motions is built up and then used throughout the book in studying some flight situations whose approach on the basis of the classical model would be considerably more difficult, perhaps even impossible. The method of the Liapunov function has allowed us to consider the unconventional problem of flight stability in an unsteady régime, permitting us to deduce the asymptotic behavior in the case of variable undisturbed flight speed and height, and to point out and estimate quantitatively the influence of the variation of these parameters of the basic motion. The effect on flight stability of the response time of the control system is made clear and estimated by resorting to some results from the theory of differential equations with time lag (with "retarded argument"). The questions of optimal programming of terrestrial flight and of optimal stabilization, by using Pontryagin's maximum principle and also techniques derived from other theories (such as Bellman's dynamic programming, etc.) are treated.

The enumeration above is only illustrative and, hence, incomplete. Without intending to tackle all the current problems in the field or all the main chapters in the study of flight stability and the control of terrestrial flight, I have endeavored to deal with those that, from my own research, I felt might contribute to the field. Within the limitations noted, I have attempted to link the problems treated so that the book forms a coherent whole.

Although the viewpoint of this book is one of theory expressed mathematically (i.e., of the mathematical model), the book is not a mathematical work. No mathematical theorems are proved in it, but the manner in which these theorems provide conclusions for the dynamics of flight is shown. Concomitantly, an endeavor is made to consider critically some settings of problems, to justify the models used; the significance of the assumptions is discussed, as well as the contents of the basic concepts. The book does not contain prescriptions for designers, nor does it describe experimental facts. Its aim is rather to stress fundamental notions, to present a way of thinking about the subject and a certain orientation in the research work in the field.

Acknowledgments. In the first place acknowledgment must be made to the Institute of Fluid Mechanics of the Academy of the Socialist Republic of Rumania; my research work in this Institute has enabled me to obtain the results that make up the framework of this book.

With great pleasure I stress the fundamental importance, in writing the mathematical topics of the book, of my association with those participating in the Seminar for Differential Equations held at the Institute of Mathematics of the Academy, particularly with Professor A. Halanay. I also thank Professor Halanay for reading the manuscript and for his critical comments.

It is an agreeable duty also to record the help given by Messrs. R. Reineck and S. Marcus with the language.

T. HACKER

Bucharest, Rumania
March 1970

Chapter I

INTRODUCTION

I.1. GENERAL REMARKS AND A FEW DEFINITIONS; BASIC MOTION AND DISTURBED MOTION

The flight of aircraft is a very complex (in principle, infinite) process; hence, hereafter, models of flight will be resorted to as the only possible approach to the subject. It will thus be assumed that motion, which, effectively, is characterized by an infinity of parameters, is satisfactorily defined by a finite, not very large number of generalized coordinates that are essential for the problem studied. The choice of these coordinates is usually the first step in conceiving the physical or theoretical model, motion being considered as a set of coherent laws to which the essential coordinates chosen are subject.

In this sense, *flight program* will mean a set of laws prescribed for the essential dynamic coordinates of the system. These laws may be established in the most varied forms, from the qualitative (intuitive) description of the succession of movements that the machine has to perform down to their mathematical expression.

The virtual motion corresponding to the given flight program† (or to a part of the given flight program) under certain set reference conditions (standard conditions) will be called *basic* or *undisturbed motion*. Hereafter, this virtual motion will often be replaced by its mathematical model, consisting of ordinary differential equations with respect to generalized coordinates and to time as the independent variable.

Although the basic motion is compatible with the characteristics of the aircraft and the medium, the actual motion of the machine generally differs from it since (i) the flight conditions practically always differ from the reference conditions, and (ii) even in the hypothesis of their coincidence, the program can only be fulfilled approximately.

† With the approximation resulting from a disregard of the parameters considered as nonessential.

The causes that induce deviations of actual flight from the basic motion will be called *disturbances*. The actual motion of the aircraft will be called *disturbed* or *induced motion*.

Just because it is a possible motion, the basic motion is far from being a mere theoretical fiction. Its main characteristic is that it allows actual flight to approximate it most closely. This is expressed by two essential properties: (1) its compatibility with the inherent characteristics of the aircraft and the medium, and with the possibilities offered by the controls, and (2) its stability. The latter property gives it the somewhat paradoxical feature of being more real than actually effected flight. Indeed, what remains essential in flight is the basic motion. When a certain flight program is carried out time after time, only the basic motion is reproduced, the faithful repetition of the disturbed motion being of a practically null probability.

The second essential property, the stability of the basic motion of aircraft, is the main object of this work. To discuss stability theoretically, some specifications regarding the notion of disturbance are required. The classifying of disturbances according to their duration as compared to the duration of the entire motion studied is of fundamental importance, at least as regards mathematical treatment. According to this criterion, disturbances are classified as *short-duration* (or instantaneous) *disturbances* and *persistent disturbances*. In dealing with problems involving short-duration disturbances, only their effect is considered, the motion being investigated from the instant when their action ends, whereas persistent disturbances are generally assumed to act during the entire motion considered.

The concepts of short-duration and persistent disturbances, as well as the manner in which they affect problems of stability, will be specified in Chapter 3.

1.2. FLIGHT AS A CONTROLLED MOTION

Any complete flight, if it is not wholly a controlled motion, includes controlled portions. In addition to somewhat a posteriori reasons supplied by flying practice, this assertion is also based on a consideration of principle. By flight are understood three-dimensional motions, in or outside the atmosphere, performed with a given purpose. Hence, any flight constitutes, in one way or another, a programmed motion. It is obvious, however, that the main means of achieving a flight program are the controls. (Since, conversely, a controlled motion can only be conceived in conditions implying the existence of a flight program, in a general sense, the ranges of the concepts of programmed flight and flight controlled in at least some portions coincide.) Therefore control, invariably present in any flight, varies from its most elementary form, consisting in subjecting the path of flight to a given inclination at the initial instant (as in the case of artillery projectiles), to the intricate forms of piloting (manually or automatically, or both) and remote control characteristic of modern flight technique of aircraft and spacecraft. It is obvious that controlled flight does not necessarily involve the permanent applying of any control. During flight there are portions of inactive controls—obvious in the case of artillery projectiles and spacecraft, and theoretically possible in the case of airplanes. Hence, hereafter, free flight will mean only that portion of the total programmed flight characterized by inactivity of the controls, but never the complete motion of the aircraft. At times, the free-flight concept will appear as a simplified model meant to make evident certain inherent characteristics of the machine that must be taken into account when, for instance, designing the control system.

As mentioned above, in given reference conditions a specific motion corresponds to a flight program. This is the basic motion (or the aggregate of basic motions, if they refer only to portions of the program), but what is actually achieved is a motion that is close to the basic one and has been called disturbed motion. Further, a *program of basic controls* (or guidance program) is linked to the basic motion. This ideal law of guidance is defined as rigorously achieving the flight program, that is, the succession of basic motions in the given reference conditions. (Failure to achieve strict correspondence that results from errors in the operation of the control devices introduces a new category of disturbances.) The existence of such an ideal law of guidance has thus been implicitly assumed. Generally, other controls are superimposed on the basic controls, their aim being to minimize the effect of the various disturbing factors. They will be called correction or stabilizing controls. As a rule, for the aircraft considered in this work, both functions are fulfilled by the same control elements.

I.3. INHERENT STABILITY AND STABILITY OF THE CONTROL SYSTEM

By *stability* will henceforth be understood a property of the basic motion by which it maintains in its vicinity the disturbed motions of the aircraft.

Hence, stability in conditions of controlled flight will refer to the basic motion (which also includes the motion law of the basic controls). However, since the actual separation between the basic controls and the correction controls is practically impossible in most aircraft, we will consider hereafter, besides the theoretical property mentioned, another stability of flight defined in the conditions of a simultaneous application of both the basic and the corrective controls.

The theoretical property, which only refers to the basic motions (including the basic controls, which may eventually be identically zero) characteristic of the machine considered, will be called its *inherent stability*. The investigation of this property is particularly useful, especially at the rough draft design stage, since it yields some information on the extent to which correction (stabilization) by controls is required, and hence on the characteristics required by the control device.

If, however, more accurate information is required, for example for final design, it becomes necessary to consider the problem of stability under flight conditions closer to the actual ones, characterized by the presence of both functions of the controls. In other words, the stability of the complete control system is to be considered.

I.4. STATE AND CONTROL VARIABLES; THE MATHEMATICAL MODEL OF CONTROLLED FLIGHT

As already mentioned, a first step in making the model of a complex process, such as flight, consists in choosing, from the (infinite) totality of parameters relating to this process, a certain number of variables that are essential for the aspect considered. Some of the variables chosen, called *state variables* or *coordinates*, will usually be generalized coordinates characterizing the mechanical motion of the airframe; the remainder, called *control functions*, *variables*, or *coordinates*, will be generalized coordinates describing the operation of the control equipment. Throughout this work use will be made of models that disregard the processes, whatever their nature (mechanical, electrical, hydraulic, psychophysiological), that occur within the control links of the system (control devices, the linkage, etc.). These processes will be replaced by hypotheses that establish the functional relationship between the input and output magnitudes corresponding to the whole control chain (cf. the concept of "mechanical model of control system as a whole" used in [**1**, Section B. **3**]). It is noteworthy that such a modeling of the functions of the control chain introduces *eo ipso* an additional group of disturbances caused by the respective simplifications and approximations. Consequently, besides the cases where processes occurring in a short time interval are considered, the resultant model of the process studied is only valid if this process is stable under persistant disturbances (see Section 3.2).

Thus, both the state and control variables will generally be quantities of the same nature (geometric, kinematic, or dynamic). The mathematical models used will be ordinary differential equations, conventional or with retarded argument, with respect to these variables and having time as the independent variable. In contrast with the state coordinates, the control parameters will sometimes appear in these equations not as dynamic variables but as time-dependent external constraints. To introduce some systematization, we present the following scheme, which shows how the mathematical models used in this book are classified according to the criterion of the use of the state and control variables.

Let x be a column vector representing the state variables considered, and u another vector whose components are control variables.

1. In some sections of the book (particularly in Chapter 11, which is devoted to the influence of time lag in the control system upon its stability), the state and control coordinates appear with an equivalent role: that of variables determining the behavior of the system considered. The motion as a whole will be defined by a system of equations of the form

$$\frac{dx}{dt} = X(t; x, u), \qquad \frac{du}{dt} = U(t; x, u), \tag{1.1}$$

and by the respective initial conditions $x(t_0) = x_0$, $u(t_0) = u_0$. In the language used in control systems theory, (1.1) represents a closed-loop (feedback) system.

2. For simplicity in a number of sections in which the inherent stability of aircraft is considered, a partial change of variables, consisting in substituting functions of the state variables assumed known for the control variables, will be resorted to. The only control variables occurring are those that refer to basic control. Having carried out this substitution in the first equation of system (1.1), we therefore consider systems of the form

$$\frac{dx}{dt} = X(t; x, u(x)) \equiv \tilde{X}(t; x), \tag{1.2}$$

hypothesizing, moreover, that the functional relationship between the basic motion and the basic control law ($u = u(x)$) is known with a sufficient degree of accuracy.

3. In other models considered, namely, the ones corresponding to the stability of partially controlled motions, by adequate hypotheses concerning the effect of controls on part of the state variables, the latter, together with the control variables, are eliminated from a *sui generis* first-approximation system, referred to as an auxiliary system. This simplified set of equations, written with respect to only part of the state variables, supplies information regarding the whole controlled motion. It is noteworthy that in contrast with case 2, where the function $u = u(x)$ is assumed known, in this case a single hypothesis is sufficient on the order of magnitude of the control force in the vicinity of the undisturbed value of the controlled state variables, for example, in the form of a Lipschitz condition

$$|R(t; y, z, u) - R(t; y, z_0, u)| < L|z - z_0|$$

where vector z represents the state variables pursued by the controls, equal to the components of z_0 in the basic motion, and y the remaining state variables. If the system of equations of the uncontrolled motion written in hypothesis 2 is

$$\frac{dy}{dt} = Y(t; y, z), \qquad \frac{dz}{dt} = Z(t; y, z),$$

then the auxiliary system will be of the form

$$\frac{dy}{dt} = Y(t; y, 0). \tag{1.3}$$

4. In Chapter 12, devoted to the problem of flight optimization, the control variables will be sought under the form of time functions that will appear in equations as external constraints on the system; that is, equations of the form

$$\frac{dx}{dt} = X(x, u), \qquad u = u(t), \tag{1.4}$$

will be considered. In contrast, for instance, with (1.1), (1.4) represents an open-loop system.

It should be noted that optimization problems do not necessarily imply consideration of open-loop systems. On the contrary, synthetized results of the form $u = u(t, x)$ would correspond to requirements connected with the programming of computing elements of automatic control systems. However, the consideration of closed-loop systems in optimization problems encounters considerable mathematical difficulties.

5. Finally, a particular case of the (1.4) system will be considered, with $u(t)$ given under the form $u(t) \equiv 0$. It is the case of so-called free flight with locked controls, studied in the sections dealing with the stability of uniform motion, the inherent stability of vertical takeoff and landing (VTOL) aircraft when hovering, and the determination of efficiency criteria for stabilization by controls.

I.5. EQUATIONS OF DISTURBED MOTION: GENERALITIES

The dynamics of flight studies mechanical motions, controlled or free, for whose description ordinary differential equations are used, with respect to state and control variables considered as essential for the problem dealt with. The stability of a given basic motion of the aircraft will be modeled by the stability of a certain solution of the respective differential equation system.

For brevity, vector representation of the equations will be used. (In connection with the validity of extending the operations used on the vectors see, e. g., [2, pages 1–2].) Let

$$\frac{d\xi}{dt} = \Xi(t; \xi) \tag{1.5}$$

be the system modeling the motions of the aircraft. The terms ξ, $d\xi/dt$, and Ξ represent column n vectors, n being equal to the number of dependent variables considered:

$$\xi = \begin{pmatrix} \xi_1 \\ \xi_2 \\ \vdots \\ \xi_n \end{pmatrix}, \qquad \frac{d\xi}{dt} = \begin{pmatrix} \dfrac{d\xi_1}{dt} \\ \dfrac{d\xi_2}{dt} \\ \vdots \\ \dfrac{d\xi_n}{dt} \end{pmatrix}, \qquad \Xi = \begin{pmatrix} \Xi_1 \\ \Xi_2 \\ \vdots \\ \Xi_n \end{pmatrix}.$$

In most of the cases treated in this book, the hypothesis of the differentiability of the functions Ξ will be sufficient. (It will be assumed that the function Ξ admits partial derivatives down to a certain order only when considering some critical cases, in Liapunov's sense, for certain stationary basic motions; see Section 6.3.)

Let, further, $\mathring{\xi}(t) = (\mathring{\xi}_1(t), \mathring{\xi}_2(t), \ldots, \mathring{\xi}_n(t))'$ (the prime denotes transpose) be a certain known solution that represents the basic motion considered, and $x(t) = (x_1(t), x_2(t), \ldots, x_n(t))$ be the deviations induced by a perturbation. The system

$$\frac{dx}{dt} = X(t, x) \tag{1.6}$$

obtained from (1.5) by the change of variables

$$\xi = x + \mathring{\xi} \tag{1.7}$$

will be called, conventionally, the equation system of the disturbed motion. Obviously, $X(t, 0) = 0$. The basic or undisturbed motion will be represented by the trivial solution $x = 0$.

Generally, the linear approximation valid in a certain domain of the variables x, according to the theorems of stability by the first approximation that will be stated in Section 3.4, will be resorted to. To this end (1.6) will be written so as to make the linear part of function X stand out. From the hypothesis of the differentiability of the function Ξ there results

$$\Xi(t; x + \mathring{\xi}) = A(t)x + N(t; x)$$

where $A(t) = \partial\Xi/\partial\xi\ (t; \mathring{\xi})$ and $N(t, x)$ satisfies a condition of the form

$$|N(t; x)| \leqslant \gamma(x)|x|$$

with $\gamma(r) \to 0$ for $r \to 0$. Hence, system (1.6) can be written

$$\frac{dx}{dt} = A(t)x + N(t; x) \tag{1.8}$$

from which, in the conditions resulting in the theory of stability by the first approximation (see Section 3.4), a linear system may be retained, namely,

$$\frac{dx}{dt} = A(t)x. \tag{1.9}$$

REFERENCES

1. G. V. Korenev, *Introduction to the Mechanics of a Controlled Body* (in Russian). Nauka, Moscow, 1964.
2. A. Halanay, *Differential Equations: Stability, Oscillations, Time Lags*. Academic Press, New York, 1966.

Chapter 2

MATHEMATICAL MODELING OF DISTURBED FLIGHT: STABILITY EQUATIONS

2.1. LONGITUDINAL MOTION OF AIRCRAFT

The simplified case of the pure longitudinal motion of aircraft will be considered first. In other words, it will be assumed that both basic and disturbed motions occur in the median longitudinal plane of the aircraft. For this, the existence of perfect symmetry about the longitudinal plane must be admitted and the aircraft must be considered as a rigid body, the latter assumption implying the absence of any gyroscopic action of the spinning masses (rotors of turbojet engines, airscrews, etc.) carried by the aircraft. (Because of the illustration provided, no details about the obtaining of the general equations considered in the theory of aircraft stability will be given. To follow the stages, reasoning, working hypotheses, etc., used in the general case, speciality books may be consulted, e.g., Chapter 4 of [1], which is noteworthy for its clearness and systematic layout.)

The essential moments and forces considered will be the aerodynamic ones, the weight of the machine, the thrust developed by engines, and the control forces and moments. The control forces and moments can, in their turn, be of an aerodynamic nature, achieved by the respective control surface (elevator), or induced by properly located gas jets.

Under these conditions, the mechanical motion of a rigid body in its plane of symmetry being studied, the state variables u, w, q, and the longitudinal attitude angle θ, the parameter of longitudinal angular control δ_Y, and that of the control of thrust (throttle control) δ_T can be considered as the essential generalized coordinates. Given the predominant importance of the aerodynamic forces and moments, the angle of attack α will be introduced as an essential variable in place of the vertical component of the velocity of the center of mass relative to the ground w, among which there is the obvious relation $\alpha = \tan^{-1}(w/u)$. Likewise, when choosing a reference frame related to the flight path (sometimes referred to as "intrinsic reference system," "stability," or "wind" axes), instead of the horizontal component of velocity, the tangential velocity of the mass center ($V = (u^2 + w^2)^{1/2}$) will be used as a variable, and sometimes, instead of the longitudinal attitude angle θ, the upward inclination of the flight path to the horizontal, γ, to which it is related by the simple equality $\theta = \gamma + \alpha$. (For symbols not defined in text, see the list of notation on page **313**.)

Hence, when making the mathematical model of the longitudinal motion of an aircraft considered as a rigid body, u, α, γ, q, δ_Y, and δ_T have been chosen as essential variables, and the time t as the independent variable. The mathematical model will be provided by the laws of classical mechanics and by additional relationships characterizing control motion. In a reference frame whose origin is at the center of mass related to the flight path (sometimes called stability axes or wind axes), the motion equations will be

$$m\frac{dV}{dt} = -\mathbf{D} - mg\sin\gamma + T\cos(\alpha - \varphi),$$

$$\frac{d\alpha}{dt} = -\frac{d\gamma}{dt} + q,$$

$$mV\frac{d\gamma}{dt} = \mathbf{L} - mg\cos\gamma + T\sin(\alpha - \varphi),$$

$$B\frac{dq}{dt} = M - Tz_T + M_c, \tag{2.1}$$

where m, $\mathbf{D}$, $\mathbf{L}$, M are known functions of the state variables considered and of t. To these should be added the equations that model the law of variation of longitudinal control and of thrust, for example, of the form

$$\begin{aligned} a\frac{d^2\delta_Y}{dt^2} + b\frac{d\delta_Y}{dt} + c\delta_Y &= f(\theta, q, \tau), \\ F(T, V, \delta_T) &= 0, \end{aligned} \tag{2.2}$$

τ being the time lag of the control system.

For simplicity, we will adopt, for attitude control, a proportional control law

$$\delta_Y(t) = \kappa\,\Delta\theta(t - \tau)$$

or a law of derivative and proportional control

$$\delta_Y(t) = \kappa_1 \Delta\theta(t - \tau_1) + \kappa_2 \Delta q(t - \tau_2);$$

κ, κ_1, and κ_2 are constant numbers; τ, τ_1, τ_2 are time lags (also assumed constant); $\Delta\theta$ and Δq denote the variation of magnitudes θ and q, respectively. Instead of parameter δ_T, the magnitude of thrust, considered as a known function of flight velocity V $(V = (u^2 + u^2)^{1/2})$, will be used.

Every motion of the aircraft in its longitudinal plane of symmetry will be represented by a special solution of the collected system (2.1) + (2.2).

As can be seen, equations (2.1) and (2.2) are nonlinear and rather complicated, being generally unapproachable in stability problems and very difficult to approach in other problems of flight dynamics. Therefore, whenever possible, linearization or simplification of the model is resorted to, by decreasing the number of variables with the aid of adequate hypotheses.

In accord with the scheme outlined in Section **1.5**, the disturbed motion equations will now be deduced. The new variables will be the deviations induced by the disturbance of the variables of system (2.1) + (2.2). If the disturbance is small in the sense to be defined hereafter (Section **3.4**), linearization is readily achieved.

To deduce the disturbed motion equations from system (2.1) it is first necessary to define the parameters on which the functions in the right-hand side of the equations depend (i.e., $\mathbf{D}$, T, $\mathbf{L}$, M, M_c), the linear part of the disturbed motion system being obtained as a variational system. Thus, the thrust T will be considered as a known function of flight velocity and independent of the remaining variables. The longitudinal control moment M_c will depend in a known manner on the induced deviations of the variables θ and q $(\theta = \alpha + \gamma,\ q = d\theta/dt)$. Regarding the aerodynamic forces $\mathbf{D}$ and $\mathbf{L}$ (drag and lift), the assumption will be made that, besides the independent variable t, they are dependent only on velocity and incidence, while the aerodynamic pitching moment M depends on the rate of pitch and rate of incidence as well.

As already noted, the motions of the machine in its plane of symmetry can be satisfactorily modeled, for the problems to be considered in this book, by a special solution of the system (2.1) + (2.2) in each case. Let $V = \tilde{V}(t)$, $\alpha = \tilde{\alpha}(t)$, $\gamma = \tilde{\gamma}(t)$, $q = \tilde{q}(t)$ be the solution corresponding to the basic motion. Then, by the change of variables

$$V = \Delta V + \tilde{V}(t), \quad \alpha = \Delta\alpha + \tilde{\alpha}(t), \quad \gamma = \Delta\gamma + \tilde{\gamma}(t), \quad q = \Delta q + \tilde{q}(t)$$

system (2.1), after a few elementary transformations, becomes

$$
\begin{aligned}
m\frac{d\Delta V}{dt} &= \left(-\frac{\partial \mathbf{D}}{\partial V}+\frac{\partial T}{\partial V}\cos(\alpha-\varphi)\right)\Delta V+\left(-\frac{\partial \mathbf{D}}{\partial \alpha}-T\sin(\alpha-\varphi)\right)\Delta\alpha \\
&\quad - mg\cos\gamma\,\Delta\gamma, \\
\frac{d\Delta\alpha}{dt} &= \Delta q-\frac{d\Delta\gamma}{dt}, \\
m\tilde{V}\frac{d\Delta\gamma}{dt} &= \left(\frac{\partial \mathbf{L}}{\partial V}+\frac{\partial T}{\partial V}\sin(\alpha-\varphi)-m\frac{d\tilde{\gamma}}{dt}\right)\Delta V+\left(\frac{\partial \mathbf{L}}{\partial \alpha}+T\cos(\alpha-\varphi)\right)\Delta\alpha \\
&\quad + mg\sin\gamma\,\Delta\gamma, \\
B\frac{d\Delta q}{dt} &= \left(\frac{\partial M}{\partial V}-\frac{\partial T}{\partial V}z_T\right)\Delta V+\left(\frac{\partial M}{\partial \alpha}+\frac{\partial M_c}{\partial \alpha}\right)\Delta\alpha+\frac{\partial M_c}{\partial \gamma}\Delta\gamma \\
&\quad +\left(\frac{\partial M}{\partial q}+\frac{\partial M_c}{\partial q}\right)\Delta q+\frac{\partial M}{\partial \dot{\alpha}}\frac{d\Delta\alpha}{dt}.
\end{aligned}
\tag{2.3}
$$

2.2. EQUATIONS OF LATERAL ASYMMETRICAL MOTION

The equations of lateral disturbed motion of aircraft are obtained similarly. For simplicity, only the steady basic motion will be considered. To facilitate the understanding of the following items, the deducing of asymmetric lateral motion equations will briefly be recalled. (For further details see, e.g., [1, Chapter 4].)

The motion will this time be referred to a reference frame fixed in the aircraft (the so-called body axes) in order to assure the invariance of the moments and products of inertia about the coordinate axes. If the aircraft—together with the reference frame—is rotating with respect to earth considered at total rest, the motion equations related to this reference frame will be the Euler equations (see Appendix, equations (3)). It is admitted that the aircraft in free (uncontrolled) flight is subject to the action of external forces, such as aerodynamic forces, its own weight, and the thrust of its engines, and as external moments to the aerodynamic moments and that of the thrust vector about its center of gravity. The latter usually only has a single component differing from zero in relation to the set of axes fixed in the aircraft (body axes) $Oxyz$, namely, the component corresponding to the lateral axis of the aircraft Oy: Tz_T. Let $(-X, Y, Z)$ and (L, M, N) be the projections on the body axes of the resultant aerodynamic force vector and of the resultant aerodynamic moment about the mass center. The total angular momentum $\mathbf{h}$ will also contain the one due to the spinning masses (rotors of jet engines, or airscrews) carried by the aircraft. Hence, $\mathbf{h} = \mathbf{h}_a + \mathbf{h}_r$. Let h_{r_x}, h_{r_y}, h_{r_z} be its projections on the body axes.†

As is known, the angular momentum of a rigid body is obtained from a summation of the angular momentum of each element of mass, $\delta \mathbf{h}_a = \mathbf{r} \times (\mathbf{V} + \boldsymbol{\omega} \times \mathbf{r})\delta m$ ($\times$ means vector product), where $\mathbf{r}$ denotes the radius vector of the mass element δm with respect to the origin. We recall that the vector $\mathbf{V}$ represents the velocity of the center of mass relative to a space-fixed system, and $\boldsymbol{\omega}$ the angular velocity of the aircraft about its center of mass (angular velocity of the rotation of the frame $Oxyz$ relative to $Ox_0y_0z_0$). The expression of $\mathbf{h}_a$,

$$\mathbf{h}_a = \int_m \mathbf{r} \times (\mathbf{V} + \boldsymbol{\omega} \times \mathbf{r})\, dm,$$

where the integral extends to the whole mass of the aircraft, is now resolved into its scalar components, namely,

$$h_{a_x} = p \int_m (y^2 + z^2)\, dm - q \int_m xy\, dm - r \int_m zx\, dm = Ap - Fq - Er,$$

$$h_{a_y} = -p \int_m xy\, dm + q \int_m (z^2 + x^2)\, dm - r \int_m yz\, dm = -Fp + Bq - Dr,$$

$$h_{a_z} = -p \int_m zx\, dm - q \int_m yz\, dm + r \int_m (x^2 + y^2)\, dm = -Ep - Dq + Cr,$$

where x, y, z are the components of vector $\mathbf{r}$ with respect to the $Oxyz$ system. Hence,

† Since the only spinning elements that can be taken into consideration in this respect in the motion equation of the aircraft are the rotors of the jet engines or the airscrews, generally, $h_{r_y} = 0$. Let σ be the angle formed by the resultant angular momentum vector $\mathbf{h}_r$ of these elements with the axis Ox; J their moment of inertia about an axis parallel to the vector $\mathbf{h}_r$; and Ω the magnitude of the angular velocity of these elements. Then, the magnitude of the two nonzero components of the vector $\mathbf{h}_r$ will be $h_{r_x} = J\Omega \cos \sigma$ and $h_{r_z} = J\Omega \sin \sigma$. The approximation can be admitted that the rotation axis of the rotors or of the airscrews coincides with the direction of thrust and hence $\sigma = \varphi$, or else the angle σ may be assumed negligible and hence $h_{r_x} = J\Omega$, $h_{r_z} = 0$.

$$h_x = h_{a_x} + h_{r_x} = Ap - Fq - Er + h_{r_x},$$

$$h_y = h_{a_y} + h_{r_y} = -Fp + Bq - Dr + h_{r_y},$$

$$h_z = h_{a_z} + h_{r_z} = -Ep - Dq + Cr + h_{r_z}.$$

The gravitational force vector has, in the $Ox_0y_0z_0$ system, a single component differing from zero: the projection upon the Oz_0 axis $(mg)_{z_0} = -mg$. The components in the directions of the $Oxyz$ frame of reference axes are readily obtained from Table A.1 (see Appendix): $(-mg\sin\theta, mg\sin\phi\cos\theta, -mg\cos\phi\cos\theta)$.

With the foregoing, the Euler equations of motion of the aircraft will be (see Appendix, equations (3))

$$m\left(\frac{du}{dt} + qw - rv\right) = -X + T\cos\varphi - mg\sin\theta, \qquad \text{(i)}$$

$$m\left(\frac{dv}{dt} + ru - pw\right) = Y + mg\sin\phi\cos\theta, \qquad \text{(ii)}$$

$$m\left(\frac{dw}{dt} + pv - qu\right) = Z + T\sin\varphi - mg\cos\phi\cos\theta, \qquad \text{(iii)}$$

$$A\frac{dp}{dt} - F\frac{dq}{dt} - E\frac{dr}{dt} + \frac{dh_{r_x}}{dt} - Dq^2 + Dr^2 - Epq + (C - B)qr + Frp + qh_{r_z} - rh_{r_y} = L, \qquad \text{(iv)}$$

$$-F\frac{dp}{dt} + B\frac{dq}{dt} + D\frac{dr}{dt} + \frac{dh_{r_y}}{dt} - Er^2 + Ep^2 + Dpq - Fqr + (A - C)rp + rh_{r_x} - ph_{r_z} = M, \qquad \text{(v)}$$

$$-E\frac{dp}{dt} - D\frac{dq}{dt} + C\frac{dr}{dt} + \frac{dh_{r_z}}{dt} - Fp^2 + Fq^2 + (B - A)pq + Eqr - Drp + ph_{r_y} - gh_{r_x} = N, \qquad \text{(vi)}$$

$$(2.4)$$

to which are added, since the variables ϕ and θ also appear in the system (as does ψ in some cases in the equations of disturbed motion) the transformation equations (see Appendix, equations (2))

$$\frac{d\phi}{dt} = p - q \sin \phi \tan \theta, \qquad \frac{d\theta}{dt} = q \cos \phi - r \sin \phi,$$

$$\frac{d\psi}{dt} = q \sin \phi \sec \theta + r \cos \phi \sec \theta. \tag{2.5}$$

Before passing on to the determination of the equations of motion induced by a disturbance, system (2.4) will be simplified on basis of the following assumptions.

1. The aircraft is perfectly symmetrical about the median plane zOx.
2. The axes Ox and Oz are assumed sufficiently close to the respective principal axes of inertia for us to disregard the product of inertia E.
3. The aircraft is assumed to be a rigid body, the gyroscopic action of the spinning masses carried, such as engine rotors or airscrews, being neglected. In other words, it is assumed that $h_{r_x} = h_{r_y} = h_{r_z} = 0$.
4. The disturbances are assumed to be sufficiently small to allow us to neglect the nonlinear terms in the disturbed motion equations, according to the theory of stability by the first approximation. (See Section 3.4.)
5. It is considered that the basic motion is steady straight level flight with the undisturbed values of the angles ϕ, ψ, and β identically equal to zero.
6. During the undisturbed motion, the components, according to the directions Oz and Ox, of the velocity field of the atmosphere relative to the earth are considered equal to zero.

From 5 and 6 it results that during the whole undisturbed motion, the atmosphere traveled by the aircraft is, locally, at rest relative to earth and, implicitly, that the undisturbed aerodynamic velocity coincides with the undisturbed velocity of the mass center relative to earth, both equal to u_0.

Hypothesis 1, according to which the median plane zOx is a symmetry plane for the aircraft, implies that Oy is a principal axis of inertia and, hence, $D = F = 0$.

From hypotheses 1 and 5 it will result that the variation of any among the variables characterizing the motion of the aircraft in its plane of symmetry (u, w or α, q) will not induce aerodynamic forces and moments in the planes xOy and yOz and hence, in the conditions of hypothesis 5 (of a symmetrical uniform basic motion) the partial derivatives, of any order, of the magnitudes Y, L, and N with respect to the variables u, w, or α and q will be equal to zero.

It is readily seen that the first-order partial derivatives of the aerodynamic forces and moments acting in the symmetry plane (X, Z, M) with respect to any of the asymmetrical variables v or β, p, r will also vanish in hypotheses 1 and 5. Indeed, because of the geometric (viz., aerodynamic) symmetry of the machine, the components in the zOx plane of the aerodynamic forces and moments (X, Z, M) induced by the variation of asymmetrical kinematic parameters β, p, or r from their zero value will be the same in sign and magnitude, whatever the sense of variation. For instance, $X(\beta) = X(-\beta)$.

As a result of the two implications of hypotheses 1 and 5 just stated, as well as of assumption 4 regarding the possibility of linearization, the aerodynamic terms coupling the symmetrical and the asymmetrical motions are eliminated from the equations. If, in addition, the fact that $D = F = 0$ according to hypothesis 1 is taken into account, and if hypothesis 3 ($h_{r_x} = h_{r_y} = h_{r_z} = 0$) is admitted, all the coupling terms are eliminated from the equations of disturbed motion, which leads to their separation into two independent sets, one of which represents the symmetrical (longitudinal) motion while the other represents the asymmetrical (lateral) motion. Assumptions 2 and 6 (as well as the one stating that the flight path is horizontal, assumption 5) were only made for convenience, and no essential change of the reasoning that follows is involved if they are neglected.

From system (2.4) only equations (ii), (iv), and (vi) will be retained. Together with equations (2.5) they will provide the equations of the asymmetrically disturbed motion of the aircraft. Admitting hypotheses 1, 2, and 3,† these equations become

$$\frac{dv}{dt} = -ru + pw + \frac{Y}{m} + g \sin\phi \cos\theta,$$

$$\frac{dp}{dt} = \frac{B-C}{A} qr + \frac{L}{A},$$

$$\frac{dr}{dt} = \frac{A-B}{C} pq + \frac{N}{C}; \tag{2.6}$$

$$\begin{aligned} \frac{d\phi}{dt} &= p - q \sin\phi \tan\theta - r \cos\phi \tan\theta, \\ \frac{d\psi}{dt} &= q \sin\phi \sec\theta + r \cos\phi \sec\theta. \end{aligned} \tag{2.7}$$

† Assumptions 4 and 5 will obviously only be used for the disturbed motion equations obtained by a change of variables of the type (1.7).

It is assumed that the aerodynamic force Y and the aerodynamic moments L and N are dependent on the variables v or β, p, and r. (Generally the variation of Y with p and r can be neglected.) Then, taking into account hypothesis 5 regarding the steadiness of the basic motion and the undisturbed value of some of the variables, and hypothesis 4 permitting the linearization of the induced motion equations, system (2.6) yields

$$\frac{d\Delta v}{dt} = \frac{1}{m}\frac{\partial Y}{\partial v}\Delta v + \frac{1}{m}\frac{\partial Y}{\partial p}\Delta p + \left(\frac{1}{m}\frac{\partial Y}{\partial r} - u\right)\Delta r + g\cos\theta\,\Delta\phi,$$

$$\frac{d\Delta p}{dt} = \frac{1}{A}\left(\frac{\partial L}{\partial v}\Delta v + \frac{\partial L}{\partial p}\Delta p + \frac{\partial L}{\partial r}\Delta r\right),$$

$$\frac{d\Delta r}{dt} = \frac{1}{C}\left(\frac{\partial N}{\partial v}\Delta v + \frac{\partial N}{\partial p}\Delta p + \frac{\partial N}{\partial r}\Delta r\right),$$

and system (2.7) yields

$$\frac{d\Delta\phi}{dt} = \Delta p - \tan\theta\,\Delta r, \qquad \frac{d\Delta\psi}{dt} = \sec\theta\,\Delta r.$$

In the last two systems u, θ, and the partial derivatives Y_v, Y_p, Y_r, L_v, L_p, L_r, N_v, N_p, and N_r correspond to the undisturbed values of the variables. For simplicity, since the undisturbed values of the quantities Δv, Δp, Δr, $\Delta\phi$, and $\Delta\psi$ are considered zero, the symbol Δ showing the respective deviation is neglected, and the symbols referring to the basic motion will have zero as a subscript. Then the last two systems, representing the asymmetrical disturbed motion, are written

$$\frac{dv}{dt} = \frac{1}{m}\left(\frac{\partial Y}{\partial v}\right)_0 v + \frac{1}{m}\left(\frac{\partial Y}{\partial p}\right)_0 p + \left(\frac{1}{m}\left(\frac{\partial Y}{\partial r}\right)_0 - u_0\right) r + (g\cos\theta_0)\phi$$

$$\frac{dp}{dt} = \frac{1}{A}\left[\left(\frac{\partial L}{\partial v}\right)_0 v + \left(\frac{\partial L}{\partial p}\right)_0 p + \left(\frac{\partial L}{\partial r}\right)_0 r\right],$$

$$\frac{dr}{dt} = \frac{1}{C}\left[\left(\frac{\partial N}{\partial v}\right)_0 v + \left(\frac{\partial N}{\partial p}\right)_0 p + \left(\frac{\partial N}{\partial r}\right)_0 r\right],$$

$$\frac{d\phi}{dt} = p - r\tan\theta_0, \qquad \frac{d\psi}{dt} = r\sec\theta_0. \tag{2.8}$$

2.3. NONDIMENSIONAL EQUATIONS

2.3.1. Introductory Remarks

In the effective solving of flight stability problems, the current usage is to express the equations of motion in nondimensional form. It should be noted from the start that, from the theoretical standpoint, nondimensional equations do not differ from dimensional equations obtained by modeling the actual process. Conventionally, a physical significance is granted to the quantities mentioned in the latter, although it is quite obvious that perfect isomorphism can never be achieved between the actual quantities and processes (approximately known and, hence, only approximately defined) and the mathematical operations and magnitudes, which are perfectly defined. (Moreover, the approximate description or representation is a basic characteristic of any model, not only of a mathematical one.) From the instant when equations (viz., a model) have been resorted to, and hence the field of mathematical operations has been entered into, until the results are interpreted, only abstract quantities are dealt with, whatever the physical nature of the actual origin of these magnitudes. Thus, in the equations, V, α, m, and so on—although they continue to be called, for convenience, velocity, incidence, mass, or whatever—are nothing but numbers.

For the mathematical treatment, using a nondimensional form may even be a source of difficulties. The nondimensional form may create some trouble, for example, in the quantitative estimations since the magnitude of a dimensionless quantity is conclusively influenced by the adopted system of dimensional units.

Nevertheless, the foregoing statements do not mean that there are no benefits to be derived from the use of nondimensional equations for most applied problems connected with flight stability. Glauert's [2] introduction of dimensionless variables and system parameters in disturbed motion equations resulted in: (i) the creation of a link between experimental findings and theory in the domain of flight stability; (ii) increased efficiency of theory by determining some invariances of the results with respect to certain design and flight régime characteristics, thus emphasizing some essential parameters from the standpoint of flight stability problems. In addition, the number of nondimensional parameters being, as will be seen, smaller than that of the initial parameters of the problem, the analysis of their influence is simplified.

The essentials in bringing the equations to a nondimensional form will be recalled and the manner in which the advantages specified are achieved will be outlined. As is known [3], the dimension of any physical quantity can be expressed in the form of a power product of dimensions, independent of one another (by dimensional independence is understood the impossibility of expressing one dimension in the form of power products of the others), adopted as fundamental units. The choice of the quantities to be used as fundamental units obviously depends on the nature of the problem considered. In problems of Newtonian mechanics the number of independent dimensions is three, the magnitudes generally used as fundamental units being time, length, and mass.

Hereunder, the dimension (derived unit) of a quantity whatever q will be denoted by $[q]$, and the nondimensional ratio $q/[q]$ by $\hat{q}$. Let x be the dynamic variables and λ_i the parameters of the system.

The system

$$\frac{dx}{dt} = X(t, x, \lambda) \tag{2.9}$$

is considered dimensionally homogeneous and therefore independent of the units chosen. (For details, rigorous definitions, statements of theorems, etc., see, e.g., [4, 5].) To bring it to a nondimensional form, the transformation

$$t = [t]\hat{t}, \qquad x = [x]\hat{x}, \qquad \lambda = [\lambda]\hat{\lambda}$$

must first be resorted to, where $[t]$, $[x]$, and $[\lambda]$ are then expressed by means of the fundamental units $[t]$, $[l]$, and $[m]$ (units of time, length, and mass). The system being dimensionally homogeneous, $[t]$, $[l]$, and $[m]$ disappear, and by simplifying, a system of the form

$$\frac{d\hat{x}}{d\hat{t}} = X_1(\hat{t}, \hat{x}, \hat{\lambda}) \tag{2.10}$$

is obtained.

This system comprises a number, for instance m, of nondimensional parameters $\hat{\lambda}_i$ equal to the number of initial dimensional parameters λ_i. If, however, k is the number of fundamental units (here $k = 3$), according to Buckingham's pi theorem [6, 7] a dimensionally homogeneous relation, independent of the choice of units among m dimensional quantities, is equivalent to a relation between $m - k$ quantities representing independent nondimensional combinations (in the form of power products) of the m parameters ($m - k$ represents, moreover, the maximum number of independent combinations formed as power products of the k fundamental units).

In other words, system (2.10) can be replaced by an equivalent system of the form

$$\frac{d\hat{x}}{d\hat{t}} = X_2(\hat{t}, \hat{x}, \hat{\nu}) \tag{2.11}$$

dependent on a number of $m - 3$ nondimensional parameters $\hat{\nu}$.

Let it now be assumed that system (2.9) represents the disturbed motion of a certain standard aircraft, corresponding to a certain standard basic motion. The parameters λ_i of this system, representing the design and aerodynamic characteristics of the aircraft and the régime of the basic motion, have been determined by experimental measurements or by approximate computations based on data determined experimentally. A priori, the conclusions yielded by this system cannot be extended to the machine studied. To ensure the validity of these results, rigorous in the framework of the model, the existence of perfect dynamic similarity is required between the standard case and the one studied. The nondimensional parameters ν_i supply the pertinent criteria of similitude. There is perfect dynamic similarity between the two motions if every nondimensional parameter ν_i coincides. Identical checking of the similarity criteria is, in most cases, impossible. On the one hand, we resort to the empirical determination of the ranges within which the values of the parameters may be considered as approximately overlapping, and on the other, we choose, from among all the criteria supplied by the nondimensional parameters, those playing an essential part in defining the process studied. Thus, the eliminating of some of these criteria is implied in the hypotheses on which the model is built.

2.3.2. Longitudinal Disturbed Motion

To illustrate the foregoing, let system (2.3) again be considered. The aerodynamic forces $\mathbf{D}$ and $\mathbf{L}$ and the longitudinal aerodynamic moment M will be expressed in the usual form, which makes the main effect of aerodynamic speed, air density, and size of the aircraft on $\mathbf{D}$, $\mathbf{L}$, and M stand out.

$$\mathbf{D} = \tfrac{1}{2}\rho V^2 S C_D, \quad \mathbf{L} = \tfrac{1}{2}\rho V^2 S C_L, \quad M = \tfrac{1}{2}\rho V^2 S l C_m. \tag{2.12}$$

Hereafter, in the expression of the moment, half the mean aerodynamic chord of the wing ($l = \bar{c}/2$) will be used as the characteristic length l. The simplifying assumption will be made that C_D and C_L are only functions of incidence (α) and the Mach number (**M**) and the pitching moment coefficient C_m is, in addition, a function of the rate of pitch q and of the derivative of incidence with respect to time $d\alpha/dt$ as well.

Let $[t]$, $[l]$, and $[m]$ be the fundamental units. Their magnitude will be shown later. All the dimensional quantities in system (2.3)† (with **D**, **L**, M replaced according to the formulas (2.12)) will be expressed as power products of fundamental units, representing the dimensions of these quantities and the nondimensional quantities‡ that characterize them, namely,

$$t = [t]\hat{t}, \quad m = [m]\mu, \quad g = \frac{[l]}{[t]^2}\hat{g} \;§$$

$$V = \frac{[l]}{[t]}\hat{V}, \quad l = [l]\hat{l},$$

$$\Delta V = \frac{[l]}{[t]}\Delta\hat{V}, \quad S = [l]^2\hat{S}, \quad T = \frac{[m][l]}{[t]^2}C_T$$

$$q = \frac{1}{[t]}\hat{q}, \quad B = [m][l]^2 i_B \qquad T_V = \frac{[m]}{[t]}C_{T_V}$$

$$\Delta q = \frac{1}{[t]}\Delta\hat{q}, \quad z_T = [l]\hat{z}_T,$$

$$M_c = \frac{[m][l]^2}{[t]^2}\hat{M}_c, \qquad M_{c_q} = \frac{\partial M_c}{\partial q} = \frac{[m][l]^2}{[t]}\frac{\partial \hat{M}_c}{\partial \hat{q}} = \frac{[m][l]^2}{[t]}\hat{M}_{c_q}.$$

The nondimensional aerodynamic rotary derivatives about the lateral axis will be noted C_{m_q} and $C_{m_{\dot{\alpha}}}$, representing the partial derivatives of the pitching moment coefficient C_m with respect to the dimensionless angular velocities $\hat{q} = q[t]$ and $\hat{\dot{\alpha}} = d\alpha/d\hat{t} = [t]\, d\alpha/dt$, respectively.

† Angles will be considered dimensionless quantities, expressed in radians.

‡ For the dimensionless quantities characterizing the aircraft mass and its longitudinal moment of inertia, the conventional notations μ and i_B have been used instead of $\hat{m}$ and $\hat{B}$, while C_T and C_{T_V} have been put for $\hat{T}$ and $\hat{T}_V$, respectively, by analogy with the symbols currently used for aerodynamic forces and aerodynamic derivatives.

§ It is readily seen that $\hat{g}$ is the reverse of Froude's number corresponding to the standard case: $\hat{g} = g[l]/[V]^2 = 1/\mathsf{NF}_{\text{stan}}$.

Taking into account the foregoing, system (2.3) becomes, after simplifying,†

$$\mu \frac{d\Delta\hat{V}}{d\hat{t}} = \left[-\hat{\rho}\hat{S}\hat{V}C_D - \frac{\hat{\rho}}{2}\hat{V}\mathbf{M}\hat{S}C_{D_\mathbf{M}} + C_{T_V}\cos(\alpha-\varphi)\right]\Delta\hat{V}$$

$$+\left[-\frac{\hat{\rho}}{2}\hat{V}^2\hat{S}C_{D_\alpha} - C_T\sin(\alpha-\varphi)\right]\Delta\alpha - \mu\hat{g}\cos\gamma\,\Delta\gamma,$$

$$\frac{d\Delta\alpha}{d\hat{t}} = \Delta\hat{q} - \frac{d\Delta\gamma}{d\hat{t}},$$

$$\mu\hat{V}\frac{d\Delta\gamma}{d\hat{t}} = \left[\hat{\rho}\hat{V}\hat{S}C_L + \frac{\hat{\rho}}{2}\hat{V}\mathbf{M}\hat{S}C_{L_\mathbf{M}} + C_{T_V}\sin(\alpha-\varphi) - \mu\frac{d\gamma}{d\hat{t}}\right]\Delta\hat{V}$$

$$+\left[\frac{\hat{\rho}}{2}\hat{V}^2\hat{S}C_{L_\alpha} + C_T\cos(\alpha-\varphi)\right]\Delta\alpha + \mu\hat{g}\sin\gamma\,\Delta\gamma,$$

$$i_B\frac{d\Delta\hat{q}}{d\hat{t}} = \left(\hat{\rho}\hat{V}\hat{S}\hat{l}C_m + \frac{\hat{\rho}}{2}\hat{V}\hat{S}\hat{l}\mathbf{M}C_{m_\mathbf{M}} - C_{T_V}\hat{z}_T\right)\Delta\hat{V} + \left(\frac{\hat{\rho}}{2}\hat{V}^2\hat{S}\hat{l}C_{m_\alpha} + \hat{M}_{c_\alpha}\right)\Delta\alpha$$

$$+\hat{M}_{c_\gamma}\Delta\gamma + \left(\frac{\hat{\rho}}{2}\hat{V}^2\hat{S}\hat{l}C_{m_q} + \hat{M}_{c_q}\right)\Delta\hat{q} + \frac{\hat{\rho}}{2}\hat{V}^2\hat{S}\hat{l}C_{m_{\dot\alpha}}\frac{d\Delta\alpha}{d\hat{t}}. \tag{2.13}$$

It can be seen that the system is dependent on a large number of dimensionless parameters, some constant ($\hat{S}$, $\hat{l}$, $\hat{g}$, or NF), but most of them generally varying with t: μ, i_B, $\hat{\rho}$, $\hat{V}$ (or $\hat{t}$), $\mathbf{M}$, α, and q (through the aerodynamic coefficients and the stability derivatives; in the case of steady fundamental motion it is obvious that these dimensionless parameters, excepting $\hat{t}$, will be constant).

According to the pi theorem, the fundamental units may be so chosen in the nondimensional system that the number of parameters is three less than the number of dimensional parameters. Indeed, let it be assumed that the aircraft mass and air density are independent of t. The decrease in the number of dimensionless parameters is equivalent to imposing additional relations on them. Let $\hat{l} = 1$ and $\hat{\rho}\hat{S} = 1$. From the first equation $[l] = l$ ($= \bar{c}/2$) and from the second

$$\rho S = [\rho][S] = \frac{[m]}{[l]} = \frac{[m]}{l} = \frac{m}{\mu l};$$

hence

$$\mu = \frac{m}{\rho S l} = \frac{2m}{\rho S\bar{c}} \tag{2.14}$$

† Recall that, because of the dimensional homogeneity of the system, the quantities $[t]$, $[l]$, and $[m]$ disappear by simplifying.

and system (2.13) can be rewritten

$$\frac{d\Delta\hat{V}}{d\hat{t}} = \frac{1}{\mu}\left[-\hat{V}C_D - \frac{1}{2}\hat{V}\mathbf{M}C_{D_{\mathbf{M}}} + C_{T_V}\cos(\alpha-\varphi)\right]\Delta\hat{V}$$
$$+\frac{1}{\mu}\left[-\frac{1}{2}\hat{V}^2C_{D_\alpha} - C_T\sin(\alpha-\varphi)\right]\Delta\alpha - \hat{g}\cos\gamma\,\Delta\gamma,$$
$$\frac{d\Delta\alpha}{d\hat{t}} = \Delta\hat{q} - \frac{d\Delta\gamma}{d\hat{t}},$$
$$\frac{d\Delta\gamma}{d\hat{t}} = \frac{1}{\mu}\left[C_L + \frac{1}{2}\mathbf{M}C_{L_{\mathbf{M}}} + \frac{1}{\hat{V}}C_{T_V}\sin(\alpha-\varphi) - \frac{\mu}{\hat{V}}\frac{d\gamma}{d\hat{t}}\right]\Delta\hat{V}$$
$$+\frac{1}{\mu}\left[\frac{1}{2}\hat{V}C_{L_\alpha} + \frac{C_T}{\hat{V}}\cos(\alpha-\varphi)\right]\Delta\alpha + \frac{\hat{g}}{\hat{V}}\sin\gamma\,\Delta\gamma,$$
$$\frac{d\Delta\hat{q}}{d\hat{t}} = \frac{1}{i_B}\left(\hat{V}C_m + \frac{1}{2}\hat{V}\mathbf{M}C_{m_{\mathbf{M}}} - C_{T_V}\hat{z}_T\right)\Delta\hat{V} + \frac{1}{i_B}\left(\frac{1}{2}\hat{V}^2C_{m_\alpha} + \hat{M}_{c_\alpha}\right)\Delta\alpha$$
$$+\frac{1}{i_B}\hat{M}_{c_\gamma}\Delta\gamma + \frac{1}{i_B}\left(\frac{1}{2}\hat{V}^2C_{m_q} + \hat{M}_{c_q}\right)\Delta\hat{q} + \frac{1}{2i_B}\hat{V}^2C_{m_{\dot{\alpha}}}\frac{d\Delta\alpha}{d\hat{t}}. \quad (2.15)$$

It may be seen that the dimensionless parameter μ, usually called relative density of aircraft, appears with considerable weight in the disturbed motion equation system. Therefore, in matters of stability, μ = const. is one of the fundamental similitude criteria; to be able to transfer results from the physical model to the aircraft studied, it is necessary for their relative density μ to be the same.

It is readily seen that, in case the condition μ = const. is fulfilled, the similitude criterion i_B = const. is equivalent to the condition that the ratio between the radius of gyration of the aircraft about its lateral axis k_y and the mean wing chord $\bar{c}$ be invariable. Indeed,

$$i_B = \frac{B}{[m][l]^2} = \frac{mk_y^2}{(m/\mu)l^2} = \mu\left(\frac{k_y}{\bar{c}/2}\right)^2 = \mu\hat{k}_y^2.$$

Usually $[t]$ is called the unit of aerodynamic time. For its definition, either the ratio m_0/ρ_0SV_0 (m_0, ρ_0, V_0 being, obviously, constants, e.g., maximum total mass of aircraft, air density at zero altitude, and a reference speed, such as the cruising speed, respectively) or the ratio l/V_0, in the given case $\bar{c}/(2V_0)$ may be used. The latter expression will here be used.

In the case of steady basic motion, the flying speed of the aircraft center of mass may be taken as reference speed, and therefore $[t] = (\bar{c}/2)V$. In this case the relation $\hat{V} = 1$ may also be added to $\hat{l} = 1$ and $\hat{\rho}\hat{S} = 1$.

Since in straight level flight the lift is at any instant approximately equal to the aircraft weight, we have

$$\hat{g} = \frac{gl}{V^2} = \frac{mgl}{mV^2} \cong \frac{(\rho/2)V^2 SC_L l}{\rho S l \mu V^2} = \frac{1}{2}\frac{C_L}{\mu}$$

and the similitude criterion connected with the maintenance of $\hat{g}$ can therefore be replaced by the one of lift coefficient identity.

Now, taking into account that for straight flight $d\gamma/dt = 0$, system (2.15) will be reduced to

$$\begin{aligned}
\frac{d\Delta\hat{V}}{d\hat{t}} &= \frac{1}{\mu}\left\{\left[-C_D - \frac{1}{2}\mathbf{M}C_{D_\mathbf{M}} + C_{T_V}\cos(\alpha-\varphi)\right]\Delta\hat{V}\right. \\
&\quad \left. + \left[-\frac{1}{2}C_{D_\alpha} - C_T\sin(\alpha-\varphi)\right]\Delta\alpha - \mu\hat{g}\cos\gamma\,\Delta\gamma\right\}, \\
\frac{d\Delta\alpha}{d\hat{t}} &= \Delta\hat{q} - \frac{d\Delta\gamma}{d\hat{t}}, \\
\frac{d\Delta\gamma}{d\hat{t}} &= \frac{1}{\mu}\left\{\left[C_L + \frac{1}{2}\mathbf{M}C_{L_\mathbf{M}} + C_{T_V}\sin(\alpha-\varphi)\right]\Delta\hat{V}\right. \\
&\quad \left. + \left[\frac{1}{2}C_{L_\alpha} + C_T\cos(\alpha-\varphi)\right]\Delta\alpha + \mu\hat{g}\sin\gamma\,\Delta\gamma\right\}, \\
\frac{d\Delta\hat{q}}{d\hat{t}} &= \frac{1}{i_B}\left\{\left(C_m + \frac{1}{2}\mathbf{M}C_{m_\mathbf{M}} - C_{T_V}\hat{z}_T\right)\Delta\hat{V} + \left(\frac{1}{2}C_{m_\alpha} + \hat{M}_{c_\alpha}\right)\Delta\alpha\right. \\
&\quad \left. + \hat{M}_{c_\gamma}\Delta\gamma + \left(\frac{1}{2}C_{m_q} + \hat{M}_{c_q}\right)\Delta\hat{q} + \frac{1}{2}C_{m_{\dot{\alpha}}}\frac{d\Delta\alpha}{d\hat{t}}\right\}.
\end{aligned} \tag{2.16}$$

2.3.3. Dimensionless Equations of Lateral Asymmetrical Disturbed Motion

To express system (2.8) in a dimensionless form, we use the dimensional units: half the wing span $b/2$ for lengths; $b/2u_0$ for time; hence, $2u_0/b$ for angular speeds. The dimensional unit of the aerodynamic force Y will be the same as in the case of the aerodynamic forces in the longitudinal plane: $(\rho/2)V^2S$; that of the aerodynamic moments L and N will be $(\rho/2)V^2S(b/2)$.

The values of the mass m of the moments of inertia A and C, as well as that of acceleration due to gravity g, will be replaced by expressions similar to those used in the case of longitudinal motion:

$$m = \mu\rho S\frac{b}{2}; \quad (A, C) = \rho S\left(\frac{b}{2}\right)^3 (i_A, i_C); \quad g = \hat{g}\,\frac{2u_0^2}{b}.$$

The variable β will be used in the dimensionless system instead of the variable v; β represents the aircraft sideslip angle, formed of the projection of the resultant aerodynamic speed on plane xOy with the axis Ox. The sideslip angle corresponds to the angle α (aircraft incidence) for the asymmetrical motion.

Taking into account that $V_0 = u_0$ and $\beta_0 = 0$, and neglecting the nonlinear terms in v and β, according to hypothesis 4 (Section 2.2), $v = u_0\beta$ is immediately obtained from the defining relation of the sideslip angle $\beta = \sin^{-1}(v/V)$, which, in the given case of motion of the aircraft following the disturbance, becomes $\beta = \sin^{-1}(v/u_0)$. We will obviously have

$$\left(\frac{\partial Y}{\partial v}\right)_0 = \left(\frac{\rho}{2}\right) u_0 S (C_{y_\beta})_0, \qquad \left[\frac{\partial}{\partial v}(L, N)\right]_0 = \left(\frac{\rho}{2}\right) u_0 S \left(\frac{b}{2}\right)(C_{l_\beta}, C_{n_\beta})_0.$$

The partial derivatives of the lateral aerodynamic force and of the aerodynamic rolling and yawing moments with respect to the angular speeds are readily obtained:

$$\left(\frac{\partial Y}{\partial p}\right)_0 = \frac{\rho}{2} u_0 S \frac{b}{2}(C_{y_p})_0, \qquad \left(\frac{\partial Y}{\partial r}\right)_0 = \frac{\rho}{2} u_0 S \frac{b}{2}(C_{y_r})_0,$$

$$\left(\frac{\partial}{\partial p}(L, N)\right)_0 = \frac{\rho}{2} u_0 S \left(\frac{b}{2}\right)^2 (C_{l_p}, C_{n_p})_0,$$

$$\left(\frac{\partial}{\partial r}(L, N)\right)_0 = \frac{\rho}{2} u_0 S \left(\frac{b}{2}\right)^2 (C_{l_r}, C_{n_r})_0.$$

After the transformation corresponding to the nondimensional system above, (2.13) becomes

$$\frac{d\beta}{d\hat{t}} = \frac{1}{2\mu} C_{y_\beta}\beta + \frac{1}{2\mu} C_{y_p}\hat{p} + \left(\frac{1}{2\mu} C_{y_r} - 1\right)\hat{r} + (\hat{g}\cos\theta_0)\phi,$$

$$\frac{d\hat{p}}{d\hat{t}} = \frac{1}{2i_A}(C_{l_\beta}\beta + C_{l_p}\hat{p} + C_{l_r}\hat{r}), \qquad \frac{d\hat{r}}{d\hat{t}} = \frac{1}{2i_C}(C_{n_\beta}\beta + C_{n_p}\hat{p} + C_{n_r}\hat{r}),$$

$$\frac{d\phi}{d\hat{t}} = \hat{p}(\tan\theta_0)\hat{r}, \qquad \frac{d\psi}{d\hat{t}} = (\sec\theta_0)\hat{r}. \tag{2.17}$$

(Subscript 0 will be neglected hereafter in the case of aerodynamic coefficients and stability derivatives, which we will consider as continuously corresponding to the undisturbed régime.) As in the case of symmetrical motion, if the basic, or undisturbed, motion is horizontal, $\hat{g}$ may be replaced by $C_L/2\mu$. The system then becomes

$$\frac{d\beta}{d\hat{t}} = \frac{1}{2\mu}\,[C_{y_\beta}\beta + C_{y_p}\hat{p} - (2\mu - C_{y_r})\hat{r} + (C_L \cos\theta_0)\phi],$$

$$\frac{d\hat{p}}{d\hat{t}} = \frac{1}{2i_A}\,(C_{l_\beta}\beta + C_{l_p}\hat{p} + C_{l_r}\hat{r}), \qquad \frac{d\hat{r}}{d\hat{t}} = \frac{1}{2i_c}\,(C_{n_\beta}\beta + C_{n_p}\hat{p} + C_{n_r}\hat{r}),$$

$$\frac{d\phi}{d\hat{t}} = \hat{p} - (\tan\theta_0)\hat{r}, \qquad \frac{d\psi}{d\hat{t}} = (\sec\theta_0)\hat{r}. \tag{2.18}$$

APPENDIX

DEFINITION OF AIRCRAFT ORIENTATION BY MEANS OF THE EULER ANGLES; EULER'S EQUATIONS

The attitude (angular position with relation to the earth) of an aircraft is usually defined by means of: angle of bank (lateral attitude) ϕ; angular pitch (longitudinal attitude) θ; and angle of azimuth ψ.

Let $Ox_0y_0z_0$ be a rectangular frame of reference in translation with respect to the earth with Oz_0 taken as vertical and its positive sense upward, and $Oxyz$ a system of axes fixed in the aircraft body (body axes). The origin of both reference systems coincides with the aircraft center of mass. The orientation of the aircraft with respect to the earth will then be given by the relative (angular) position of the two systems, defined by the three Euler angles mentioned. For their definition the semifixed rectangular reference frame $Ox_0y_0z_0$ (see Figure 2.1) is rotated around point O until it is superposed on the fixed reference trihedron $Oxyz$ as follows (the sequence of rotations is not indifferent).

1. The reference frame $Ox_0y_0z_0$ is rotated by ψ about the axis Oz_0; the new position of the axis will be shown by subscript 1; Oz_1 coincides with Oz_0 and the plane Ox_1y_1 with the plane Ox_0y_0; the azimuth of the axis Ox (and implicitly that of the aircraft) is thus definitely fixed.

2. By further applying to the trihedron $Ox_1y_1z_1$ a rotation θ around the axis Oy_1, a position $Ox_2y_2z_2$ is reached that definitely fixes, together with its elevation, the orientation of the axis Ox.

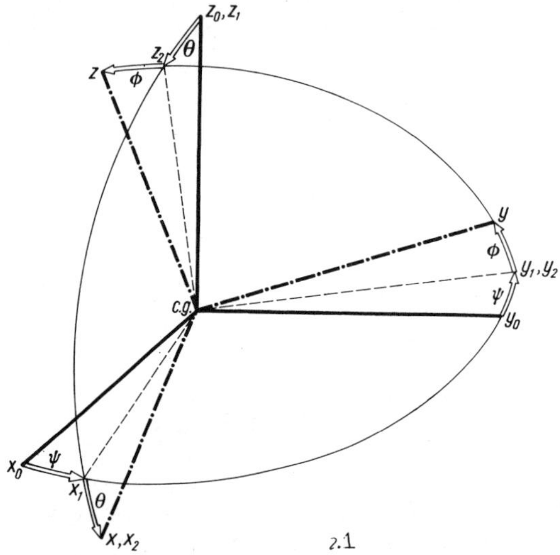

Figure 2.1.

3. Finally, the trihedron $Ox_2y_2z_2$ is rotated by an angle ϕ around the axis Ox_2 in order to obtain its final position, coinciding with the orientation of the system of body axes $Oxyz$.

Table A.1 shows the expressions of the direction cosines of the aircraft axes as a function of the angles ϕ, θ, and ψ that allow passing from one system of coordinates to the other (from the semifixed to the fixed, or vice versa).

Table A.I

	x_0	y_0	z_0
x	$\cos\theta\cos\psi$	$-\cos\theta\sin\psi$	$\sin\theta$
y	$\sin\phi\sin\theta\cos\psi$ $+\cos\phi\sin\psi$	$-\sin\phi\sin\theta\sin\psi$ $+\cos\phi\cos\psi$	$-\sin\phi\cos\theta$
z	$\sin\phi\sin\psi$ $-\cos\phi\sin\theta\cos\psi$	$\sin\phi\cos\psi$ $+\cos\phi\sin\theta\sin\psi$	$\cos\phi\cos\theta$

The relationships will now be established between the time derivatives $\dot{\phi}$, $\dot{\theta}$, and $\dot{\psi}$, on the one hand, and the scalar components p, q, and r along the body axes $Oxyz$ of the aircraft angular speed $\boldsymbol{\omega}$ on the other, the earth being assumed at an absolute rest. This relationship can readily be deduced by examining Figure 2.1, and the following linear expressions are obtained with respect to p, q, r, $\dot{\phi}$, $\dot{\theta}$, $\dot{\psi}$.

$$
\begin{aligned}
p &= \dot{\phi} + \dot{\psi} \sin \theta, \\
q &= \dot{\psi} \cos \theta \sin \phi + \dot{\theta} \cos \phi, \\
r &= \dot{\psi} \cos \theta \cos \phi - \dot{\theta} \sin \phi,
\end{aligned}
\tag{1}
$$

or

$$
\begin{aligned}
\dot{\phi} &= p - q \sin \phi \tan \theta - r \cos \phi \tan \theta, \\
\dot{\theta} &= q \cos \phi - r \sin \Phi, \\
\dot{\psi} &= q \sin \phi \sec \theta + r \cos \phi \sec \theta,
\end{aligned}
\tag{2}
$$

respectively.

Euler's equations. Referring the aircraft motion to the body axis $Oxyz$, in rotation versus the earth considered at rest, with an angular velocity $\boldsymbol{\omega}$, the scalar equations of motion will be written in the form

$$
\begin{aligned}
m\left(\frac{du}{dt} - rv + qw\right) &= F_x, \\
m\left(\frac{dv}{dt} - pw + ru\right) &= F_y, \\
m\left(\frac{dw}{dt} - qu + pv\right) &= F_z, \\
\frac{dh_x}{dt} - rh_y + qh_z &= M_x, \\
\frac{dh_y}{dt} - ph_z + rh_x &= M_y, \\
\frac{dh_z}{dt} - qh_x + ph_y &= M_z,
\end{aligned}
\tag{3}
$$

where (p, q, r), (u, v, w), (F_x, F_y, F_z), (M_x, M_y, M_z), and (h_x, h_y, h_z) denote the scalar components of the vectors $\boldsymbol{\omega}$ (angular velocity about the center of mass O), $\mathbf{V}$ (speed of mass center relative to the earth), $\mathbf{F}$ (resultant of external forces), $\mathcal{M}$ (resultant of external moment), and $\mathbf{h}$ (total angular momentum), respectively.

REFERENCES

1. B. Etkin, *Dynamics of Flight: Stability and Control.* Wiley, New York, 1959.
2. H. Glauert, "A Non-Dimensional Form of the Stability Equations of an Aeroplane," *Aeronaut. Res. Council Rept. & Memo.* **1093**, 1927.
3. J. B. J. Fourier, *Théorie analytique de la chaleur*, 1822.
4. L. I. Sedov, *Similarity and Dimensional Methods in Mechanics* (English transl. from the 4th Russian ed., 1957). Academic Press, New York, 1960.
5. G. Birkhoff, *Hydrodynamics: A Study in Logic, Fact and Similitude* (rev. ed.). Princeton Univ. Press, Princeton, New Jersey, and Oxford Univ. Press, London, 1960.
6. E. Buckingham, "On Physically Similar Systems," *Phys. Rev.* **4**(1914), 354–376.
7. E. Buckingham, "Model Experiments and the Forms of Empirical Equations," *Trans. Amer. Soc. Mech. Eng.* **37**(1915), 263–296.

Chapter 3

GENERALITIES AND SOME MATHEMATICAL AIDS

3.1. THE LIAPUNOV STABILITY: SHORT-DURATION DISTURBANCES

The general theory developed along the line of Liapunov's decisive contribution [1] will be resorted to in this book for the problems of flight stability. In this theory as well as in other mathematical theories of the stability of motion, a fundamental role is assigned to disturbances of limited duration, whose effect (i.e., the induced motion) is investigated from the moment their action ceases. In this way, in the chosen mathematical model the disturbance will appear only as a modification of the initial conditions of the solution, and not of the equations themselves, the disturbed motions being described by the solutions of the same system, for instance (1.6), that are different from the trivial one $x = 0$. The initial condition x_0 of such a solution will represent the instantaneous disturbance suffered by the basic motion. The basic motion $x = 0$ will be stable if deviations x are kept within admissible limits over the whole interval of time subsequent to the action of the disturbance for sufficiently small values of the initial conditions x_0. The basic motion $x = 0$ will be asymptotically stable if it is stable and if, in addition, the deviations will die out in time.

The respective definitions will be stated in accordance with the theory of Liapunov stability [1]. (The concepts further defined here were introduced by Liapunov and his followers. For general information, in addition to Liapunov's memoir [1], see [2–4] and Chapter 1 of [5].)

Let the system (1.6) be considered. If t_0 is the initial moment coinciding with the instant when the action of the disturbance ceases, then a certain solution of system (1.6) corresponding to an initial condition x_0 will be denoted by $x(t; t_0, x_0)$.

DEFINITION 1. The trivial solution $x(t; t_0, 0) \equiv 0$ describing the basic motion is stable whenever for any $\varepsilon > 0$ there exists an $\eta(t_0, \varepsilon) > 0$ such that the inequality $|x_0| < \eta$ implies $|x(t; t_0, x_0)| < \varepsilon$ for $t \geqslant t_0$.

DEFINITION 2. The stability is said to be uniform if η is independent of t_0.

The requirement of resorting to the concept of uniform stability is connected with the practical necessity of providing a sufficiently large region of attraction (η), whatever the instant t_0 when the disturbance occurs, since this instant is evidently random in relation to the basic motion.

DEFINITION 3. If the solution $x = 0$ is stable, and if for sufficiently small initial conditions

$$\lim_{t \to \infty} x(t; t_0, x_0) = 0,$$

it is said to be asymptotically stable.

To provide a region of attraction and a damping that are acceptable no matter what the initial instant t_0, the concept of uniform asymptotic stability of the trivial solution of the system (1.6) will be resorted to.

DEFINITION 4. The solution $x = 0$ is uniformly asymptotically stable if it is uniformly stable, and if, in addition, for sufficiently small initial conditions, $x(t; t_0, x_0)$ tends to zero uniformly with respect to t_0, as $t \to \infty$.

To emphasize the main quantitative characteristics of asymptotic stability, that is, the size of the region of attraction and the velocity of damping, which are sometimes required to be estimated, we give the following definition, equivalent to Definition 4.

DEFINITION 4′. The solution $x = 0$ is uniformly asymptotically stable if there exist a number δ and two functions $\eta(\varepsilon)$ and $T(\varepsilon)$ such that $|x_0| < \eta$ implies $|x| < \varepsilon$ for $t \geqslant t_0$, and the inequalities $|x_0| < \delta$ and $t \geqslant t_0 + T(\varepsilon)$ entail $|x(t; t_0, x_0)| < \varepsilon$.

The fundamental problem of proving whether a given solution is stable, uniformly stable, and so on, according to the foregoing definitions will be considered in Section 3.4.

3.2. STABILITY UNDER PERSISTENT DISTURBANCES

The foregoing section defines the concepts of stability, asymptotic stability, and uniform asymptotic stability of a given solution of a system of ordinary differential equations when the initial conditions are perturbed as these concepts are used today in the theory of Liapunov stability. In the general (i.e., mathematical) theory of motion stability this case, which models the actual case of perturbations of short duration, is considered fundamental. Many other cases of perturbation reduce or are referred to this one.

As mentioned in Chapter 1, the disturbances defined as causes inducing the modification of the basic motion by changing the flight conditions prescribed by the program effectively act during the whole motion.

Let it be noted that the distinction between the given external forces and the forces of perturbation is not a rigid one. Actually, the latter category includes all those external forces whose laws of variation are not sufficiently well known at a given stage to be included in the program, and to be modeled mathematically in dynamic terms. Perturbation forces are thus considered to include those due to: atmospheric turbulence, in the case of an atmospheric flight; irregularities of engine operation; errors of piloting; and the like. However, attempts were made to treat some of these forces dynamically, as for instance the forces induced by gusts, for which some adequate modeling was sought (see [6, Chapter 10]). For this reason it is fitting to define persistent disturbances with reference to the adopted model as the aggregate magnitudes neglected in making the model. Now, it is obvious that a fundamental criterion for the validity of the model is the condition that the omissions should not affect the results the model helps to obtain. In the given case, particularly, because the mathematical model is a system of differential equations, it is necessary that the deviations inherent in writing the system not induce inordinately high deviations from the actual situation. In this way we are led to consider the stability of a given solution of the system for perturbations not only of the initial conditions, but also of the system itself, and therefore to consider the stability under persistent perturbations or disturbances.

For this, the perturbed system

$$\frac{dx}{dt} = \bar{X}(t, x),$$

which will be written in the form

$$\frac{dx}{dt} = X(t, x) + R(t, x) \tag{3.1}$$

where

$$R(t, x) = \bar{X}(t, x) - X(t, x),$$

will be considered in addition to system (1.6). Let $\bar{x}(t; t_0, x_0)$ be the solution of this system, corresponding to the initial condition $x(t_0; t_0, x_0) = x_0$.

DEFINITION 5. The solution $x = 0$ of system (1.6) is said to be uniformly stable under persistent disturbances if there exist two functions $\rho(\varepsilon)$ and $\eta(\varepsilon)$ such that for any $R(t, x)$ satisfying the inequality $|R(t, x)| < \rho$ for $|x| < \varepsilon$ and $t \geqslant t_0$, and for any initial condition x_0 in the region $|x_0| < \eta$, the inequality $|\bar{x}(t; t_0, x_0)| < \varepsilon$ will hold for $t \geqslant t_0$.

DEFINITION 6. The solution $x = 0$ of system (1.6) is uniformly asymptotically stable under persistent disturbances if it is uniformly stable under persistent disturbances in accordance with the preceding definition and if, in addition,

$$\lim_{t \to \infty} \bar{x}(t; t_0, x_0) = 0.$$

This means that there is a positive number δ and a positive function $T(\varepsilon)$ for which the inequalities $|x_0| < \delta$ and $t \geqslant t_0 + T(\varepsilon)$ imply $|\bar{x}(t; t_0, x_0)| < \varepsilon$.

3.3. SOME THEORETICAL ASPECTS OF THE RELATION BETWEEN STABILITY AND CONTROL

It was mentioned in Chapter 1 that any flight program is achieved by means of controls, and that a basic control program corresponds to the basic motion prescribed by the program. Let us now assume a set of stationary trimmed motions of the aircraft. Each one of these motions will be described by two sets of constant magnitudes, one representing the values of the state variables, usually referred to as *flight régime*, and the other the values of the control variables (particularly the trim positions of the controls) under standard reference conditions, both corresponding to the given stationary trimmed motion.

Let x_i and u_j be constant column vectors representing the flight régime and the trim positions of the controls, respectively. The passage from a flight régime x_i to another flight régime x_{i+1} is obviously achieved by a modification of the setting of the controls. We will consider first the case of two neighboring régimes.

1. Let x_1 and x_2 be two close flight régimes, the machine having to pass from x_1 to x_2. In this case, that is, if the values of the difference $x_2 - x_1$ are sufficiently small and if the stationary trimmed motion described by x_2 is in addition asymptotically stable, the passage from x_1 to x_2 can be

achieved by a sudden, that is, discontinuous, modification of the control setting from u_1 to u_2. As for the mathematical model, the instantaneous passage from the control setting u_1 to u_2 is tantamount to a modification of the system of equations. The system admitting $x = x_1$ as a singular point is superseded by a new system having $x = x_2$ as the singular point. For the latter system, x_1 becomes the initial condition, and $x_2 - x_1$ the instantaneous disturbance. If the stationary trimmed motion $x = x_2$ is asymptotically stable, the disturbed motion induced by the disturbance $x_2 - x_1$ represents the transitory régime leading from the steady state x_1 to the steady state x_2 when $x_2 - x_1$ is small enough. The characteristics of the control will depend on those of the disturbed motion: heavy damping of the disturbances signifies a ready response to controls and so on. The connection between stability and handling characteristics appears thus to be obvious: good dynamic stability characteristics are required not only for minimizing the necessity of stabilizing intervention of the controls but also for providing an acceptable response to them.

Consider an illustrative example concerning aircraft stability. A requirement imposed upon the longitudinal stability characteristics of an aircraft is to secure in free flight a sufficiently rapid damping of the disturbance of the incidence angle ("rapid incidence adjustment") by the inherent qualities of the machine. The requirement is connected with the fact that in machines with a sufficiently great static stability (i.e., a high enough value of the coefficient of static stability, or a large enough stick-fixed static margin), the variation of the deviation of the angle of incidence induced by a disturbance in all the usual régimes is a high-frequency oscillation, which renders it difficult if not impossible to achieve stabilization by controls. Damping of the disturbance must thus be secured by the inherent qualities of the uncontrolled aircraft. The same requirement for sufficiently heavy damping of the induced deviation of incidence in free flight is also imposed on grounds of handling qualities connected with the longitudinal response of the aircraft to elevator control. Essential for the longitudinal response is the motion about the center of mass of the aircraft; the incidence must tend rapidly enough toward an equilibrium (trimmed) value corresponding to the new position of the stick (i.e., the new elevator setting) so that the transitory process does not affect the flight speed. The quantitative aspect of the problem describing the rate of damping obviously pertains to the field of flight practice. A compromise has to be made since a too-rapid response can cause piloting difficulties by rendering the aircraft oversensitive.

2. Let us now consider the case of the transition from a steady state x_1 to another x_2 that is not in the vicinity of x_1. In this case a continuous control $u(t)$ will have to be introduced by means of which to reach the régime x_2 from x_1. In the first place we have to settle the issue of the existence of a control function $u(t)$ that will achieve the transition to x_2. The existence of $u(t)$ with the stated property is not a priori obvious. It is, indeed, possible that the controls at hand would not permit actually reaching the régime x_2, but only approaching it (i.e., getting arbitrarily close to x_2).

In general, the function $u(t)$ performing the transition from x_1 to x_2, or close to x_2, is not uniquely determined. The problem consists, then, in choosing a certain function $u(t)$ that should achieve this transition under the most favorable conditions. It is a problem of finding the optimal control policy.

Under the conditions of the transition from a steady state x_1 to another distant one x_2, the problem of stability takes the following two aspects.

(i) From the instant when the machine, constrained by the control $u(t)$, has reached close to the régime x_2, let us say the régime x_3, the problem reduces to a case similar to that of the neighboring régimes discussed above. The difference, which in the given case is not essential, is that the initial régime x_3 is not necessarily stationary. The essential condition is that $x_2 - x_3$ be bounded with a sufficiently small upper bound. The uniform asymptotic stability of the stationary motion corresponding to this régime then appears as a necessity for reaching the régime x_2.

(ii) To ensure the accuracy of maneuver, and to minimize the correction controls superposed on the basic control $u(t)$, it is obviously necessary to have the stability of the basic motion throughout, that is, including the transitory interval from x_1 to x_2 (or x_3). The unconventional problem of investigating the stability of an unsteady motion is thus posed (see Chapters 8–10). Naturally, certain stability characteristics (e.g., stability in "fixed controls" or in "partial control" conditions) are required from the stationary motion x_2 also, even if the control $u(t)$ achieves its object.

3.4. HOW TO INVESTIGATE STABILITY

3.4.1. Introductory Remarks

The fundamental concepts of stability in the sense of Liapunov have been discussed in Sections 3.1 and 3.2. As is but natural, in the course of the history of mechanics and mathematics from Aristotle to our own times numerous theoretical models of the empirically known property of stability of a state of motion (and of equilibrium in particular) have been developed. (By "stability of a state of motion" is understood either only a weak receptivity to the actions of fortuitous disturbances, or the damping of their effect as well.) Many mathematical definitions and schemes are still being put forward as covering most exactly the various physical aspects of stability. At present, however, stability as defined by Liapunov enjoys increasing attention. Important mathematical schools are concerned with the development of this theory, which was created more than 70 years ago. The outstanding viability of Liapunov's concept of stability is attributable, on the one hand, to its inherent properties as a model fitting number of meaningful problems (with all the reservations that will be stated in subsequent sections) and, on the other, to its highly general techniques, which make it possible to evaluate the stability.

The essential problem consists in the possibility of investigating stability (or to quote Etkin [6, page 191], the "'yes' or 'no' evaluation of the stability."). If we are not able to do this, any definition, however adequate, will be meaningless. This is not only a significant but also a particularly difficult problem. The systems of linear equations with constant coefficients are the only models for which it was solved completely. (It is noteworthy that in this case the various definitions of stability generally coincide.) For this simple case it was possible to give the necessary and sufficient stability conditions in the form of easy to handle practical criteria (see Sections 5.1 and 5.2, as well as the Appendix to Chapter 5). These criteria do not imply the imposition of certain restrictive conditions relating to the magnitude of the disturbances: their satisfaction secures the stability *in the large* (global stability). Moreover, the investigation of solutions of linear systems with constant coefficients under all aspects including that of stability is always possible since these systems can be integrated in a closed form. The criteria mentioned above permit us to make certain deductions about the stability of a particular solution without requiring that we integrate the system (i.e., in a less laborious way). As will be seen, for instance, in Section 5.3, they thus enable us to analyze the effect of various parameters on stability by comparatively simple procedures.

3.4.2. Stability by the First Approximation

To extend the validity of results obtained by using the linear model with constant coefficients, methods were developed for investigating the stability by the first approximation. According to these methods, from the point of view of stability systems (1.8) can be replaced by systems (1.9). The more general problem of stability by the first approximation of the solution of the systems depending explicitly on the independent variable was subsequently considered. In what follows, the fundamental theorem of stability by the first approximation will be stated for the general case of nonautonomous systems (depending explicitly on t) when the first-approximation system is linear.† In Chapter 5, Section 5.2, the theorem will be applied to the particular case of linear first-approximation systems with constant coefficients, and a theorem of instability by the first approximation will be stated.

Consider system (1.8) with matrix $A(t)$ bounded and $|N(t, x)| \leqslant \lambda |x|$, $\lambda =$ const., and the first-approximation system linear with respect to x (1.9). The theorem is as here.

THEOREM. *If the solution $x = 0$ of system* (1.9) *is uniformly asymptotically stable, then for a sufficiently small number λ the zero solution of system* (1.8) *is uniformly asymptotically stable also.*

For general systems of the form (1.9), no practical stability criteria are known, as when matrix A is constant. Nevertheless, linear systems with respect to x of the type (1.9) are as a rule easier to handle than systems like (1.8), since certain general properties of systems linear with respect to x with useful implications in practice are known. This assertion will be illustrated in Chapter 8, which is devoted to a discussion of certain problems of flight stability in a nonstationary régime.

The fact will finally be emphasized that the theory of stability by the first approximation secures only local stability; that is, it is valid only for sufficiently small instantaneous disturbances (initial conditions). In conclusion, under these conditions the uniform asymptotic stability of the first-approximation system implies the uniform asymptotic stability of the nonlinear system.

3.4.3. Method of the Liapunov Function

Essentially significant for studying stability is the method of the Liapunov function, which consists in seeking for each problem a function with certain

† It is noteworthy that a system by the first approximation is not necessarily linear, and the terms that can be ignored are not necessarily of a higher degree as compared to the first-approximation system. For the applications this book is concerned with, it is enough to consider the linear first-approximation systems.

properties. Its more detailed account, which is beyond the scope of this book, can be found in Liapunov's classical memoir [1] and in every treatise or monograph devoted to the general theory of stability (e.g., Chapters 2, 5, and 6 of [2]; Sections 8–12 of [4]; Chapter 1 of [5]). Certain elements will be given here that are prerequisite to the reading of those sections of the book in which this method is used.

At first, definitions will be given for the concepts of functions of definite and semidefinite signs, and for the concept of total derivative with respect to the independent variable of a function along the integral curves of a system.

The functions $\mathscr{V}(x)$ and $\mathscr{W}(t, x)$ defined in a certain neighborhood $|x| < h =$ const. of the origin on the semiaxis $t \geqslant 0$ of the independent variable, continuous together with their first partial derivatives in the range of definition, will be considered in the sequel. (Henceforth, by range of definition will be understood either the conditions $|x| < h =$ const., $t \geqslant 0$, or only one of them, according to the circumstances.)

A function $\mathscr{V}(x)$ having the foregoing properties is said to be positive definite if it can take only positive values within the definition range, and vanishes exclusively for $x = 0$. The function $\mathscr{V}(x)$ is positive semidefinite if, without changing its sign, and vanishing for $x = 0$, it can also be canceled when $x \neq 0$. Similar definitions apply for the negative definite and semidefinite functions, respectively.

Let $\mathscr{V}_1(x)$ be a positive definite function in the region $|x| < h =$ const. The function $\mathscr{W}(t, x)$, ($\mathscr{W}(t, 0) = 0$), is said to be positive definite if in the respective definition range $\mathscr{W}(t, x) \geqslant \mathscr{V}_1(x)$, and is said to be negative definite if $\mathscr{W}(t, x) \leqslant -\mathscr{V}_1(x)$.

Consider now the system

$$\frac{dx}{dt} = X(t, x). \tag{3.2}$$

By the total time derivative of the function $\mathscr{W}(t, x)$ along the integral curves of system (3.2) is meant the function whose expression is

$$\left(\frac{d\mathscr{W}}{dt}\right)_{(3.2)} = \frac{\partial \mathscr{W}}{\partial t} + \frac{\partial \mathscr{W}}{\partial x} X(t, x).$$

The fundamental stability and instability theorems of the method of the Liapunov function used in this book will now be stated.

THEOREM 1. *If there exists a positive definite function* $\mathscr{W}(t, x)$ *defined for* $t \geqslant 0$ *and* $|x| \leqslant h = \text{const.}$ *such that its total time derivative along the integral curves of system* (3.2) *is negative semidefinite or identically zero, then the zero solution of* (3.2) *is stable.*

If in addition $\mathscr{W}(t, x)$ *is dominated for* $t \geqslant 0$ *and* $|x| < h = \text{const.}$ *by a positive definite function, independent of t,* $\mathscr{V}_2(x)$, $\mathscr{V}_2(x) \geqslant \mathscr{V}_1(x)$, *the stability is uniform.*†

THEOREM 2. *If there exists a positive definite function* $\mathscr{W}(t, x)$ *defined as in Theorem* 1 *such that its total time derivative along the integral curves of system* (3.2) *is negative definite, then the trivial solution of* (3.2) *is asymptotically stable.*

Again, if in addition to the properties above, $\mathscr{W}(t, x) \leqslant \mathscr{V}_2(x)$, *the zero solution of* (3.2) *is uniformly asymptotically stable.*

THEOREM 3. *If there exists a function* $\mathscr{W}(t, x)$ *such that its total time derivative along the integral curves of system* (3.2) *is negative* (*or positive*) *definite and the function* $\mathscr{W}(t, x)$ *itself is not positive* (*or negative*) *semidefinite, the trivial solution of system* (3.2) *is unstable.*

Thus, to investigate the stability of a system we seek to work out a function $\mathscr{W}(t, x)$ having the properties required by Theorem 1 or 2, which is used to derive practical criteria giving sufficient stability conditions from the conditions $d\mathscr{W}/dt \leqslant 0$ or $d\mathscr{W}/dt \leqslant -\mathscr{V}(x) < 0$. (By $d\mathscr{W}/dt$ is denoted the derivative of $\mathscr{W}$ along the integral curves of the system considered, and by $\mathscr{V}(x)$ a positive definite function.) The method is general, but difficulties are encountered in its application for finding such Liapunov functions as will be most convenient. The seeking of functions that will give criteria of the least restrictive character requires a certain ingenuity, but mostly a certain skill that comes from experience. A more intimate knowledge of the modeled physical process can facilitate the finding of the Liapunov function that is closest to the optimum one.

3.4.4. Theorems of Stability under Persistent Disturbances

So far, the fundamental problem of studying Liapunov stability has been considered for the case of instantaneous disturbances. The following two

† In Liapunov's terminology it is said that the function $\mathscr{W}(t, x)$ dominated by a positive definite function $\mathscr{V}_2(x)$: $\mathscr{W}(t, x) \leqslant \mathscr{V}_2(x)$ admits an "infinitely small upper bound." In this case $\mathscr{W}(t, x)$ tends toward zero together with x uniformly with respect to t. In other words, whatever $\varepsilon > 0$ might be, there exists a function $\nu(\varepsilon)$ such that $t \geqslant 0$ and $|x| < \nu$ imply $|\mathscr{W}(t, x)| < \varepsilon$.

theorems, originally proved by I. G. Malkin [7], establish the relation between stability under persistent disturbances and stability when initial conditions are perturbed (disturbances of short duration).

The disturbed system (3.1) is considered now in addition to system (3.2).

THEOREM 1. *If the zero solution of system* (3.2) *is uniformly asymptotically stable in accordance with Definition* 4 *or* 4′, *it is stable under persistent disturbances.*

THEOREM 2. *It is further assumed that the function* $R(t, x)$ *satisfies a Lipschitz condition in the vicinity of* $x = 0$, *and that* R *tends to zero as* $t \to \infty$ *uniformly with respect to* x, *when* x *is sufficiently small. Under these conditions the uniform asymptotic stability of the trivial solution of system* (3.2) *implies its uniform asymptotic stability under persistent disturbances* (*see* [5, *page* 104]).

Several implications of these significant theorems will be considered in this book.

3.5. THE NECESSITY OF NUMERICAL ESTIMATES; THE CONCEPT OF PRACTICAL STABILITY

In certain problems of flight dynamics it is sometimes enough to prove the stability (particularly the asymptotic one) or instability of given motions. Stability criteria deduced from the general theory will be used to this end. Most frequently, however, the simple recognition of the properties defined in the preceding sections is not sufficient for a satisfactory knowledge of the flying qualities of a machine.

For practical purposes, indeed, it is not sufficient to make sure that there exists a region of the initial conditions corresponding to an arbitrary upper bound of the disturbance-induced deviations, or that the effect of the disturbance finally dies out. It is generally interesting to know how great is the magnitude of the region of the initial conditions corresponding to a given upper bound of the value of deviations (viz., if this region is not somehow exceeded by current disturbances) or how long it takes for the induced deviation to decrease below a given negligible value. Consideration of the quantitative aspects is significant for the case of instability also. It is a well-known fact that not every instability of free flight is dangerous, and it would be interesting to estimate the rate of increase of deviation amplitude, the oscillation frequency, and the like. In the light of these facts we will now examine the definitions given in the previous sections.

From the first definition it follows that motion is stable toward instantaneous disturbances if to an arbitrary upper bound ε of induced deviations there corresponds an upper bound η, depending on ε, of the initial disturbance implying the inequality $|x(t)| < \varepsilon$ secure for $t \geqslant t_0$. According to Definition 5, to any upper bound ε of the value of deviations x for persistent disturbances there corresponds, in addition to a maximum admissible region of initial conditions, a region for the variation suffered by the equations, that is, an admissible region of the value of the persistent disturbance, $|R(t, x)| < \rho(\varepsilon)$. According to Definitions 4 and 6, uniform asymptotic stability implies, moreover, the existence of a time interval $[t_0, T(\varepsilon)]$ after the elapsing of which the value of deviations should become smaller than ε if the initial conditions are sufficiently small ($|x_0| < \delta$). Thus, the defined properties of stability, asymptotic stability, and so on refer to the possibility of imposing an arbitrary upper bound ε to the value of deviations. However, in applications from the field of flight dynamics as in other practical applications, the upper bound of deviations is not "however small," but is set according to the specific character of the problem. Various operation requirements, including flight safety, comfort conditions, the accuracy of programmed motions, maneuverability, and others, impose certain more or less precise values on the upper bound ε. To be able to foresee whether for a given flight machine under set operating conditions it is possible to keep the disturbance-induced deviations within these limits, or to select aerodynamical and structural design characteristics securing the fulfillment of this condition, it becomes necessary to estimate the quantities η, ρ, T, corresponding to the given values of ε. Thus, for instance, $\eta(\varepsilon)$ and $\rho(\varepsilon)$ are estimated in order to infer a safety limit of gust gradient. (ε represents admissible limits, from the point of view of safety requirements, of the magnitude of deviations of kinematic parameters from their undisturbed values. An estimate of the difference $T(\varepsilon) - t_0$ leads to a lower bound of the time interval after the elapsing of which the effect of disturbance can be considered practically damped. In this case the components of ε represent magnitude orders of the "perceptible" lower bound of deviations.) General techniques were developed for such estimates. Some of them will be outlined in the following section.

In consideration of the foregoing statements, certain doubts can arise concerning the practical value of the concepts defined in the previous sections. That these definitions require the fulfillment of certain properties for any positive number ε might give rise to the question whether it does not imply conditions that are too restrictive, taking into account that the respective properties are necessary only for certain given values, or at most for a certain range of positive numbers.

There is no single answer valid for all cases. For certain cases it is, indeed, possible to weaken the conditions by taking into account that for practical purposes ε need not be however small (and even zero, particularly). This has induced certain authors to define a concept of practical stability, or ε_0-stability (see [5, page 216]) and asymptotic ε_0-stability. Such concepts differ from the corresponding conventional ones presented in the preceding sections in that the functions $\eta(\varepsilon)$ and $T(\varepsilon)$ are defined for $\varepsilon \geqslant \varepsilon_0 > 0$.† However, no general method was developed for constructing certain criteria of practical stability. (Theorems concerning practical stability on the basis of which criteria of practical stability might be inferred were given only for sets of equations of a particular form [8].) This gap seems not to affect to any great extent the theory of flight dynamics. Actually, in almost all the mathematical models to which this discipline has recourse, consideration of ε_0-stability instead of stability in the conventional sense would not result in a weakening of the respective conditions. Indeed, in the case of the sets of equations utilized, as for instance the particular one of the linear and quasilinear systems, the existence of ε_0-stability necessarily implies that of conventional stability (ε_0-stability can occur only when stability in the sense of Liapunov is secured, i.e., by considering ε arbitrary).

3.6. ESTIMATING TECHNIQUES

3.6.1. Preliminary Remarks

It is evident from Section 3.4 that for the stability of the solution of some system to secure the stability of the solution of another system, obtained by a modification (such a modification being obtained by addition of either certain terms of a higher degree in relation to the variables, or certain terms representing persistent disturbances) of the first, the said system must have in a certain sense a margin of dynamic stability. Such a margin is furnished

† In [2], a simple illustrative example is given: the trivial solution of the equation $dx/dt = \alpha^2 x - x^3$, with constant α, is obviously unstable in the Liapunov sense, but it is ε_0-stable even in the large. Indeed, if $x(t)$ is a solution, $\lim_{t\to\infty} x(t) = \pm\alpha$. Hence, however small the initial value of x, $|x(t_0)|$ might be, after a sufficiently large lapse of time, $|x(t)|$ becomes greater than ε if $\varepsilon < |\alpha|$. In exchange, for $\varepsilon_0 \geqslant \alpha$ the trivial solution $x = 0$ is ε_0-stable, however large the initial disturbance $x(t_0)$ might be.

by the property of uniform asymptotic stability of the solution. It was shown in the preceding section that for practical purposes, however, it is not sufficient to prove the existence of a margin of dynamic stability alone. Its amount also has to be estimated. Thus, for a given system it is useful to estimate the allowable upper bound of the value of additional terms (e.g., of nonlinear terms or persistent disturbances), depending on the given margin of dynamic stability of the linear system, while the required margin of dynamic stability is sought for given additional terms or for their given order of magnitude. The persistent disturbances expressed by additional terms, which may or may not vanish in the origin, can be of different natures. For instance, they can represent the effects on the system of certain random processes, such as atmospheric turbulence, the secondary effects of controls (see Appendix I to [9] as well as Sections 4.3 and 5.7 of this book), the errors implied by the mathematical model used, small actual variations, depending on t, of the parameters of the system modeled as constants (see Chapter 8) and others. Two techniques will be given in this section for their estimation, one based on Lagrange's method of variation of parameters, and the other on the method of the Liapunov function.

3.6.2. Estimate of Damping Velocity of Disturbances

The damping velocity of the disturbance can be used, for instance, as a measure of the dynamic stability margin. (By the conventional designation of damping velocity of the deviation x_1 is meant the ratio of the instantaneous value of the derivative dx_1/dt to the value of the deviation itself.)

For linear systems with constant coefficients this is expressed by means of the greatest real part of all their characteristic roots (see Chapter 5).

In the case of nonautonomous systems, if an adequate Liapunov function is known, the damping velocity can be estimated as shown in the following.

Let us assume that a Liapunov function $\mathscr{W}(t, x)$ was found for the system (1.6) by means of which the uniform asymptotic stability of the trivial solution of the system was proved in consideration of Theorem 2 of Section 3.4. The function $\mathscr{W}(t, x)$ therefore satisfies the double inequality $\mathscr{V}_1(x) \leqslant \mathscr{W}(t, x) \leqslant \mathscr{V}_2(x)$ within the range of definition, and the total time derivative of $\mathscr{W}$ along the integral curves will be dominated by a negative definite function: $(d\mathscr{W}/dt)_{(1.6)} \leqslant -\mathscr{V}_3(x)$; $\mathscr{V}_1$, $\mathscr{V}_2$, and $\mathscr{V}_3$ being positive definite functions. Obviously, there exists a positive constant ν such that in the definition range ($|x| < h$) the inequality $\mathscr{V}_3 \geqslant 2\nu\mathscr{V}_2$ holds for any x. It can be seen at once that the damping velocity of the disturbance can be estimated by means of the number ν. Indeed, by assuming that such a constant has been found, the following estimate may be stated.

$$\frac{d\mathscr{W}}{dt} \leqslant -2\nu\mathscr{V}_2$$

and therefore

$$\frac{d\mathscr{W}}{dt} \leqslant -2\nu\mathscr{W}$$

or

$$\mathscr{W} \leqslant \mathscr{W}(t_0) \exp[-2\nu(t - t_0)].$$

Let x_i, $i = 1, 2, \ldots, n$, denote the components of the vector x. In the hyperspace x_i, $\mathscr{W}(t, x_1, \ldots, x_n) = C$ represents a set of closed moving hypersurfaces containing the origin, tending toward the origin as $x_1 \to 0$, $x_2 \to 0, \ldots, x_n \to 0$ uniformly with respect to t. As a result of asymptotic stability the integral curve of the system will intersect these surfaces from the exterior toward the interior, approaching the origin at least with the velocity imposed by the last inequality. Let $\mathscr{W}$ denote a quadratic form with time-dependent coefficients. Therefore, $\mathscr{W} = C$ will represent a moving hyperellipsoid. In this case the moduli of x_i are bounded for all $t \geqslant 0$ by the semiaxes of the hyperellipsoid decreasing with a velocity at least equal to the exponential function $\exp[-\nu(t - t_0)]$. Hence it will be possible to estimate the damping velocity of the induced deviations x by means of the number ν.

3.6.3. Estimate of Additional Terms: First Procedure

System (3.1) is written in the form

$$\frac{dx}{dt} = Ax + N(t, x) + R(t, x) \tag{3.3}$$

defined for $t \geqslant t_0$ and $|x| < h$, where h denotes a positive constant. It is assumed that within the range of definition the function $N(t, x)$ satisfies a Lipschitz condition $|N(t, x)| < l|x|$, $l =$ const., and vanishes if $x = 0$, for any $t \geqslant t_0$. The function $R(t, x)$ is assumed bounded in the same region $|R(t, x)| < \rho =$ const., $t \geqslant t_0$, $|x| < h$. Generally, the function $R(t, x)$ representing persistent disturbances does not vanish if $x = 0$.

Let ν_i, $i = 1, 2, \ldots, n$ (n standing for the order of the system), denote the real parts of the eigenvalues of constant matrix A, and let ν be the greatest of the magnitudes ν_i. It is assumed that $\nu < 0$ (ν represents the dynamic stability margin in the sense shown above), and therefore the solution $y = 0$ of the system

$$\frac{dy}{dt} = Ay \tag{3.4}$$

obviously is uniformly asymptotically stable. Then, according to the stability theorem by the first approximation (see Section 5.2) for a sufficiently small Lipschitz constant l, the zero solution of the system

$$\frac{dx}{dt} = Ax + N(t, x)$$

will also be uniformly asymptotically stable; whence, according to Malkin's first theorem, stated in Section 3.4, for a sufficiently small ρ it will be stable under persistent disturbances.

Henceforth, estimating relations will be established among the initial condition (x_0), l, ρ, and ν for which:

(i) the deviations induced by the initial disturbance x_0 and by the persistent disturbances R, $|R| < \rho$, will remain smaller than a given quantity ε along the whole semiaxis $t \geqslant t_0$; or

(ii) these deviations remain smaller than ε within a certain interval of time $[t_0, T]$.

The first procedure to be described is based on Lagrange's method of the variation of parameters. (A similar technique is used by Bellman within the framework of the theory of stability by the first approximation [10].)

At first Gronwall's well-known lemma [11] will be stated in a more general form (see, e.g., [12, page 10]):

LEMMA. *If the functions* $x(t)$, $\varphi(t) \geqslant 0$, *and* $k(t) \geqslant 0$, *continuous in the interval* $[t_0, t_1]$, *satisfy within this interval the inequality*

$$x(t) \leqslant \varphi(t) + \int_{t_0}^{t} k(s)x(s)\, ds,$$

then the relation

$$x(t) \leqslant \varphi(t) + \int_{t_0}^{t} \varphi(s)k(s) \exp\left(\int_{s}^{t} k(\tau)\, d\tau\right) ds$$

will hold within the same interval.

Now, according to [10], the differential equation (3.3) will be transformed into an integral equation by using the method of variation of parameters. The solution of the vector equation (3.4) can be expressed in a form analogous to that of the scalar equations

$$y = e^{AT}y(0)$$

where e^{AT} represents the solution of the matrix equation

$$\frac{dY}{dt} = AY$$

corresponding to the initial condition $Y(0) = I$ (I being the unit matrix). For the validity of operations to be performed on matrix e^{At} by analogy with the scalar exponential function $e^{\lambda t}$, see [10, Chapter 1, Section 6].

Let $y(0) = x(0) = x_0$. Hence

$$y = e^{AT}x_0.$$

According to the method of variation of parameters, equation (3.3), by the substitution $x = e^{At}u(t)$, yields

$$x = \exp(At)x_0 + \int_0^t \exp[A(t-s)][f(s, x(s)) + R(s, x(s))]ds. \tag{3.5}$$

Equation (3.5) is the integral equation that will serve as a point of departure for estimations. We readily have the relation

$$|x| \leqslant ||\exp(At)||\, |x_0| + \int_0^t ||\exp[A(t-s)]||(|f(s, x(s))| + |R(s, x(s))|)\, ds.$$

The norm of matrix e^{At} can be dominated thus

$$||e^{At}|| \leqslant Ke^{\nu t}$$

where K is a constant depending on the system parameters through the intermission of the elements of matrix A. Then, by considering also the conditions imposed on the functions N and R, the relation

$$|x| \leqslant K \exp(\nu t)|x_0| + \int_0^t K \exp[\nu(t-s)](l|x| + \rho)\, ds,$$

holds. Hence

$$|x| \leqslant \frac{K\rho}{-\nu} + K \exp(\nu t)\left(|x_0| - \frac{\rho}{-\nu}\right) + Kl \int_0^t \exp[\nu(t-s)]\, |x|\, ds.$$

For application of the above-stated lemma to be possible, both members of the inequality are multiplied by $e^{-\nu t}$, with the result

$$e^{-\nu}|x| \leqslant \frac{K\rho}{-\nu} e^{-\nu t} + K\left(|x_0| - \frac{\rho}{-\nu}\right) + \int_0^t Kle^{-\nu s}|x|\, ds.$$

Then, in consideration of the above lemma, the relation

$$\exp(-\nu t)|x| \leqslant \frac{K\rho}{-\nu} \exp(-\nu t) + K\left(|x_0| - \frac{\rho}{-\nu}\right)$$

$$+ \int_0^t \left[\frac{K\rho}{-\nu} \exp(-\nu s) + K\left(|x_0| - \frac{\rho}{-\nu}\right)\right] Kl \exp[Nl(t-s)]\, ds,$$

holds, and hence, by certain simple transformations, the estimate of x is obtained in the form

$$|x| \leqslant -\frac{K\rho}{\nu + Nl} + K\left(|x_0| + \frac{\rho}{\nu + Kl}\right) \exp[(\nu + Nl)t] \cdot (\leqslant K|x_0|). \tag{3.6}$$

(i) If

$$\nu + Kl < 0, \qquad K|x_0| \leqslant \varepsilon, \tag{3.7}$$

then $|x(t)| \leqslant \varepsilon$ for any $t \geqslant t_0$.

(ii) A sufficient condition for $|x| \leqslant \varepsilon$ is given by the inequality

$$\frac{-K\rho}{\nu + Kl} + K\left(|x_0| + \frac{\rho}{\nu + Kl}\right) \exp[(\nu + Nl)T] \leqslant \varepsilon, \tag{3.8}$$

and the time interval T in which the inequality $x(t)$ is secured can be estimated by means of the relation

$$T \leqslant \frac{1}{\nu + Kl} \ln \left\{ \frac{\varepsilon + \dfrac{K\rho}{\nu + Kl}}{N\left(|x_0| + \dfrac{\rho}{\nu + Kl}\right)} \right\}. \tag{3.9}$$

3.6.4. Estimate of Additional Terms: Second Procedure

We will now outline an estimating technique derived from the method of the Liapunov function (cf. the technique used by Malkin [2] to prove the theorem of stability under persistent disturbances) that is conveniently applied also to nonautonomous systems for which an appropriate Liapunov function is known.

Consider the system

$$\frac{dx}{dt} = X(t, x) + R(t, x) \tag{3.10}$$

defined for

$$t \geqslant t_0, \qquad |x| < h \tag{3.11}$$

with function X continuous, and vanishing for $x = 0$. Function $R(t, x)$ has the properties enumerated earlier (see Section 3.4.4).

It is assumed that a function $\mathscr{V}(t, x)$ has been found, defined in the domain (3.11), vanishing for $x = 0$, and having a continuous partial derivative with respect to x, which in this domain satisfies the relations

$$ax^2 \leqslant \mathscr{V}(t, x) \leqslant bx^2 \tag{3.12}$$

and

$$\frac{\partial \mathscr{V}}{\partial t} + \frac{\partial \mathscr{V}}{\partial x} X(t, x) \leqslant -cx^2 \tag{3.13}$$

where a, b, and c are certain positive constants. Then, the solution $x = 0$ of the system

$$\frac{dx}{dt} = X(t, x) \tag{3.14}$$

is stable under persistent disturbances. The additional term $R(t, x)$ representing these disturbances will be estimated depending on the parameters of the system (3.14) and on the initial conditions such that the solution $x(t)$

of system (3.10) satisfies the inequality $|x(t)| < \varepsilon$ for any $t \geqslant t_0$, ε being a given constant. In other words, an estimate is sought that will ensure that the deviations induced by disturbances will not exceed a certain value imposed by the respective technical requirements (such as the safety and comfort of flight and the accuracy of evolution).

Let $\varepsilon < h$ be a given positive constant, and let k be another constant satisfying the inequality $k < a\varepsilon^2$. Then, by taking into consideration (3.12), for $\mathscr{V}(t, x) = k$ we have $(ax^2)_{\mathscr{V}=k} < k < a\varepsilon^2$ and hence $|x|_{\mathscr{V}=k} < \varepsilon$ and $k < \mathscr{V}(t, \varepsilon)$. Let us further assume that the initial disturbance satisfies the relation $|x_0| < \min(\varepsilon, (k/b)^{1/2})$. (By min is meant the smallest of the two bracketed values.) By taking into consideration the second inequality (3.12), we obtain $\mathscr{V}(t_0, x_0) < k$. Summing up, we have

$$\mathscr{V}(t_0, x_0) < k < \mathscr{V}(t, \varepsilon).$$

Let us consider the integral curve of system (3.10) with its origin in the point $x(t_0) = x_0$, $|x_0| < \min(\varepsilon, (k/b)^{1/2}) \leqslant \varepsilon$. For $|x(t)|$ to become equal to the given constant ε along the integral curve, it is necessary that the function $\mathscr{V}(t, x(t))$ starting from the value $\mathscr{V}(t_0, x_0)$ reach $\mathscr{V}(t, \varepsilon)$ in passing through the value k, so that for $\mathscr{V} = k$ the derivative $d\mathscr{V}/dt$ be positive along the integral curve of system (3.10). Therefore, the inequality

$$\left[\frac{\partial \mathscr{V}}{\partial t} + \frac{\partial \mathscr{V}}{\partial x}(X(t, x) + R(t, x))\right]_{\mathscr{V}=k} < 0 \tag{3.15}$$

is a sufficient condition for $|x(t)| < \varepsilon$ for any $t \geqslant t_0$. It can readily be seen that under the given conditions there exists a $\rho > 0$ such that for $|R(t, x)| < \rho$ the inequality (3.15) is possible. Indeed, from the 2nd inequality (3.12) it follows that for $\mathscr{V}(t, x) = k$, $(x^2)_{\mathscr{V}=k} \geqslant k/b$, and therefore (3.13) becomes

$$\left[\frac{\partial \mathscr{V}}{\partial t} + \frac{\partial \mathscr{V}}{\partial x} X(t, x)\right]_{\mathscr{V}=k} \leqslant -\frac{ck}{b} \neq 0,$$

whence the existence of a number ρ with the stated property follows at once in consideration of the continuity of the partial derivative $\partial\mathscr{V}/\partial x$ for $|x| \leqslant h$ and $t \geqslant t_0$.

Since (3.15) must hold whatever the sign of the additional term $R(t, x)$, and by considering the relation $|R| < \rho$, the condition can be written in the form

$$\left[\frac{\partial \mathscr{V}}{\partial t} + \frac{\partial \mathscr{V}}{\partial x}(X(t, x) + \rho)\right]_{\mathscr{V}=k} < 0, \tag{3.16}$$

or, if no more refined estimate is aimed at, in the form

$$-\frac{ck}{b} + \rho\left|\frac{\partial \mathscr{V}}{\partial x}\right|_{\mathscr{V}=k} < 0. \tag{3.17}$$

3.6.5. Estimate of Allowable Velocity of Repeated Vertical Gusts for Securing Longitudinal Stability with Respect to the Incidence

Consider the effect of persistent disturbances on the flight in an atmosphere characterized by a nonhomogeneous velocity field exerted on the quantitative characteristics of the stability of the machine [13, 14]. The purpose of the example is purely illustrative, and therefore certain simplifications will be resorted to.

Atmospheric turbulence will appear as a persistent disturbance of which it is necessary only to know an upper bound of its intensity. (Unlike such a problem setting limited to the particular aspect of stability, there exists an interesting literature devoted to the problem of the response of flight machines to the action of atmospheric disturbance described in a statistical sense. It includes the studies carried out at the Institute of Aerospace Studies of the University of Toronto, and more particularly the works of Etkin, see [6, Chapter 10], and the quoted works, as well as [15] and [16] published subsequently, and others.)

The basic motion will be assumed horizontal rectilinear and uniform (undisturbed flight velocity constant). The horizontal component ΔV_x of the induced deviation of flight velocity will be neglected, by considering only the rapid mode of the longitudinal disturbed motion. The value of the static stability coefficient is supposed to be sufficiently great for this mode to represent an oscillation of comparatively high frequency, and heavily damped owing to the comparatively great values of the rotary derivatives C_{m_q} and $C_{m_{\dot{\alpha}}}$ (as well as of the derivative C_{L_α}).

Consider the disturbed motion in "fixed controls" condition. Hypotheses conditioning the separation of the longitudinal disturbed motion from the lateral one (see Section 2.2) are supposed valid in the presence of persistent disturbances also. In particular, the local atmospheric velocity V_g is considered sufficiently small to permit the linearization of the expression of induced forces with respect to this argument.

It is admitted that the gradient of the velocity field in the atmosphere is sufficiently small for the aircraft to be treated as a point. In keeping with this supposition, it will be possible to express the forces induced by atmospheric turbulence only depending on the components of the local atmospheric velocity. By choosing a system of reference fixed to the trajectory, as in establishing the equations of motion in the longitudinal plane in Section 2.1, the scalar components of the velocity vector $\mathbf{V}_g$ include the horizontal one V_{g_x}, parallel to the path of the center of mass and hence to the plane of symmetry of the machine as well; the vertical one V_{g_z}; and the perpendicular one V_{g_y}, to the plane of symmetry of the machine. Of these, the lateral component V_{g_y} will not appear in the equations of the longitudinal disturbed motion as a result of the assumption with respect to the absence of coupling terms. The components V_{g_x} are also supposed negligible. In other words, the flight will be considered under the conditions of repeated vertical gusts. By adopting the same rule of signs for V_{g_z} as for the vertical component of the velocity of the center of mass (and not of the aerodynamical velocity, equal to it but of opposite sign), that is, $V_{g_z} > 0$ for upward vertical gusts, we define the incidence induced by the atmospheric disturbance having the vertical velocity V_{g_z} by the relation

$$\alpha_g = \tan^{-1} \frac{-V_{g_z}}{V}$$

and in consideration of the hypothesis of small disturbances

$$\alpha_g = -\frac{V_{g_z}}{V}.$$

Since the undisturbed flight velocity V is constant, we can use either $\alpha_g(t)$ or $V_{g_z}(t)$ as a measure of the intensity of atmospheric turbulence.

Aerodynamical forces and moments will suffer modifications corresponding to the modifications of the value of incidence:

$$\alpha_{\text{tot}} = \alpha + \Delta\alpha + \alpha_g$$

where α is the undisturbed incidence and $\Delta\alpha$ the incidence induced by an initial disturbance of short duration. Small disturbances permitting a linear approximation with respect to $\Delta\alpha$ and α_g being assumed, the share of the total aerodynamical force F due to persistent (continuously acting) perturbation of the incidence by the presence of an additional incidence $\alpha_g(t)$ different from zero can be written in the form of a separate term

$$F_{\text{tot}} = F + \Delta F + F_g = F + \frac{\partial F}{\partial \alpha}(\Delta\alpha + \alpha_g).$$

Then, if we neglect the deviation of the flight velocity $\Delta\hat{V}$, thus eliminating the equation corresponding to the tangential component of the longitudinal disturbed motion, and if we note that for the horizontal flight with locked controls $\Delta\gamma$ is not included in the remaining equations, we obtain from (2.16), by addition of the nonlinear terms† and of the terms corresponding to the persistent disturbance, the system

$$\frac{d\Delta\alpha}{d\hat{t}} = -\frac{1}{\mu}(\tfrac{1}{2}C_{L_\alpha} + C_T\cos(\alpha - \varphi))\,\Delta\alpha + \Delta\hat{q} + A(\Delta\alpha, \Delta\hat{q}, \ldots) - \frac{1}{V}C_{L_\alpha}V_{g_z}(\hat{t}),$$

$$\frac{d\Delta\hat{q}}{d\hat{t}} = \frac{1}{2i_B}\left[C_{m_\alpha} - \frac{1}{\mu}(\tfrac{1}{2}C_{L_\alpha} + C_T\cos(\alpha - \varphi))\right]\Delta\alpha$$

$$+ \frac{1}{2i_B}(C_{m_q} + C_{m_{\dot\alpha}})\,\Delta\hat{q} + Q(\Delta\alpha, \Delta\hat{q}, \ldots) - \frac{1}{V}C_{m_\alpha}V_{g_z}(\hat{t}). \qquad (3.18)$$

For convenience, we adopt the notation $\Delta\alpha = x$, $\Delta\hat{q} = y$,

$$-\frac{1}{\mu}(\tfrac{1}{2}C_{L_\alpha} + C_T\cos(\alpha - \varphi)) = -a,$$

$$\frac{1}{2i_B}\left[C_{m_\alpha} - \frac{1}{\mu}C_{m_{\dot\alpha}}(\tfrac{1}{2}C_{L_\alpha} + C_T\cos(\alpha - \varphi))\right] = -b,$$

$$\frac{1}{2i_B}(C_{m_q} + C_{m_{\dot\alpha}}) = -c, \quad A = X, \quad Q = Y, \quad \frac{1}{V}C_{L_\alpha} = k_1,$$

$$-\frac{1}{V}C_{m_\alpha} = k_2,$$

$$-\frac{1}{V}C_{L_\alpha}V_{g_z}(\hat{t}) = -k_1V_{g_z} = R_1,$$

$$-\frac{1}{V}C_{m_\alpha}V_{g_z}(\hat{t}) = k_2V_{g_z} = R_2.$$

† It is somewhat arbitrarily assumed that the nonlinear terms depend only on the two variables of the system, $\Delta\alpha$ and $\Delta\hat{q}$, also.

The nondimensional time will be denoted t (without a circumflex). The total derivative with respect to t will be denoted by an overdot. With these notations, system (3.18) becomes

$$\dot{x} = -ax + y + X(x, y) + R_1(V_{g_z}(t)),$$

$$\dot{y} = -bx - cy + Y(x, y) + R_2(V_{g_z}(t)). \tag{3.19}$$

By the assumptions made, a, b, c, k_1, and k_2 are positive constants.

The velocity of gusts $V_{g_z}(t)$ is considered bounded in the modulus: $|V_{g_z}(t)| \leqslant w = \text{const.}$; hence $|R_1| \leqslant \rho_1 = k_1 w = \text{const.}$, $|R_2| \leqslant \rho_2 = \text{const.}$ It is also admitted that the nonlinear parts satisfy the conditions

$$|X(x, y)| \leqslant l_1(|x| + |y|), \qquad |Y(x, y)| \leqslant l_2(|x| + |y|)$$

with known Lipschitz constants.

First estimating technique. In accordance with the foregoing hypotheses, the characteristic equation of the linear system

$$\dot{x} = -ax + y, \qquad \dot{y} = -bx - cy \tag{3.20}$$

will have a pair of complex roots $\lambda_{1,2} = \nu \pm i\omega$, $\nu = -(a + c)/2$, $\omega = (b + ac - \nu^2)^{1/2}$, with negative real part ($\nu < 0$).

By seeking for (3.20) solutions in the form $x = \gamma_1 e^{\lambda t}$, $y = \gamma_2 e^{\lambda t}$, the general solution can be written

$$x = e^{\nu t}(D_1 \cos \omega t + D_2 \sin \omega t),$$

$$y = e^{\nu t}\{D_1[(\nu + a) \cos \omega t - \omega \sin \omega t] + D_2[(\nu + a) \sin \omega t + \omega \cos \omega t]\}. \tag{3.21}$$

In order to transform (3.19) into a system of integral equations of type (3.5) we will have recourse to the procedure of the variation of parameters by seeking the solution of the system with additional terms (3.19) in the form

$$x = e^{\nu t}(u_1(t) \cos \omega t + u_2(t) \sin \omega t),$$

$$y = e^{\nu t}\{u_1(t)[(\nu + a) \cos \omega t - \omega \sin \omega t] + u_2(t)[(\nu + a) \sin \omega t + \omega \cos \omega t]\}. \tag{3.22}$$

Functions u_1 and u_2 are determined by replacing x and y in (3.19) with the expressions (3.22) and taking into consideration that (3.21) with D_1, D_2 arbitrary represents the solution of the linear system (3.20),

$$
\begin{aligned}
u_1(t) &= u_1(0) + \int_0^t e^{-\nu s}\{(X+R_1)[(\nu+a)\sin\omega s + \omega\cos\omega s] \\
&\quad - (Y+R_2)\sin\omega s\}\,ds, \\
u_2(t) &= u_2(0) + \int_0^t e^{-\nu s}\{-(X+R_1)[(\nu+a)\cos\omega s - \omega\sin\omega s] \\
&\quad + (Y+R_2)\cos\omega s\}\,ds.
\end{aligned}
\tag{3.23}
$$

$u_1(0)$ and $u_2(0)$ are obtained from (3.22) by taking $t = 0$.

$$u_1(0) = x(0) = x_0, \quad u_2(0) = -\frac{\nu+a}{\omega}x_0 + \frac{1}{\omega}y_0, \qquad y_0 = y(0).$$

Estimations will be deduced by starting from the integral equations (3.22) in which $u_1(t)$ and $u_2(t)$ have the expressions (3.23). We readily obtain

$$
\begin{aligned}
|x| + |y| &\leqslant e^{\nu t}\{1 + [(\nu+a)^2 + \omega^2]^{1/2}\}(|u_1(t)| + |u_2(t)|) \\
&= e^{\nu t}(1+\sqrt{b})(|u_1(t)| + |u_2(t)|)
\end{aligned}
$$

and further

$$
\begin{aligned}
|x| + |y| \leqslant e^{\nu t}(1+\sqrt{b})\Bigg[\left(1 + \frac{|\nu+a|}{\omega}\right)|x_0| \\
+ \frac{1}{\omega}|y_0| + 2\int_0^t e^{-\nu s}(\sqrt{b}|X+R_1| + |Y+R_2|)\,ds\Bigg]
\end{aligned}
$$

or, by taking into consideration the conditions imposed upon the functions R_1, R_2, X, and Y, simple transformations yield

$$e^{-\nu t}(|x| + |y|) \leqslant m_1 e^{-\nu t} + m_2 + m_0\int_0^t e^{-\nu s}(|x| + |y|)\,ds,$$

where

$$m_1 = m_1(\rho_1, \rho_2) = \frac{2(\sqrt{b}+1)}{-\nu}(\sqrt{b}\rho_1 + \rho_2),$$

$$m_2 = m_2(x_0, y_0, \rho_1, \rho_2) = \omega(\sqrt{b}+1)[(|\nu+a| + \omega)|x_0| + |y_0|] - m_1,$$
$$m_0 = m(l_1, l_2) = 2(\sqrt{b}+1)(l_1 + l_2).$$

Then, by the lemma stated earlier, the inequality

$$\exp(-\nu t)(|x|+|y|) \leqslant m_1 \exp(-\nu t) + m_2 + m_0 \int_0^t (m_1 \exp(-\nu s) + m_2) \exp[m_0(t-s)]\, ds$$

holds, and hence

$$|x| + |y| \leqslant \kappa_0 + \kappa \exp[(\nu + m_0)t]$$

where

$$\kappa_0 = \frac{m_1 m_0}{-(\nu + m_0)}, \qquad \kappa = \frac{-\nu m_2 - m_0(m_1 + m_2)}{-(\nu + m_0)}.$$

If the margin of dynamic stability expressed by the value of the real part ν of the characteristic roots of the linear system is sufficiently large in comparison with the magnitude of the nonlinear terms expressed by the Lipschitz constants, namely, if

$$|\nu| > m_0 = 2(\sqrt{b} + 1)(l_1 + l_2),$$

the sum of the value of the deviations is dominated by an exponentially decreasing time function bounded by the quantity $\kappa_0 + \kappa = m_2$.

Therefore a sufficient condition to secure the fulfillment of the inequality $|x| + |y| < \varepsilon$ for any $t \geqslant t_0$ is given by the satisfaction of the last inequality simultaneously with

$$m_2(x_0, y_0, \rho_1, \rho_2) < \varepsilon.$$

The inequality $|x| + |y| < \varepsilon$ may no longer hold at a given instant, if this last condition is not fulfilled. The time interval $[0, T]$ during which this condition is secured may then be estimated by means of the relation (cf. (3.9))

$$T \leqslant \frac{1}{\nu + m_0} \ln \frac{\varepsilon - \kappa_0}{\kappa}.$$

If the nonlinear terms are negligible ($l_1 \approx 0$, $l_2 \approx 0$), it can readily be seen that the following estimates are obtained for $|x|$ and $|y|$.

$$|x| = |\Delta\alpha| \leqslant e^{\nu t}(|u_1(t)| + |u_2(t)|) \leqslant \frac{2}{-\nu}(\sqrt{b}\rho_1 + \rho_2)(1 - e^{\nu t})$$

$$\leqslant \frac{2}{-\nu}(\sqrt{b}\rho_1 + \rho_2), \tag{a}$$

$$|y| = |\Delta \hat{q}| \leqslant e^{\nu t}(|u_1(t)| + |u_2(t)|) \leqslant \frac{2\sqrt{b}}{-\nu}(\sqrt{b}\rho_1 + \rho_2)(1 - e^{\nu t})$$

$$\leqslant \frac{2\sqrt{b}}{-\nu}(\sqrt{b}\rho_1 + \rho_2). \tag{b}$$

Let ε be an upper bound of the deviation of the incidence imposed by service requirements. For instance, $\varepsilon = \alpha_{st} - \alpha$ where α_{st} is the stalling incidence and α is the actual undisturbed incidence, or $\varepsilon = \alpha$ (if the deviation $\Delta\alpha$ were to become equal to $-\alpha$, the lift would vanish). Then, a sufficient condition for the value of the deviation of incidence to remain always smaller than ε is given by the relation

$$\frac{2(\sqrt{b}\rho_1 + \rho_2)}{-\nu} < \varepsilon, \tag{3.24}$$

whence follows an estimate of the allowable vertical gust velocity for a given dynamic stability margin ν:

$$V_{g_z(\mathrm{ad})} < \frac{-\nu\varepsilon V}{C_{L_\alpha}\sqrt{b} - C_{m_\alpha}}.$$

In ν the dominating term usually is $(C_{m_q} + C_{m_{\dot\alpha}})/4i_B$, characterizing the aerodynamic damping, while in b it is $-C_{m_\alpha}/2i_B$, equal to the coefficient of static stability, by approximation of the factor $-1/2i_B$. Hence in this estimate the allowable velocity of vertical gusts is bounded by a quantity that increases with the aerodynamic damping and decreases with the static stability. Such an effect of static stability is explained by the fact that the additional term R_2 is proportional to the value of the parameter C_{m_α}.

If $V_{g_z(\mathrm{ef})}$ denotes the greatest effective value of the velocity of vertical gusts within the whole space in which the flight is performed, from (3.24) we have an estimate of the dynamic stability margin:

$$|\nu| > \frac{C_{L_\alpha}\sqrt{b} - C_{m_\alpha}}{\varepsilon}\,\frac{V_{g_z(\mathrm{ef})}}{V}.$$

Second estimating technique. A Liapunov function is sought for the system (3.20) in the form

$$\mathscr{V}(x, y) = \beta_{11}x^2 + 2\beta_{12}xy + \beta_{22}y^2$$

whose total time derivative along the integral curves of the system is

$$\left(\frac{d\mathscr{V}}{dt}\right)_{(3.20)} = -2(x^2 + y^2).$$

The coefficients of the quadratic form $\mathscr{V}(x, y)$ are determined from the identity

$$[(\beta_{11}x + \beta_{12}y)(-ax + y) + (\beta_{12}x + \beta_{22}y)(-bx - cy)] \equiv -x^2 - y^2.$$

Their expressions, depending on the coefficients of system (3.20), will be

$$\beta_{11} = \frac{b(b+1) - c(a+c)}{\varDelta}, \qquad \beta_{12} = \frac{c - ab}{\varDelta},$$

$$\beta_{22} = \frac{b + 1 + a(a+c)}{\varDelta}, \qquad \varDelta = (a+c)(b+ac).$$

As it was assumed that the characteristic equation of the system has a pair of complex roots $\nu \pm i\omega$, with the real part negative, it follows, according to one of Liapunov's theorems [1, Section 20, Theorem II], that the uniquely determined quadratic form $\mathscr{V}$ will be positive definite. Hence $\beta_{11} > 0$, $\beta_{22} > 0$, $\beta_{11}\beta_{22} - \beta_{12}^2 > 0$, $\nu < 0$ imply also for $|x| < h$, $|y| < h$, with h sufficiently small, a uniform asymptotic stability and consequently stability under persistent disturbances of the zero solution of the nonlinear system of equations

$$\dot{x} = -ax + y + X(t, x, y),$$

$$\dot{y} = -bx - cy + Y(t, x, y).$$

To estimate the safety limit of the velocity V_{g_z} of vertical gusts, proceed according to the scheme already mentioned.

If μ_1 and μ_2 are the eigenvalues of the matrix of the quadratic form $\mathscr{V}(x, y)$, that is, of

$$\begin{pmatrix} \beta_{11} & \beta_{12} \\ \beta_{12} & \beta_{22} \end{pmatrix}$$

equal to $\frac{1}{2}\{\beta_{11} + \beta_{22} - [(\beta_{11} - \beta_{22})^2 + \beta_{12}^2]^{1/2}\}$ and $\frac{1}{2}\{\beta_{11} + \beta_{22} + [(\beta_{11} - \beta_{22})^2 + \beta_{12}^2]^{1/2}\}$, respectively, then the double inequality

$$\mu_1(x^2 + y^2) \leqslant \mathscr{V}(x, y) \leqslant \mu_2(x^2 + y^2)$$

holds for any x and y.

We seek to determine an upper bound of the value of the velocity V_{g_z} such that $|x| = |\Delta\alpha| \leqslant \varepsilon_1$ and $|y| = |\Delta\hat{q}| \leqslant \varepsilon_2$, where ε_1 and ε_2 are chosen according to some service requirements (e.g., for ε_1 see previous case). Let ε_1' and ε_2' be two numbers smaller than ε_1 and ε_2, respectively (which for calculation purposes can be taken as equal in practice), and let $k = \mu_1\varepsilon'^2$, $\varepsilon'^2 = \varepsilon_1'^2 + \varepsilon_2'^2$.

The behavior of the integral curves of system (3.20) will be investigated by starting at the initial moment t_0 from the point (x_0, y_0), $|x_0| < \min(\varepsilon_1, \varepsilon'(\mu_1/\mu_2)^{1/2})$, $|y_0| < \min(\varepsilon_2, \varepsilon'(\mu_1/\mu_2)^{1/2})$, at its intersection with the ellipse $\mathscr{V}(x, y) = k$. If this integral curve will intersect the ellipse $\mathscr{V} = k$ from the outside toward the inside, the constraint of bounds imposed on x and y is satisfied. Therefore, the conditions will have to be established under which the derivative $d\mathscr{V}/dt$ is negative along the solutions of systems (3.20) in any point of the curve $\mathscr{V} = k$; that is,

$$[-(x^2 + y^2) + (\beta_{11}x + \beta_{12}y)(X + R_1) + (\beta_{12}x + \beta_{22}y)(Y + R_2)]_{\mathscr{V}=k} < 0.$$

By substituting in this inequality the respective estimates for X, Y, R_1, and R_2, $k_iV_{g_{z(\mathrm{ad})}}$ for R_i in particular, $V_{g_{z(\mathrm{ad})}}$ being the allowable upper bound of the velocity V_{g_z} of vertical gusts, an estimate of this upper bound depending on the characteristics of the machine, and on the undisturbed flight régime is readily obtained. (For details of computation, see [14].)

3.7. STABILITY AS A PROPERTY OVER AN UNLIMITED INTERVAL OF TIME

It was shown in the previous sections that for the Liapunov stability to have a practical sense it sometimes appears necessary besides establishing whether the system is stable, to provide some quantitative estimates. Another characteristic of the concepts defined in this chapter will be pointed out in the sequel, which might also lead to a restriction of the validity of this mathematical theory. It is the question that the property of stability is defined for an unlimited interval of time.

It has to be specified that the validity over an unlimited interval of time is an essential and practically necessary characteristic of the property of stability. Indeed, the existence of a certain finite region of initial disturbances ensuring that induced deviations are kept within allowable (theoretically arbitrary) limits over a finite interval of time is provided in a trivial manner by the continuity of the solution with respect to the initial conditions in every case, including those of the unstable solution. Indeed the continuity implies for any $\varepsilon > 0$ and for any interval of time T, the existence of a function $\eta(\varepsilon, T)$ with the property that $|x_0| < \eta$ entails $|x(t)| < \varepsilon$ for $t_0 \leqslant t \leqslant t_0 + T$. But although any examined motion, and any evolution of the flying machine in particular, occurs over a finite and possibly known interval of time, the duration T of the interval considered from an instant defined by the appearance of the disturbances is a random quantity, which evidently cannot be found among the data of the problem. Thus, the question is raised how to secure the stated property for any interval; how to find, therefore, the conditions under which η is independent of T. On this basis the concept of stability in the sense of Liapunov compels recognition.

In handling the general theory according to Liapunov it is important to take this fact into account with sufficient consistency; all the hypotheses and reasoning, including those that concern the basic motion, will have to be compatible with the problem considered over an infinite interval of time. In this connection, difficulties may appear in the case of certain unsteady basic motions that are not defined over an infinite interval of time within the framework of the model used (e.g., the uniformly decelerated motion, or even the uniformly accelerated one). The mathematical model must be conceived so that finite intervals that may appear even "short" by our scale might be subjectively assimilated to intervals of time however long. This problem will come up again in Chapter 9.

3.8. STABILITY WITH RESPECT TO A PART OF THE VARIABLES (INCOMPLETE STABILITY)

In Chapter 1 we noted that stability is a property of the basic motion. For this reason it is not possible to talk about the stability of a flying machine as about the stability of any motion that it can perform. The checking of stability on the basis of certain criteria is usually carried out for some motions only, the choice of which depends on the destination of the machine. On the other hand, the practical criteria for checking the flying properties do not generally refer to an unconditional stability of free flight. Most frequently, however, an incomplete stability even of the most common motions (such as the steady straight level cruising flight of airplanes) is enough. Thus, for instance, it appears that none of the aircraft being produced at present possesses complete inherent stability in all the basic régimes utilized. A slight oscillatory instability in the slow stage of the longitudinally disturbed motion, which can readily be corrected through the controls for the common régimes, is generally admitted. Greater freedom is thus obtained in securing satisfactory flight properties of another kind (better maneuverability, a larger range of régimes with satisfactory stability properties, etc.).

To define the concept of incomplete stability more accurately, we will have recourse to the mathematical model. Let us assume that the n components of the vector x in system (1.6) represent the dynamic variables describing the uncontrolled or free disturbed motion of an aircraft. It will be recalled that in the case of a system of the form (1.6) the dynamic variables represent the deviations of the generalized coordinates that describe the basic motion corresponding to the trivial solution $x(t; x_0, 0) = 0$ of the system. Now, two categories of dynamic variables will be considered, which will be represented by two column vectors y and z, k- and m-dimensional, respectively ($k + m = n$ if x is an n vector).

Let us further assume that during the whole course of the disturbed motion only the deviations y have to remain in a certain given neighborhood of the zero value, meaning in a way that the property of stability refers only to the components y.

It was sought to frame a general theory of stability with respect to a part of the variables, without, however, succeeding in establishing sufficiently effective procedures. The question of achieving an efficient approach is still open, so that a definition of stability with respect to a part of the variables, and a theorem providing a sufficient condition for this stability according to [17], are given hereunder to emphasize this fact.

Let us consider the system

$$\frac{dy}{dt} = Y(t, y, z), \qquad \frac{dz}{dt} = Z(t, y, z) \tag{3.25}$$

in which the functions Y and Z are defined for $t \geqslant t_0$, $|x| \leqslant h = \text{const.}$ and y arbitrary and vanishing for $y = z = 0$, and let $(y(t; t_0, y_0, z_0), z(t; t_0, y_0, z_0))$ be the solution corresponding to the initial conditions (y_0, z_0).

Definition. The solution $y(t; t_0, 0, 0) = 0$, $z(t; t_0, 0, 0) = 0$ is stable with respect to the variables y if there exists an $\eta(\varepsilon, t_0)$ for which $|y_0| + |z_0| < \eta$ and $t \geqslant t_0$ imply $|y(t; t_0, x_0, y_0)| < \varepsilon$.

Stability is uniform if η is independent of t_0.

The concept of positive definite function is further extended for this case. Let $\mathscr{V}(y, z)$ be a function defined for $|y| < h = \text{const.}$, positive for y and z arbitrary and vanishing only for $y = 0$. The function $\mathscr{W}(t, y, z)$ is said to be positive definite if for $t \geqslant t_0$, $|y| < h$ and z arbitrary, the inequality $\mathscr{W}(t, y, z) \geqslant \mathscr{V}(y, z)$ holds uniformly with respect to t. The function $\mathscr{W}(t, y, z)$ is negative definite if under the same conditions $\mathscr{W}(t, y, z) \leqslant -\mathscr{V}(y, z)$.

The following theorem can be stated.

THEOREM. *If there exists a function $\mathscr{W}(t, y, z)$ positive definite with respect to y, for which the total time derivative along the integral curves of* (3.25), *that is, the function $\partial\mathscr{W}/\partial t + \partial\mathscr{W}/\partial y\, Y(t, y, z) + \partial\mathscr{W}/\partial z\, Z(t, y, z)$, is positive semi-definite, then the solution $y(t; t_0, 0, 0) \equiv 0$, $z(t; t_0, 0, 0) \equiv 0$ is stable with respect to y.*

The sufficient condition of uniform stability with respect to y, moreover, contains the inequality $\mathscr{W}(t, y, z) \leqslant \mathscr{U}(y, z)$, where $\mathscr{U}(y, z)$ is a function positive definite with respect to y and z in the usual sense.

The main difficulty in applying these theorems consists in finding the function $\mathscr{W}$ with the required properties, for which so far no adequate practical procedures have been established.

Also left open is the matter of establishing the conditions under which incomplete stability is possible. This amounts to the different behavior with respect to stability of variables related to each other through a system of equations (not separated into independent systems). It appears that this question has not so far been the subject of some general theory. It can be assumed, at least for the case of linear systems with constant coefficients, that such a situation is possible only by separation into independent systems.

In addition to the question of incomplete stability of flight, the more general question of the different behavior of the components of disturbance from the point of view of stability is also of interest. For instance, it is possible for the system to present asymptotic stability with respect to a share of the variables, and nonasymptotic stability with respect to the remaining variables.

The possibility of such a differentiated behavior of the solution is shown in [5, page 79] for a system of the form

$$\frac{dy}{dt} = A(t)y + Y(t, y, z), \qquad \frac{dz}{dt} = Z(t, y, z) \tag{3.26}$$

with A denoting a time-dependent matrix, $|Y(t, x, y)| \leqslant k|y|^\beta$ for $\beta > 0$, $|y| \leqslant h$, $|z| \leqslant h$, $h =$ const., k sufficiently small, $|Z(t, y, z)| \leqslant K|y|$, by means of the following proposition.

If the trivial solution of the linear system $dy/dt = A(t)y$ is uniformly asymptotically stable, the trivial solution of system (3.26) is uniformly stable; in addition, for sufficiently small initial values, any solution satisfies the equalities $\lim_{t\to\infty} y(t; t_0, y_0, z_0) = 0$, $\lim_{t\to\infty} z(t; t_0, y_0, z_0) = l(y_0, z_0)$.

Let it be noted that for current values of the disturbances (i.e., initial values) such a stability with respect to z can be insufficient for flight practice. In other words, there are cases when, to secure an ε (according to Definitions 1, 2 in Section 3.1) sufficiently small from the point of view of the practical purpose aimed at (reasons of security, comfort, accuracy of evolution, etc.), the value of the function $\eta(\varepsilon)$ is exceeded by that of current disturbances. It is also possible that $l(y_0, z_0)$ from the stated proposition might be too large in comparison, for instance, with ε for usual initial values y_0, z_0. In such cases it can be said that we have to deal with a practical instability with respect to z (although the system on the whole is stable in the sense of Liapunov).

A particular case of the systems (3.26) is represented by systems of the form

$$\frac{dy}{dt} = A(t)y + Y(t, y, z), \qquad \frac{dz}{dt} = u \tag{3.27}$$

where the components of the vector z stand for the angular coordinates and the components of the vector u, belonging to y, the respective angular velocities.

A classic example of the latter system is provided by the simplified equations of short-period longitudinal oscillations (for the aspects discussed hereafter see Section 6.1) written in B. Melvill Jones's hypotheses [18]. As is known, they serve to determine approximately the behavior of incidence (rapid incidence adjustment) and the pitching velocity of an aircraft in a first short interval of time following a disturbance. Added to these equations is the kinematic relationship between the longitudinal attitude and the rate of pitch. It will be recalled that in addition to the general hypotheses that are at the basis of the conventional equations of the longitudinal disturbed motion of the aircraft, B. Melvill Jones also admitted that the flight velocity is constant during the interval of time considered, since no sensible deviation of it is induced in the first seconds. Then, if the steady straight level flight is considered to be the basic motion for which $\Delta\alpha = \Delta\hat{q} = \Delta\theta = 0$, the motion of the aircraft in a stick-fixed condition can be described by a system of the form

$$\frac{d\Delta\alpha}{d\hat{t}} = a\,\Delta\alpha + \Delta\hat{q},$$

$$\frac{d\Delta\hat{q}}{d\hat{t}} = b\,\Delta\alpha + c\,\Delta\hat{q},$$

$$\frac{d\Delta\theta}{d\hat{t}} = \Delta\hat{q}$$

where $\Delta\alpha$, $\Delta\hat{q}$, and $\Delta\theta$ denote the deviations of the incidence, angular velocity of pitch, and attitude angle, respectively, induced by the disturbance (in nondimensional form, $\hat{t}$ standing for the nondimensional time), while a, b, and c are dimensionless constants depending on certain aerodynamic and other design characteristics of the aircraft as well as on the undisturbed flight régime. It can readily be seen that, given the values of the variables $\Delta\alpha$ and $\Delta\hat{q}$ at the initial instant, their time variation law is completely determined by the first two equations of the system. In all cases of practical interest the parameters a, b, and c are such that the zero solution of the system formed by the first two equations is uniformly asymptotically stable, namely, exponentially with a comparatively large rate of decay. At the same time, on the basis of the third equation, the value of the angular deviation of pitch increases, tending in general asymptotically toward a constant magnitude depending linearly on the initial conditions. Hence, the system is stable with respect to $\Delta\theta$ but not asymptotically stable.

NOTE. The situation of an asymptotic stability with respect to $\Delta\theta$ is obviously excluded. It still has to be proved that it is nevertheless not a question of instability in the sense of Liapunov.

Let $\lambda_{1,2} = -\mu \pm i\nu$ (μ and ν are positive) be the characteristic roots of the system in $\Delta\alpha$ and $\Delta\hat{q}$. Then, the component $\Delta\hat{q}$ of the solution is of the form $\Delta\hat{q} = C\exp[-\mu(\hat{t}-\hat{t}_0)]\sin(\nu\hat{t}+\varphi)$, the constant C depending linearly on $\Delta\alpha_0$ and $\Delta\hat{q}_0$, and $\Delta\theta = \Delta\theta_0 + \int_{\hat{t}_0}^{\hat{t}} C\exp[-\mu(s-\hat{t}_0)]\sin(\nu s+\varphi)\,ds$ whence $|\Delta\theta| \leqslant |\Delta\theta_0| + |C\{1-\exp[-\mu(\hat{t}-\hat{t}_0)]\}|/\mu \leqslant |\Delta\theta_0| + l(\Delta\alpha_0, \Delta\hat{q}_0)$ where $l = c/\mu$ is a linear function of $\Delta\alpha_0$ and $\Delta\hat{q}_0$. For $k > 0$ large enough we evidently have $|\Delta\alpha_0| + l \leqslant k(|\Delta\alpha_0| + |\Delta\hat{q}_0| + |\Delta\theta_0|)$ and if, for instance, $\eta(\varepsilon) = \varepsilon/k$, the inequality $|\Delta\alpha_0| + |\Delta\hat{q}_0| + |\Delta\theta_0| < \eta$ entails $|\Delta\theta| < \varepsilon$ no matter what the value of ε. In practice, however, the region of actual initial conditions can exceed any η corresponding to a required ε, and consequently the stability with respect to $\Delta\theta$ in the sense of Liapunov does not secure a satisfactory behavior of the aircraft with respect to this variable. Therefore, this theoretical stability of the system considered with respect to $\Delta\theta$, unlike the asymptotic stability with respect to $\Delta\alpha$ and $\Delta\hat{q}$, has no other practical significance than to emphasize that $\Delta\theta$ behaves differently than $\Delta\alpha$ and $\Delta\hat{q}$. From this result, when the validity of B. Melvill Jones's hypotheses ceases, we can further expect a situation of instability to develop with respect to all the variables of the system ($\Delta\alpha, \Delta\hat{q}, \Delta\theta$, to which also $\Delta\hat{V}$ is added), but with stronger effects on the variable $\Delta\theta$ (and $\Delta\hat{V}$). Thus, the results obtained on the basis of the simplified model of short-period oscillations furnish certain qualitative information concerning the margin of stability with respect to the different dynamic variables, whereby this model is justified.† But as the deviation of the flight velocity develops, the separation in the system by the first approximation of the equations in $\Delta\alpha$ and $\Delta\hat{q}$ from the rest is no longer acceptable. It becomes necessary to consider a more complete system in which $\Delta\hat{V}$ is not neglected. Such a system would be of the form

$$\frac{d\Delta\hat{V}}{d\hat{t}} = A\,\Delta\hat{V} + B\,\Delta\alpha + C\,\Delta\theta,$$

$$\frac{d\Delta\alpha}{d\hat{t}} = D\,\Delta\hat{V} + E\,\Delta\alpha + \Delta\hat{q},$$

$$\frac{d\Delta\theta}{d\hat{t}} = \Delta\hat{q},$$

$$\frac{d\Delta\hat{q}}{d\hat{t}} = F\,\Delta\hat{V} + G\,\Delta\alpha + H\,\Delta\hat{q},$$

where A, B, C, D, E, F, G, H are parameters depending on the aircraft and on the flight régime appropriate for the basic motion.

The zero solution of this system can be (and as a rule is) slightly unstable, naturally with respect not only to $\Delta\hat{V}$ and $\Delta\theta$ but also to $\Delta\alpha$ and $\Delta\hat{q}$. For this reason, the known hypothesis adopted along the line of Lanchester's simplified theory of idealized phugoid paths (see [23, Chapters 2 and 5]) for the equations corresponding to the slow mode phase of the disturbed motion, namely, that the incidence remains constant during the whole course of this phase (see, e.g., [18, page 86; 24, page 541; or 25, page 29]) is, strictly

† The various mathematical models of stability within a finite interval of time find application and justification in such cases as this (see, e.g., [19–21]; in [22], the general theory given in these references is also applied to, among others, the mentioned case of stability during the rapid stage of the disturbed motion for unsteady basic motion).

speaking, groundless. Moreover, for setting up a criterion of stability of free motion and not for estimating the characteristics of the transitory régime induced by the disturbances over a finite interval of time, neglecting the variation of flight velocity from the simplified theory of the rapid stage of the longitudinal disturbed motion (rapid incidence adjustment) is, rigorously, not valid for the very reason that it eliminates the possibility of evidencing a slight instability of the basic motion on the whole (including with respect to the incidence) if such be the case. Nevertheless, when the real motion is studied in the first seconds following upon the discontinuation of the action of the disturbing cause, such neglecting is considered allowable, since the results obtained when so doing are in keeping with the reality. The apparent contradiction is attributable to the fact that the typical real motion is not generally free flight. The pilot, human or automatic, actuates the controls continually in order to secure flight stability. Thus, for instance, the human pilot counteracts permanently by slight reflex movements of the controls the effect of various disturbances that appear even during flight in the generally still atmosphere. Such disturbances result from: momentary asymmetry in working of the engines; shifts in the center of mass due to mass displacements within the machine; gusts of small intensity; slight errors in piloting; and so on. Because of the reflex character of these movements, the pilot, moreover, is usually unaware of the existence of the slight instability of the free flight. Thus, the stabilizing intervention of the controls converts the slightly unstable motion into a stable one, somehow justifying a posteriori the admitted omissions (e.g., the neglect of the variation of flight velocity induced by disturbance in studying the behavior of incidence and pitching velocity). Indeed, the variables neglected in the simplified models remain small during motion even under the conditions of a slight instability of free flight, because of the stabilizing intervention of the controls. Thus, the conditions under which the flight actually proceeds make for a consideration of the stability of partially controlled motions instead of the partial stability of flight. Chapter 4 of this book is devoted to the general theory of the stability of partially controlled motions.

REFERENCES

1. A. M. Liapunov, *Obshchaya zadacha ob ustoychivosti dviženiya*. Published by Khar'kov Mathematical Society, Khar'kov, 1892. English transl. in *Stability of Motion* by A. M. Liapunov, with a contribution by V. A. Pliss. Academic Press, New York, 1966.

2. I. G. Malkin, *Theory of Stability of Motion*. English transl. from Russian by U.S. Atomic Energy Commission, AEC-tr-3352, pp. 19–20, 188–194, 221, 224.
3. J. L. Massera, "Contribution to Stability Theory," *Ann. of Math.* **65**(1956), 182–206.
4. J. La Salle and S. Lefschetz, *Stability by Liapunov's Direct Method, With Applications*. Academic Press, New York, 1961.
5. A. Halanay, *Differential Equations: Stability, Oscillations, Time Lags*. Academic Press, New York, 1966.
6. B. Etkin, *Dynamics of Flight: Stability and Control*. Wiley, New York, 1959.
7. I. G. Malkin, "On the Reversibility of Liapunov's Theorem on Asymptotic Stability" (in Russian), *Prikl. Mat. Mekh.* **18**(1954), 129–138.
8. V. M. Popov, "On the Question of Practical Stability of Automatic Control Systems Including an Element with Nonuniform Nonlinearity" (in Russian), *Rev. Electrotechn. Energ. Acad. R. P. R.* **6**(1959), 81–101.
9. T. Hacker, "Stability of Partially Controlled Motion of an Aircraft," *J. Aerospace Sci.* **28**(1961), 15–26.
10. R. Bellman, *Stability Theory of Differential Equations*. McGraw-Hill, New York, 1954.
11. T. H. Gronwall, "Note on the Derivatives with Respect to a Parameter of the Solutions of a System of Differential Equations," *Ann. Math.* **20**(1918), 292–296.
12. N. Bourbaki, *Fonctions d'une variable réelle*, Livre IV, Actualités scientifiques et industrielles. Hermann, Paris, 1951.
13. T. Hacker, "Gust Influence on the Longitudinal Motion of an Aircraft," *Rev. Méc. Appl. Acad. R. P. R.* **3**(1958), 17–24.
14. T. Hacker, "On the Longitudinal Stability of an Aircraft under Repeated Disturbances," *Rev. Méc. Appl. Acad. R. P. R.* **4**(1959), 229–236.
15. B. Etkin, "A Theory of the Response of Airplanes to Random Atmospheric Turbulence," *J. Aerospace Sci.* **26**(1959), 409–420.
16. B. Etkin, "Theory of the Flight of Airplanes in Isotropic Turbulence: Review and Extension," *UTIA Rept.* **72**(February 1961).
17. V. V. Rumyantsev, "On the Stability Relative to a Part of the Variables" (in Russian), *Vestn. Mosk. Univ. Ser. Mat. Mekh. Astron. Fiz. i Khim.* **4**(1957), 9–16.
18. B. Melvill Jones, *Dynamics of the Airplane* (Volume 5 of the series *Aerodynamic Theory*, W. F. Durand, ed.), div. N, pp. 1–222. Springer, Berlin, 1935.
19. G. V. Kamenkov, "On the stability of motion over a finite time interval" (in Russian), *Prikl. Mat. Mekh.* **17**(1953), 529–540.
20. A. A. Lebedev, "Contribution to the question of motion stability over a finite time interval" (in Russian), *Prikl. Mat. Mekh.* **18**(1954), 75–94.
21. A. A. Lebedev, "On the stability of motion over a given time interval" (in Russian), *Prikl. Mat. Mekh.* **18**(1954), 75–94.
22. T. Hacker, "Longitudinal stability of unsteady motion of aircraft" (in Russian), Dissertation abstract, Moskow Aviation Institute, 1954.
23. F. W. Lanchester, *Aerial Flight II: Aerodonetics*. Constable, London, 1908.
24. R. von Mises, *Theory of Flight*. McGraw-Hill, New York, 1945.
25. W. J. Duncan, *The Principles of the Control and Stability of Aircraft*. Cambridge Univ. Press, London and New York, 1952.

Chapter 4

STABILITY OF PARTIALLY CONTROLLED MOTIONS

The first three sections of this chapter are based on the author's papers published between the years 1959 and 1961 (see [1]). Section 4.4 presents the general framework within which Neumark's theory of "constrained stability" [2] is placed. This theory is the first attempt to consider flight under conditions of controlling a share of the state variables.

4.1. INTRODUCTORY CONSIDERATIONS

As mentioned earlier, at its present stage the theory of flight stability must take into account, in several problems, the effect of manual or automatic controls or of both. Indeed, it is useless to try to obtain complete inherent stability conditions and it is increasingly difficult and unrealistic to find *a priori* simplified models in which adequate hypotheses could be substituted for the effect of the controls, as had been possible, for instance, in the simplified theory of short-period oscillations. Present flight technique implies a complication of the problems, and at the same time that piloting and flight practice cease to be an inexhaustible source of valid hypotheses.

Depending on the end in view, on the one hand, and on the quality and amount of available information about the operation of the control system, on the other, the theory of flight stability accounts for the effect of controls in two ways:

1. If we are mainly concerned with the control system, for instance, with choosing the characteristics of the automatic pilot or automatic stabilizer, or if our chief concern is the behavior of an automatically controlled aircraft, and the equations of motion for the control system are known with sufficient accuracy, the aircraft together with the control system will be considered as a single dynamic system (the aggregate being referred to as the "control system").

2. We might, however, be mainly concerned only with the mechanical behavior of the controlled object (i.e., of the airframe), in which case, in order to facilitate the theoretical investigation, we would avoid introducing an additional number of degrees of freedom when considering the operation of the control devices. Moreover, the operation of the control system might not be known with sufficient accuracy to permit its description by means of dynamic equations. Such would be the case of the human pilot, an element of the control system whose response characteristics vary not only from person to person, but also in the same individual according to his degree of fatigue, psychological and physiological condition, and so on. In either case the effect of the corrections is considered as external forces having the nature of persistent disturbances.† In this chapter we will treat only case 2, in the form of a model of partially controlled systems. Certain aspects of the first case, in which the control devices and the airframe are considered as a single dynamic system, will make up the subject matter of Chapter 11.

4.2. MATHEMATICAL STATEMENT OF THE PROBLEM OF PARTIALLY CONTROLLED MOTIONS

Let

$$\frac{dy}{dt} = Y(t, y, z), \qquad \frac{dz}{dt} = Z(t, y, z), \tag{4.1}$$

where y and z denote k- and m-column vectors, respectively ($k + m = n$), be the system of equations of the uncontrolled motion. Functions Y and Z

† A study of the pilot–aircraft combination as a single dynamic system that, in addition to the inherent qualities of the machine, has as a starting point the response characteristics of the pilot as well as certain data pertaining to the control system linkage is of interest for certain purposes. Such a study turns out to be necessary because the pilot's intervention, when the control characteristics are unsatisfactory, can have a negative effect on the flying qualities. Thus, for instance, it can lead to instability of the pilot–aircraft system for régimes in which the uncontrolled aircraft is stable (e.g., see [3, 4]). The relation between the pilot's response characteristics, such as his time of reaction, and the aircraft's response characteristics plays a decisive part in the efficiency of the pilot's intervention. This problem will be resumed in the sequel as well as in Chapter **11** in connection with a discussion of the influence of time lags on stability. As regards the human pilot, any theoretical study must be completed with measurements carried out by means of ground simulators or flight tests.

are defined for $t \geqslant t_0$, $|y| \leqslant h_1 = \text{const.}$, $|z| \leqslant h_2 = \text{const.}$, and vanish for $y = z = 0$. Moreover Y and Z must generally have properties similar to those stated for the right-hand side of the system (1.6).

Let us assume that the pilot or the automatic pilot are concerned only with the behavior of the variables z, seeking to maintain them within certain limits or to cancel them by adequate control movements. It will be agreed to call these variables *controlled* or *constrained*, as against the remaining variables, components of the vector y, which will be conventionally designated *uncontrolled* or *free*.†

As regards the equations of motion, the intervention of the controls generally finds expression in a modification of the right-hand sides of all the equations of system (4.1). (See Section **4.3**. An obvious analogy with the problem of the behavior of the solution in the case of persistent disturbances will be noted in the sequel. This analogy will lead to the results obtained in the present section).

Let

$$\frac{dy}{dt} = Y_1(t, y, z), \qquad \frac{dz}{dt} = Z_1(t, y, z) \tag{4.2}$$

be the system of equations of partially controlled motion (corresponding to the constraint through controls of the variables z).

It is assumed that the properties of the generally unknown functions Y_1 and Z_1 are those mentioned for Y and Z, and that in addition

$$Y_1(t, y, 0) \equiv Y(t, y, 0), \qquad Z_1(t, y, 0) \equiv Z(t, y, 0). \tag{4.3}$$

The additional relations (4.3) express the hypothesis by which the stabilizing effect of the controls released by the appearance of the deviations z cease rigorously at the instant when deviations z vanish. In other words, they cease when all the constrained kinematic parameters recover their

† These traditional designations are undoubtedly misnomers. Indeed, by gearing the controls, we generally seek to change the whole picture of the motion characterized by the aggregate of the variables y and z, which is achieved on the basis of the dynamic relationships existing among all the variables described by the equations of motion. It can readily be seen that any constraint imposed on a part of the variables (the additional relationships finding expression in a modification of the equations) implies a modification of the solution, hence of the law of variation in general of all the variables of the system. Rigorously, therefore, we cannot speak of the free or uncontrolled variables when actuating a control. A more apt, but also more obscure, mode of expression would be a classification into directly controlled and indirectly controlled variables.

undisturbed value. The limit of validity of this basic hypothesis depends on the rate of variation of the variables z, as well as on the magnitude of the time lag of the whole control system, including the time lag due to the aircraft response.

Now a definition will be given of the stability of the partially controlled basic motion represented by the trivial solution of the system (4.2) suggested by the definition of stability under persistent disturbances.

DEFINITION 1. The solution $y = 0$, $z = 0$ of the system of equations of the partially controlled motion corresponding to the constraint of the variables z (4.2) (the abbreviated expression "the partially controlled system (4.2)" will also be used in the sequel) is uniformly stable with respect to the variables y if for any $\varepsilon > 0$ there are two functions $\eta(\varepsilon)$ and $\delta(\varepsilon)$ such that $|y_0| < \eta$ and $|z(t)| < \delta$ for $t \geqslant t_0$ imply $|y(t; t_0, z_0)| < \varepsilon$ for $t \geqslant t_0$ and z_0 arbitrary (within the limits $|z_0| < \delta$).

The expression $y(t; t_0, z_0)$ represents the component y of the solution of the system (4.2) corresponding to the initial conditions $y(t_0) = y_0$, $z(t_0) = z_0$.

DEFINITION 2. The solution $y = 0$, $z = 0$ of the partially controlled system is uniformly asymptotically stable with respect to the variables y if it is uniformly stable with respect to these variables according to Definition 1, and moreover, $\lim_{t\to\infty} y(t; t_0, y_0, z_0) = 0$.

Noting that

$$R(t, y, z) = Y_1(t, y, z) - Y_1(t, y, 0)$$

and taking into account (4.3), we can write the first vector equation of the system (4.2) in the form

$$\frac{dy}{dt} = Y(t, y, 0) + R(t, y, z). \tag{4.4}$$

Obviously $R(t, y, 0) \equiv 0$, and as there is continuity in z, the function R will be small at the same time as z. It is assumed that the generally unknown function $R(t, y, z)$ can be estimated in the vicinity of $z = 0$.

Now the following auxiliary system will be considered in addition to (4.2).

$$\frac{dy}{dt} = Y(t, y, 0). \tag{4.5}$$

The uniform stability and the uniform asymptotic stability of the trivial solution of a partially controlled system, defined above, can be evidenced by means of the following two propositions.

PROPOSITION 1. If the solution $\tilde{y}(t; t_0, y_0) = 0$ of the auxiliary system (4.5) is uniformly asymptotically stable (according to Definition 4 or 4′, Section 3.1), then the solution $y = 0$, $z = 0$ of the partially controlled system (4.2) is uniformly stable with respect to the variables y (in the sense of Definition 1 above).

PROPOSITION 2. If the solution $\tilde{y}(t; t_0, y_0) = 0$ of the auxiliary system is uniformly asymptotically stable, and $\lim_{t\to\infty} z(t; t, y, z) = 0$, then the trivial solution of the partially controlled system is uniformly asymptotically stable with respect to the variable y (in the sense of Definition 2 above).

It can readily be seen that these propositions are immediate consequences of the theorems of uniform stability and uniform asymptotic stability under persistent disturbances, respectively, stated in Section 3.3 and of the continuity in z of the function $R(t, y, z)$, which is used as a basis in seeking to secure sufficiently small values for these functions by partial control.

It is therefore possible to draw a valid conclusion as to the stability of partially controlled systems in the aggregate from the behavior of the solution of the auxiliary system, for sufficiently small initial values of the free variables, if we can secure the inequality $|z| < \delta$ throughout the considered interval of time by watching the behavior of the variables z; that is, under such conditions the uniform asymptotic stability of the trivial solution of the auxiliary system implies the uniform stability of the partially controlled motion in the sense of Definition 1. The behavior of the trivial solution of the auxiliary system obviously depends on the inherent dynamic characteristics of the flight machine and on the régime of the basic motion on which the parameters of the system depend.

Under standard flying conditions (except when crossing a zone of intense atmospheric turbulence), the pilot or automatic pilot usually achieves a correction through the controls that is even more efficient. He (or it) succeeds in practice in completely dampening the deviations of the controlled variables induced by the disturbances, achieving with a satisfactory approximation the supplementary condition required by the second proposition, $z \to 0$. Thus, most frequently, the uniform asymptotic stability of the trivial solution of the auxiliary system is equivalent to that of the actual partially controlled motion.

Therefore, in studying the stability of the aircraft under the conditions put forth, it is essential to consider the auxiliary system (4.5). Thus, the problem of finding certain criteria of stability of the real motion considered is solved by considering the solution of the auxiliary system. (This system being smaller by an order k than the initial system of the motion equations, a simplification of the numerical computation also follows implicitly.)

But in numerous flight situations, that is, in the most common ones, investigation of the solution of the auxiliary system offers more than simple criteria of stability. Under common flight conditions when in a small-disturbance condition, as a result of the action of controls, the constrained variables z become negligible, the auxiliary system, although a mathematical fiction, can represent a valid simplified model of the machine's real motion. This model can thus serve to obtain information concerning the (oscillating, monotone, or other) character of the transitory motion induced by a disturbance, to estimate with a satisfactory approximation the main quantitative characteristics of this motion (the damping velocity of the free variables y, their oscillation frequency, etc.).

In order to find the validity limit of this rough model, we can proceed to estimate the quantity δ (see Definition 1) depending on the allowable value ε of the deviations of the free dynamic variables, and to compare it with the magnitude of the constrained variables resulting from the calculations of the response of the aircraft to actuation of the controls, depending on the intensity of the disturbance, or read off the instrument board.

NOTE. In [5] Halanay gives a mathematical formulation of the problem of the stability of partially controlled motions that differs slightly from, but is generally equivalent (with regard to the theoretical implications) to, the one given above. It also results in a consideration of the same auxiliary system for identifying the stability of the system.

The formulation given in [5, pages 89–92] will now be reproduced. Consider the system (4.2) with y, z, Y_1, Z_1 as above. Suppose that the trivial solution of this system is stable with respect to the constrained variables z (in the sense of the definition given for incomplete stability in Section 3.8) and that, moreover, the trivial solution of the auxiliary system (4.5) is uniformly asymptotically stable. Then the trivial solution of system (4.2) is uniformly stable. If an incomplete uniform asymptotic stability with respect to these variables is admitted instead of the incomplete stability, the trivial solution of system (4.2) will be uniformly asymptotically stable.

As can be seen, this formulation offers the advantage of operating with known concepts, namely, the stability in a classical sense of the trivial solution of the complete system (4.2), including the operation of the controls acting on the variables z. In agreement with this formulation, the uniform asymptotic stability of the trivial solution of the auxiliary system implies the stability and the uniform asymptotic stability, respectively, in the sense of the classical definitions (see Definitions 1 and 4, Section 3.1), of the trivial solution of system (4.2). For the scope of the present work, however, the previously given formulation was preferred although it requires an additional definition of the stability of the partially controlled motion with respect to the free variables. In this formulation the difference between the constrained variables and the free variables appears more clearly, and more intuitively from the point of view of the piloting process (see Section 4.3). But the essential advantage for the proposed object is emphasized from the moment the requirement appears for the quantitative estimates mentioned above. Indeed, methods have been developed for estimating the function $\delta(\varepsilon)$ that are sufficiently accurate, and capable of being improved, whereas no such methods yet exist for investigating the incomplete stability required by the latter formulation, and even less for obtaining estimates.

4.3. SOME COMMENTS

A discussion of certain ideas and concepts introduced in the preceding sections will now be given. The discussion will concern the concept of partial control and the classification of the dynamic variables into constrained and free, as well as the physical and mathematical meaning of the two concepts.

In the presented model of partially controlled motions it was assumed that the pilot or the automatic pilot concerned with the behavior of certain state variables would maintain within allowable limits the deviations of those variables induced by the disturbance. Two essential aspects would have to be examined in connection with this assumption, namely:

(1) whether it is possible and in keeping with common piloting practice to watch selectively the disturbance components; and

(2) what is the response efficiency to the partial control from the point of view of the stability of the system.

The latter category includes considerations involving the time of response, the selection of the most adequate controls with regard to the direct effect, and the problem of secondary effects.

The first aspect will be discussed here very briefly. The answer to the two questions is straightforward in the case of an automatic control. Indeed, the automatic pilot is conceived so that it should react to certain input signals representing well-defined components of the motion of the machine registered by the detecting instruments of the automatic pilot. Therefore, under all circumstances an automatic pilot watches only certain selected components of the motion, components that are represented in the given case by the variables z from the preceding section.

Generally, a human pilot controls the flight by responding to the motion of the aircraft perceived by his sense organs both directly and through the flight instruments. The present trend is to use instrument flying even with conventional aircraft. This method of piloting has come to be the only possible one, for instance, in the case of piloted space vehicles. But even in the case of piloting without instruments, up to a point, the pilot's reflexes are selective with respect to the components of motion. The control in this case is exerted over the sufficiently slow modes of the motion induced by the disturbance, and in the rest the stability of flight is to be secured through the inherent properties of the machine. In other words, a case of partial control in the sense of the preceding section is spontaneously achieved.

The second aspect will be now examined. Suppose that the constrained variables z were selected so that the corresponding auxiliary system (4.5) should be uniformly asymptotically stable. According to the presented theory, in this case the basic motion considered is uniformly stable or uniformly asymptotically stable (if the variables controlled tend toward zero), respectively, in relation to the free variables. In what case does such a formal result correspond to a physical reality, or what are the additional conditions that have to be fulfilled in every case considered in order that the stability be satisfactory from a practical point of view? The answer is implied by the hypotheses and conditions of the theory, which will have to serve as a starting point. In other words, a number of conditions required by the theory must be fulfilled and transformed into effective quantitative relations by estimations in order to check if they are satisfied. The most significant of these conditions include (i) an acceptable approximation of the relations (4.3); moreover, (ii) the magnitudes of the functions $\eta(\varepsilon)$ and $\delta(\varepsilon)$ (see Definition 1) estimated for securing an adequate value for ε should not be too small as compared to current actual values of the initial disturbance of the free dynamic variables, or of the magnitude at each moment of the constrained variables, respectively.

Without going into detail for the time being, certain objectives to be aimed at for obtaining the most favorable conditions will result immediately, yielding finally as efficient as possible an artificial stabilization.

A first obvious aim to be achieved on the basis of the inherent properties of the flying machine by conveniently selecting the constrained variables is to secure a sufficient "dynamic stability margin" with respect to the free variables, hence a highly stable auxiliary system.

Another aim is to secure through the controls a most convenient effective upper bound of the norm of the vector z. (If certain contradictory conditions implicated by other requirements (e.g., maneuverability) do not interfere, "the most convenient" effective upper bound of the value of the variables z coincides with the smallest possible one.) The smaller the effective upper bound, the easier it will be to satisfy the required condition $|z(t)| < \delta(\varepsilon)$. On the other hand, the smallest possible upper bound of z will lead, on the basis of the relation $R(t, y, 0) = 0$ and of the continuity of the function R with respect to z, to the smallest possible values of this function, hence for a given ε to the highest possible admissible values of η and δ.

The main aim of the control is to maintain the value of the variables z within some given bounds. But the secondary effects of the controls also influence the efficiency of stabilization through partial control by the agency of the additional terms $|Y_1(t, y, z) - Y(t, y, z)|$. The secondary effects are meant here to include those consequences of applying the controls that are not desired or intended by the pilot. In the equations of motion, if not neglected, the secondary effects can appear in the form of certain unknown terms that are estimable at the most, instead of appearing in the form of dynamically defined terms. In that form they will enter into the expressions of the functions Y_1 and Z_1, hence also of the function R. A simple example of a secondary effect attending the control of the longitudinal attitude or of the incidence by actuating the stick will be given. In this case, the main effect consists in the generation of a longitudinal control moment achieved by a modification of the nondimensional aerodynamic coefficient C_m of the machine. But the deflection of the elevator causes a modification of the aerodynamical properties of the machine in the aggregate, and in particular as a secondary effect also a change of the value of the lift and drag coefficients directly, that is, not through the agency of the incidence or other state variables.

In computations, the secondary effects are most frequently neglected. Theory points out rigorously the limits within which the neglect is valid (cf. Malkin's theorems concerning stability under persistent disturbances, the theorems by the first approximation, etc.). For certain applications it is not always required to define quantitatively the validity range for each case considered. This category includes the problem of investigating the stability property of certain conventional machines operating under standard conditions, by means of certain stability criteria confirmed through repeated use. However, as soon as a qualitatively new element interferes, either as a design formula or in the form of operating conditions, and so on, the effective quantitative estimation of the validity range in each particular case considered appears first and foremost, frequently implying also the necessity to take into account the secondary effects of the controls. It is to be noted, however, that some of these secondary effects prove on actual estimations to be always negligible.

Finally, the efficiency of the controls (in stabilizing the flight through partial control, too) depends on the relation between the dynamic characteristics of the airframe and those of the control system, particularly on the length of the time lag. Indeed, a certain interval of time elapses between the instant a disturbance appears and the instant the corresponding control moment or force becomes manifest as a result of the control applied. The processes occurring in this interval of time include:

(i) sensing or detecting of the disturbance by the pilot or automatic pilot as a command signal;

(ii) the response of the pilot, or automatic pilot, including such moments as the computing one, which consists in comparing the signal at the input†

† The command signal does not generally reduce itself to a deviation from the undisturbed value of the state variable considered. The pilot reasons on the basis of the aggregate information received (not necessarily simultaneously) about the controlled variable, relying on his flying experience. In this respect the possibilities of the computing element of the automatic pilot are inevitably more limited. For practical purposes they are, however, in many cases sufficient. Thus, if a position coordinate, e.g., an angular attitude, is followed up, detecting by the automatic pilot of the first derivatives with respect to time (and moreso if the higher-order derivatives are added) of the respective deviation amounts to forecasting the behavior of that deviation during the interval of time following immediately. Alternatively, detecting the respective integral term (the integrated value of the deviation from the instant when the disturbance appears till the instant considered) somehow amounts to taking into account the previous history of the deviation induced by the disturbance (the memory of the computing element).

with the known potentialities of the controls of the machine; selecting the dynamic variables that will be acted upon; choosing the controls considered to be the most efficient as well as the manner in which they will be acted upon;

(iii) the performance by the pilot or servomotor of the control motion in keeping with the decision taken;

(iv) further transmission of the controls through the respective control system linkage (cables, rods, beams, etc.) to the output, that is, the elements performing the control motion (aerodynamical control surfaces, engine throttle, additional control jet nozzles, etc.);

(v) the aerodynamic (Wagner effect) or gas-dynamic transition process (to which the mechanical transition process is added in the case of motors with spinning masses) until a steady state is reached corresponding to the new setting of the control elements, hence until the desired control force or moment appears.

During the interval of time that the processes mentioned last, the motion induced by the disturbance develops under more or less free conditions. (The reservation refers to stage (v), in which the transition process of development of the control forces takes place. As a rule this stage is short in relation to the whole interval of time corresponding to processes (i)–(iv) and can be neglected, while the disturbed motion is considered to be free during this interval.)

For stabilization through controls to achieve its object (i.e., to be sufficient) it is necessary that during this interval of time the disturbance-induced deviations of the values of the kinematic parameters subject to constraint (the variables z) should not have acquired dangerous values, or changed their direction of variation. In the latter case, the controls would emphasize the effect of the disturbance instead of countering it. We are thus led to consider the two basic quantitative characteristics of the free disturbed motion: the increment velocity (or, conversely, the decrement or damping velocity) of the induced deviations, and their frequency of oscillation. The deviations of the various dynamic parameters generally vary in different ways. The magnitudes of the two quantitative characteristics mentioned above compared by means of an adequate relation with the length of the time interval required for the completion of the previously described processes (i)–(iv) constitutes one of the significant criteria for selecting from the aggregate of the main kinematic parameters those that are to be constrained under partial control. An inadequate relation would lead to high effective values of the variables z having the above-mentioned implications.

To conclude, the efficiency of partial control to secure flight stability is directly connected with the selection of the constrained variables (from the aggregate of the essential state variables), as well as of the corresponding controls, implying:

(a) a uniformly asymptotic stability of the auxiliary system with the heaviest possible damping, failing a contradictory requirement of another nature than that of stability (e.g., of maneuverability);

(b) the smallest possible damaging secondary effects, such that the resulting values of the functions $\eta(\varepsilon)$ and $\delta(\varepsilon)$ should not be too low;

(c) time lags of the control system small enough to permit sufficiently small effective values of the controlled variables, or their vanishing at a sufficiently rapid rate.

4.4. CONSTRAINED STABILITY ACCORDING TO NEUMARK

The general mathematical framework within which Neumark's theory of constrained stability is placed will be set forth first, and the efficiency of this method will then be commented on. For the subject matter treated in this chapter, Neumark's method of constrained stability is of documentary significance. Indeed, [2] furnished the notion of treating the stability, when only a share of the dynamic variables are constrained, without adding an additional number of degrees of freedom (corresponding to the control variables). This approach results in considering a simplified system in which the constrained variables do not appear instead of the initial complete system.

The general idea as defined in [2, page 21] reads: "It is assumed that one of the controls is operated so as to keep one of the usual variables or a linear combination of them, equal to 0 (or constant), while other elements of the disturbance are still free to vary according to the (suitably modified) system of dynamic equations. The modification of the system will consist in one equation (that directly affected by the control in operation) being removed, while the remaining equations simplify in a manner consistent with the assumption." (The assumption is taken into consideration that the controlled variable is identically zero. Note that the equation "directly affected" by the control applied does not necessarily coincide with the equation referring to the controlled variable from the initial system completely written in the standard form.)

The corresponding mathematical model will be presented in a more general form by considering a certain (nonlinear and nonautonomous) system of differential equations containing a certain number of constrained variables.

Let (cf. system (4.1))

$$\frac{dy}{dt} = Y(t, y, z), \qquad \frac{dz}{dt} = Z(t, y, z) \tag{4.6}$$

where y and z denote k- and m-column vectors ($k + m = n$), and Y and Z are as in (4.1), the system of equations of uncontrolled motion. Let the variables y be designated *free*, and the variables z *constrained*, in keeping with Neumark's usage. The content of the latter concept differs from that of the concept of controlled variable defined in the preceding sections of this chapter; in compliance with the general idea in [2] quoted above, it will be assumed that by operating the controls, the constrained variables z are canceled out and kept zero. (Recall that the preceding sections admitted only that the value of the controlled variables is kept, through the controls, within a certain (sufficiently narrow) limit, and is occasionally in addition subject to the requirement of damping in time.)

It will be further assumed that actuating the controls modifies only a part of the system's equations, and that the number of equations affected by the controls is equal to the number of constrained variables. (System (4.6) being written in a standard form, each of its (scalar) equations corresponds to a certain variable. The last hypothesis is not completed with specifications as to some correspondence between the equations written for the constrained variables and the equations affected by the controls.)

Then, provided certain mathematical conditions are satisfied (see the Appendix), the behavior of the free variables will be perfectly determined through the equations that are unaffected by the control, in which zero was substituted for the constrained variables and their time derivatives. The equations affected by the control can serve subsequently to estimate the corresponding control variables.

In this way, a simplified system of differential equations the order of which is at the most equal to $k = n - m$ (the number of free variables) will come to be considered instead of the system (4.6). Let

$$\frac{du}{dt} = U(t, u) \tag{4.7}$$

be this system, where u is a p-dimensional column vector with $n - 2m \leqslant p \leqslant n - m = k$ (for details concerning the deduction and form of system (4.7), see the Appendix to this chapter).

Unlike the auxiliary system (4.5) considered in the preceding sections, which (rigorously speaking) represents a mathematical fiction, system (4.7) constitutes a (highly simplified) direct mathematical model describing real flight with the approximation of the hypotheses made in describing system (4.6) and of the additional hypotheses enumerated. In the previously presented theory it was necessary to give a mathematical proof of the equivalence of the results obtained on the basis of the auxiliary system (4.5) with those corresponding to the model considered to represent the physical phenomenon (4.2). For the present theory, if the validity of the results furnished by the system (4.6) is admitted, then the validity of the results obtained on the basis of the system (4.7) is conditioned only by the validity of the additional hypotheses, that is, their correspondence with the real situations. We sum up these hypotheses as follows.

(i) Through the controls, the constrained variables are kept identically zero ($z = 0$, $dz/dt = 0$, $d^2z/dt^2 = 0$, etc.);

(ii) The controls affect only a share of the equations, and the number of the affected equations is equal to the number of the constrained variables.

Another hypothesis, mathematical in nature, is to be added to the foregoing, as will be seen in the Appendix.

Let us now examine these two hypotheses. In compliance with the requirement expressed in the theory set forth in the preceding section about the main effect of the controls (maintaining the value of the controlled variables within acceptable limits, viz., $|z(t)| < \delta(\varepsilon)$, during the whole course of the disturbed motion, or in addition their tendency to damping in time: $\lim_{t\to\infty} z(t) = 0$), the partially controlled disturbed motion is considered starting from some arbitrary moment t_0, for instance, from the moment when the disturbance occurred. As a result, the duration of the transitory process induced by actuating the controls, or in other words, the time lag of the control system, including the response of the machine, plays a part only in determining the efficiency of the partial control without affecting the validity of the model as such.

In Neumark's theory of constrained stability, however, the part played by this lag becomes decisive in determining the validity of the (simplified) model used. Indeed, the conditions of the first hypothesis, that not only the constrained variables z but also their derivatives dz/dt be equal to zero, can on no account even approximately be fulfilled during the transitory process that takes place following the disturbance and operation of the controls. Hence, a moment subsequent to the practically complete expiration (damping) of this transitory process must be considered as the initial moment t_0. Hence, the first hypothesis implies also the acceptance of another assumption, namely, that the duration of the transitory process of response to the controls is small compared to the lapse of time between two successive disturbances. But for the process to be useful in practice, it is necessary that the transitory period of response to the controls also be short in comparison to the whole disturbed motion. In the language of the respective model, this means that the time lag of the control system, including the response to the controls, must be as small as possible in comparison to the duration of the response to the disturbance of the free variables. This condition obviously cannot be fulfilled unless it is deliberately taken into account when selecting the constrained variables and the controls.

The case of keeping the flight velocity constant under throttle control, considered in [2], is an example. The simplified equation (4.7) coincides in this case with the well-known system of equations of short-period oscillations (see [2, page 26]). It is, however, doubtful that such a system could be obtained consistent with the hypotheses of the constrained stability, since in most of the usual cases the damping of these oscillations is practically exhausted long before it could have been possible to cancel the deviation of the flight velocity through the mentioned control. Note that the objection refers to the hypotheses underlying the theory exposed, and not to the possibility of considering the system of equations of the short-period oscillations under the conditions of a partial control of flight velocity (and of the longitudinal attitude). On the contrary, as will be seen in Section 6.2, considering this system under the conditions described is rigorously justified within the framework of the theory set forth in the preceding sections.

Some reservations could also be expressed in connection with the second hypothesis, and with its implications concerning the practical process of inferring the simplified system (4.7) through removal of the equations affected by the controls.

As is known (see the preceding section), in order to secure the highest efficiency of the control so as to determine in a given case a most energetic counteraction to the deviation of a certain variable induced by the disturbance, the pilot generally resorts to several controls simultaneously. But a single control generally also affects the quasi-totality of the equations of motion (it generally leads to a modification of all the equations of the system). Under these conditions, the hypothesis concerning the equality of the number of constrained variables with that of the equations affected seems artificial.

Also to be examined are the conditions under which it is justified to neglect the equations "directly affected" by the controls. If the neglected equations are those corresponding (in the system written in standard form) to the constrained variables, their omission is, at least formally, compensated by the hypothesis $z \equiv dz/dt \equiv 0$. However, if among the removed equations are some written for the free variables, the validity of the procedure, being subject to discussion, does not appear of itself. Indeed, this case implies the admission that the effect sought by operating the controls (canceling of the constrained variables) does not coincide with their main (direct) effect, bent immediately on affecting the behavior of the ("free") variables, but only through their agency, and by a concatenation of processes on constraining the variables z. It is, therefore, probable that the hypothesis postulating the last consequences of the chain of processes, that is, the identical vanishing of the vector z, might not be generally sufficient, and that it would be necessary also to consider these processes in themselves, hence the equations corresponding to the variables y, which were left out for being directly affected by the controls. In such cases, the removal of these equations can lead to invalid results. It seems that this category is to include the case of maintaining a constant velocity under elevator control, in which there results a similar aperiodical (viz., exponential) very rapid decrease for the incidence and the angle of pitch considered as free variables (see [2, page 25]).

Moreover, omitting none but the equations directly affected by the controls (to diminish the order of the system in keeping with the decrease of the number of variables, by postulating the identical vanishing of the variables z) appears somehow arbitrary. It is recollected that hypothesis (i) mentioned earlier in this section is acceptable only during the interval of time between the consummation of the transitory process and the appearance of a new disturbance causing z to become different from zero. Thus, during the whole interval in which the problem of constrained stability is considered, the control force is zero with the same approximation as z, or dz/dt. (Otherwise, instead of stabilizing the flight, the controls would act as disturbing forces, which could be contrary to the aim.) Therefore, in the absence of some control we cannot speak of equations affected by controls, making a distinction between these and the ones that are not affected.

To conclude, a decision has to be reached for each case separately as to whether the selection of the equation removed by the method of the constrained stability is justified or not. For its simplicity and the practical methods of computation that it implies, however, this method is too attractive to be given up without first testing the validity of its hypotheses for the case considered.

APPENDIX

DEDUCTION OF THE SYSTEM (4.7)

Consider the system (4.6) to start with, namely,

$$\frac{dy}{dt} = Y(t, y, z), \qquad \frac{dz}{dt} = Z(t, y, z)$$

with

$$y = \begin{pmatrix} y_1 \\ \vdots \\ y_k \end{pmatrix}, \qquad z = \begin{pmatrix} z_1 \\ \vdots \\ z_m \end{pmatrix}$$

where y_i and z_j stand for the free variables and constrained variables of the system, respectively. It is assumed that a number of p equations out of the k scalar equations $dy_i/dt = Y_i(t, y, z)$, and a number of q equations out of the m scalar equations $dz_j/dt = Z_j(t, y, z)$ are not affected by the controls. By hypothesis, $p + q = k$. It is also admitted that z vanishes identically as a result of actuating the controls, and no other disturbances appear subsequently. Hence, for any $t \geqslant t_0$, $z(t) = 0$ and $dz/dt = 0$.

Let therefore

$$\begin{aligned} \frac{dy'}{dt} &= Y'(t, y', y'', 0, 0), \\ 0 &= Z'(t, y', y'', 0, 0), \end{aligned} \tag{1}$$

with

$$y' = \begin{pmatrix} y_1 \\ \vdots \\ y_p \end{pmatrix}, \qquad Y' = \begin{pmatrix} Y_1(t, y, 0) \\ \vdots \\ Y_p(t, y, 0) \end{pmatrix},$$

$$z' = \begin{pmatrix} z_1 \\ \vdots \\ z_q \end{pmatrix}, \qquad Z' = \begin{pmatrix} Z_1(t, y, 0) \\ \vdots \\ Z_q(t, y, 0) \end{pmatrix},$$

be the equations unaffected by the controls, and

$$\begin{aligned} \frac{dy''}{dt} &= Y''(t, y', y'', 0, 0, \sigma), \\ 0 &= Z''(t, y', y'', 0, 0, \sigma), \end{aligned} \tag{2}$$

with

$$y'' = \begin{pmatrix} y_{p+1} \\ \vdots \\ y_k \end{pmatrix}, \qquad Y'' = \begin{pmatrix} Y_{p+1}(t, y, 0, \sigma) \\ \vdots \\ Y_k(t, y, 0, \sigma) \end{pmatrix},$$

$$z'' = \begin{pmatrix} z_{q+1} \\ \vdots \\ z_m \end{pmatrix}, \qquad Z'' = \begin{pmatrix} Z_{q+1}(t, y, 0, \sigma) \\ \vdots \\ Z_m(t, y, 0, \sigma) \end{pmatrix},$$

the equations "affected" by the controls. The m vector σ represents the control variables, and $Y_i(t, y, 0, \sigma)$ and $Z_j(t, y, 0, \sigma)$ the right-hand sides of the respective equations over the interval within which the controls operate. Moreover, as was mentioned in Section 4.4, over the semiaxis $t \geqslant t_0$ in which $z \equiv 0$, $\sigma \equiv 0$ ought also to obtain, and therefore it would be more accurate to write the equations (2) without introducing the new function σ, in the form

$$\frac{dy''}{dt} = Y''(t, y', y'', 0, 0), \qquad \frac{dz''}{dt} = Z''(t, y', y'', 0, 0)$$

with

$$Y'' = \begin{pmatrix} Y_{p+1}(t, y, 0) \\ \vdots \\ Y_k(t, y, 0) \end{pmatrix}, \qquad Z'' = \begin{pmatrix} Z_{q+1}(t, y, 0) \\ \vdots \\ Z_m(t, y, 0) \end{pmatrix}$$

while y'' and z'' are as above. However, this system would generally be incompatible with (1). Indeed, it can be seen that for determining the free variables y (i.e., the unknown functions y_i, $i = 1, 2, \ldots, k$) the system (1) ($p + q = k$) is sufficient without any other supplementary relations. Only this system will be considered in the sequel. Then, admitting at last that for the considered values of the components of the vector y'', the Jacobian

$$\det \frac{\partial Z'(t, y', y'', 0, 0)}{\partial y''} \neq 0,$$

the equation $Z'(t, y', y'', 0, 0) = 0$ can be solved with respect to the unknown quantities y'', and we have $y'' = y''(t, y')$. Noting that $y' = u$ and $Y'(t, u, y''(t, u), 0, 0) = U(t, u)$, we have the simplified system (4.7)

$$\frac{du}{dt} = U(t, u).$$

This system is of the order p (i.e., the same order as that of the system $dy'/dt = Y'$) and p varies between the values $n - 2m$, corresponding to the situation in which none of the equations written for the constrained variables is "directly affected" by the controls ($Z = Z'$), and $n - m$, corresponding to the situation in which all the equations $dz/dt = Z$ are affected by the controls ($Z = Z''$).

REFERENCES

1. T. Hacker, "Stability of partially controlled motions of an aircraft," *J. Aerospace Sci.* 28(1961), 15–26.
2. S. Neumark, "Problems of Longitudinal Stability below Minimum Drag Speed and Theory of Stability under Constraint," *Aeronaut. Res. Council Rept. & Memo.* 2983(1957).
3. C. D. Perkins, *Flight Test Manual*, Vol. 2, *Stability and Control.* AGARD.
4. W. H. Phillips, B. P. Brown, and J. T. Matthews, *Review and Investigation of Unsatisfactory Control Characteristics Involving Instability of Pilot-Airplane Combination and Methods for Predicting these Difficulties from Ground Tests.* NACA TN 4064(1957).
5. A. Halanay, *Differential Equations: Stability, Oscillations, Time-Lags.* Academic Press, New York, 1966.

Chapter 5

LINEAR SYSTEMS WITH CONSTANT COEFFICIENTS: STEADY BASIC MOTIONS

5.1. INTRODUCTORY REMARKS

The model of systems of linear differential equations with constant coefficients covers, as far as the mathematical approach is concerned, the entire field of what has so far existed in the domain of flight stability as a unitarian theory as well as techniques having taken root in current practice, in both design and investigation. Indeed, starting with the use of small-disturbance methods and the application of Routh's theory to the problem of flight stability by Bryan and Williams in 1903,† the linearized model of equations in variations has remained the basis of the methods used in practically every work in this area. Moreover, it is the only method firmly implanted in the great works of synthesis, treatises and textbooks dealing with the stability of flying machines. The reason for this uncommon vitality should first be sought in the general mathematical properties of the linear model, particularly of the linear systems with constant coefficients.

As regards linear systems with constant coefficients, the methods are as effective as they are universal: (i) for such systems the solution is obtained in terms of elementary functions by means of simple algebraic operations; (ii) there exist general simple stability criteria.

This model, so easy to handle and convenient, has been blended with the conventional theory of the dynamic stability of aircraft. A new element

† The first communication regarding the application of Routh's theory to the problems of gliding flight stability was made by Bryan and Williams in 1903, even before the first flight of the Wright brothers, and was published in 1904 [1]. The first theory of aircraft stability based on a general theory (Routh's) was discussed in a consistent form in [2]. Routh's theory [3, 4] is essentially one of stability by the first approximation, based on the linearization of the disturbed motion equations. Only later did Liapunov, stating the conditions of stability by the first approximation, substantiate rigorously the use and indicate the limits of applicability of this class of models (see Section 5.2).

has appeared relatively recently, coming to strengthen a tradition that was already strong. This element has issued from the contact of flight dynamics, particularly the theory of flight stability, with the linear theory of control systems mentioned in the preceding chapters. As a result of this association, the theory of flight stability acquired some techniques that had already proved their efficiency in several domains of applied science (chiefly of engineering). Thus, operational technique is currently resorted to in order to solve some problems of flight stability: instead of equations, the so-called transfer characteristics (transfer functions and frequency characteristics), which can be directly determined experimentally, are used to describe the operation of the single control system, including the airframe (but sometimes even only the mechanical motion of the machine); instead of stability criteria of the Routh–Hurwitz type, other equivalents are considered, which can, like the former, be deduced from Cauchy's theorem of the variation of argument, adapted, however, to the specifics of operational methods (Nyquist-type criteria and the like). It should be stressed that all these new techniques do not, actually, introduce a new mathematical content, but remain within the boundaries of the theory of linear systems with constant coefficients.

The prevailing use of this model in the theory of flight stability has conclusively influenced the definition of the fundamental concepts, beginning with the very definition of the property of stability, and since, as underlined in Chapters 1 and 2, stability is a property of the basic motion, in this theory stability is defined only for the steady basic motion,† particularly for steady straight level flight, to which, by the first approximation,

† "Our book only discusses the theory of the stability of steady straight flight. It should not be thought that the notion of stability only applies to stationary motions, it can be applied to any motion of the aircraft. Nevertheless, we abstain from formulating the general notion of stability, since it will not be required in our book" [5, page **11**].

"It must also be emphasized that the concept of stability only applies to systems in equilibrium, at rest or in some regular motion, either free or with prescribed forces. We cannot, for example, discuss the stability of an aircraft which is not trimmed to fly at the speed and in the attitude postulated" [**6**, page **113**].

"The equilibrium of an airplane in flight ... is of the second kind: that is, uniform motion .... Stability, or the lack of it, is a property of an equilibrium state" [**7**, page **4**].

"A dynamical system is said to be stable, or to possess stability if, when slightly disturbed from a state of equilibrium, it tends to return to and remain in that state, the disturbance acting only for a finite time" [8, page **36**].

there corresponds a system of linear equations with constant coefficients for disturbed motion.

The requirements in actual flight practice are (i) to exclude any possibility for the deviations induced by disturbances to be amplified, at any stage or instant of the disturbed motion; and (ii) to finally restore (and, practically, in a reasonably short lapse of time) the undisturbed (basic) motion. These two requirements are expressed, with the reservation of some numerical estimates, by the mathematical concept of asymptotic stability in the sense of Liapunov, defined in Section 3.1. In the current literature devoted to the flying qualities of aircraft, stability is generally defined in compliance with the second requirement (of complete damping out of the disturbances),† the first one, in the case of linear systems with

† "If, in the solution obtained, the variations with time *tend towards zero*, the aircraft is considered stable" [5, page 13].

"Stability of motion of a body is the property of the kinematic parameters which characterizes this motion, viz., *to return to their initial values corresponding to the basic motion*, after a deviation of the body from its initial motion and the discontinuance of the action resulting from the cause which determined this deviation" [9, page 215].

"If the subsequent motion [following any disturbance of the equilibrium] finally *restores the equilibrium*, the system is termed dynamically stable" [10, page 6].

"The equilibrium [in the sense of a uniform motion, see footnote at preceding page] is *stable* if, when the body is slightly disturbed in any of its degrees of freedom, it returns ultimately *to its initial state*" [7, page 4].

"A dynamical system is said to be stable when the constituents of the disturbed motion *ultimately become vanishingly small*" [8, page 126].

All the italics and notes in brackets are ours. Some of the definitions found in the literature even point to the fact that stability implies *only* this asymptotic behavior and not the maintenance of the disturbed motion all the time in the neighborhood of the basic motion:

"It must be emphasized that stability is a term concerned with the ultimate consequence of a disturbance. Thus, it might happen that a disturbance became largely amplified in the early stages of its history in a completely stable system" [6, page 113].

In other words, the exclusiveness of the requirement regarding asymptotic behavior is also extended to cases where this behavior does not also necessarily imply excluding the possibility of an increase beyond the admissible bounds of deviations during the whole time interval following a disturbance and hence also to the unsteady basic motion (for which, moreover, the stability in [6] is not defined).

constant coefficients, being implied. Indeed, in this case the two properties of the solution, stability and asymptotic behavior, are invariably correlated: any stable solution is implicitly exponentially stable too; conversely, whenever the solutions of the system, differing from the trivial one, tend asymptotically toward zero, then the trivial solution is necessarily stable.

A concept that arose under the influence of the adoption as a mathematical model of the linear system with constant coefficients, and of the fact that Routh's general theory had been considered satisfactory for all the steady basic motions, is that of neutral stability.

Postulating the validity of the linear model with constant coefficients for all cases corresponding to steady straight level undisturbed flight (for a discussion of the limits of this model's validity, see Section 5.2), the conventional theory considers possible, as regards stability, three situations, according as (i) the real part of all the characteristic roots of the system is negative; (ii) at least one of them is positive; and (iii) one at least is zero, and the nonzero real parts are negative, namely those of stability, instability, and neutral stability†). As will be mentioned in the following section, in the case of a zero characteristic root, usually it is not the basic motion of the machine that has a "neutral" behavior, but the solution of the linear system with constant coefficients. Indeed, for this system, in the case of any zero or imaginary characteristic root, there corresponds to every initial condition a certain value, different from zero, toward which the respective solution (or its amplitude) tends asymptotically. This limit value depends continuously on the initial conditions, generally being small together with them. It therefore comes to a case of nonasymptotic stability of the trivial solution, in Liapunov's sense. Actually, in the case of characteristic roots having their real part zero or small in value, the linear model ceases to represent the effective motion adequately. The theoretical problem becomes

† "In the intermediary case it may happen for the variation to tend towards a constant value whatever, different from zero; in this case, the aircraft is neutral" [**5**, page **13**]. "The concept of stability of the flying machine is used to characterize the general properties of free disturbed motion. This concept is related to three kinds of variation of the deviations [state coordinates] in free motion. In the first case, for an unlimited increase of time the deviations of all the motion parameters ... tend towards zero: they damp out. It is then said that the aircraft is stable. In the second case, the deviations do not damp out, but they do not increase, either. It is said of such an aircraft that it is *neutral*. In the latter case all deviations with time increase infinitely and the aircraft is unstable" [**11**, pages **447–448**].

very involved, and it becomes necessary to take into account terms of higher degree. The problem of neutral dynamic stability will be discussed in Sections 5.5 and 5.6. For the time being, it will only be noted that a situation where the deviations—strictly speaking—do not either damp out or increase is scarcely likely to appear in flight practice, neutral behavior meaning, in most cases, very slow variations of these deviations. (Cf. the remark in [12, page 53] regarding the "spiral" indifference. Defining the spiral indifference by the constancy of sideslip angle deviation, we are reminded that the behavior of conventional aircraft is usually *neighboring* to spiral indifference. It is believed that, *mutatis mutandis*, the same remark is valid for all the motions to which "neutral stability" is attributed: they are in the neighborhood of the ideal case corresponding to the linear model.)

Some arguments of a mathematical nature that explain the use of linear systems with constant coefficients as major models in the theory of flight stability have been reviewed above. Nevertheless, the justifying of the model's efficiency would be neither complete nor convincing if some aspects of the practical importance of the actual motions for which this model is adequate with a sufficiently high degree of accuracy were not also added to the argument.

As already recalled, the system of disturbed motion equations and hence, particularly, matrix A in (1.9) as well is dependent on some design characteristics of the machine and on the régime of undisturbed flight. Hence, in order for matrix A to be constant, it should be assumed that both the characteristics of the aircraft and those of the flight régime are constant. Hence, strictly speaking, the linear model with constant coefficients corresponds, at most,[†] only to stationary basic motions. The conventional theory of aircraft dynamic stability deals only with such basic motions. In this book the physical model of stationary motions will constitute the object of Chapters 5 to 7. The case will be represented in the framework of the theory, expounded in Chapter 4, of the stability of partially controlled motions. The well-known results of the theory of free control stability will not be related here, while the considerations connected with the hypothetical case of fixed controls will be reduced to the minimum necessary from

[†] The reservation refers to the fact that, generally, from the linear model we can deduce only local stability. As mentioned, the stationary basic motion is itself a model. The motions, which will be called quasi-stationary or quasi-autonomous, that constitute the originals of this model will be dealt with further in Chapter 8.

the standpoint of the material presented. We will also discuss, in the framework of the autonomous linear model: in Section 6.4, the problem of height perturbation on uniform steady level flight (according to Neumark), and in Section 7.3, some aspects of the stability of VTOL aircraft in hovering flight.

5.2. LIMITS OF VALIDITY FOR LINEAR SYSTEMS OF EQUATIONS WITH CONSTANT COEFFICIENTS: QUASILINEAR SYSTEMS

As already mentioned, since Bryan's and William's paper [1], and down to the present day, the small-disturbance method has made up the base for almost all models of disturbed motion used in problems of flight stability, the linear approximation being *a priori* admitted. It should be noted that the general theory justifies linearization in the most frequent cases of flight practice. Moreover, in the cases of asymptotic stability and instability linearization gives clear and accurate results. On the other hand, the general theory invalidates some conclusions implied by linear approximation for cases at the boundary between asymptotic stability and instability, finding them either unjustified theoretically but actually proper, with a sufficient degree of approximation, or else erroneous.

Indeed, as already seen in Section 3.4, the main point in the theory of stability by the first approximation is that, under rather general conditions imposed on the nonlinear terms, neglect of these terms is rigorously justified if the trivial solution of the corresponding linear system is uniformly asymptotically stable, whereas taking into account the nonlinear terms is essential when the trivial solution of the linear system is merely stable (but not asymptotically).

Now, the theory set forth in Section 3.4, giving, as stated by Liapunov ([13, Chapter 2]; see also [14, Chapter 4, Sections 4–7]), the theorems of stability and instability by the first approximation when the system of first approximation is linear with constant coefficients, will be applied to a special case. To this end it will be assumed that matrix $A(t)$ in the right-hand side of system (1.8) is constant. The following system will therefore be considered

$$\frac{dx}{dt} = Ax + N(t, x) \tag{5.1}$$

where A is constant matrix and $|N(t, x)| < l|x|$ at any instant t, for $|x| < h$, h being a given constant, and l a sufficiently small constant. Moreover, $N(t, 0) \equiv 0$. Then the theorem of stability by the first approximation may be stated as follows.

THEOREM 1. *If all the eigenvalues of matrix A (which, according to current use, will be called in all that follows the characteristic roots of the system) have negative real parts, then the zero solution of system* (5.1) *is uniformly asymptotically stable.*

The theorem of instability by the first approximation will be:

THEOREM 2. *If, among the characteristic roots of system* (5.1) *there is at least one having a positive real part, then the trivial solution of this system is unstable.*

If matrix A has no eigenvalues with positive real parts, but it has at least one zero root, or a pair of pure imaginary roots, then the trivial solution of system (5.1) is either uniformly asymptotically stable, or uniformly stable or unstable according to the form of function N.

Thus, if system (5.1) has, besides characteristic roots having negative real parts, at least one having its real part equal to zero, then consideration of the system $dy/dt = Ay$ alone yields no valid conclusions regarding the stability applicable to the basic motion described by the zero solution of nonlinear system (5.1), and it also becomes necessary to account for the nonlinear terms. If not, the conclusion concerning the neutral behavior of the trivial solution for the entire nonlinear system is not rigorously justified, except, perhaps, for some particular cases. (Such a case will be considered in Section 6.3. Likewise, it may also be admitted that the basic motions properly modeled by systems of the type (3.26) or (3.27)—a case often met hereafter, have neutral stability. Further details will be found in Section 5.6.) An investigation is therefore useful, from the standpoint of the mathematical model, of what, in flight tests, seems to the pilot to be a neutral behavior. This problem will be discussed in Sections 5.5 and 5.6.

5.3. STABILITY REGION IN SPACE OF THE PARAMETERS

The term "stability region in space of parameters" seems self-explanatory. The importance of determining this domain in problems of flight dynamics proceeds from the possibility of the existence of unstable basic motions in the case of certain design characteristics. They can, however, become stable by appropriate modifications or by adequate operation of the controls. This is connected with the problem of dangerous and safe stability boundaries and of the mathematical modeling of neutral stability, which will be discussed in subsequent sections.

The system parameters for the models used in this book for disturbed motion, will, on the one hand, characterize the basic motion (flight régime) and, on the other, the geometric (i.e., aerodynamic) properties, the aircraft's weight and center-of-mass location, and so on.

Let m be the number of parameters considered essential for the given problem. Then the m-dimensional vector p defines a point in the hyperspace of the parameters. A domain is sought in this space (hereafter the terms space, surface, etc. will be used instead of hyperspace, hypersurface, etc., even when $m > 3$) in which the trivial solution of the system taken into consideration, for example,

$$\frac{dx}{dt} = X(t, x; p), \tag{5.2}$$

is everywhere uniformly or uniformly asymptotically stable. This domain will be referred to as the *stability region* in the space of the parameters of system (5.2). The boundary separating it from the domain of instability will be called the stability boundary of the system (5.2).

The region of stability or some of its parts is determined by means of the criteria deduced from the methods for investigating stability mentioned in Chapter 3. Thus, for example, if a Liapunov function $\mathscr{W}_1(t, x; p)$ has been found, positive definite for any point p of the space of parameters, then, if for any t and x

$$\left(\frac{d\mathscr{W}_1(t, x; p)}{dt}\right)_{(5.2)} < 0 \tag{5.3}$$

where the left-hand side of the inequality represents the total time derivative of $\mathscr{W}_1$ along the integral curves of system (5.2), (5.3) defines a set of points in the space p belonging to the stability region. If a function $\mathscr{W}_2(t, x; p)$ is found such that its time derivative along the integral curves of system (5.2) be negative definite, then the points in space p where $\mathscr{W}_2$ vanishes for any t and x define the stability boundary.

Hereafter, the stability region and the stability boundary will be defined on the basis of the theory of stability by the first approximation. Since this chapter deals with autonomous systems, their connection to the conventional definition of the stability boundary will be achieved in this manner. Indeed, the knowledge of certain criteria indicating the sign of the real parts of the characteristic roots of linear systems with constant coefficients without actual integration (Routh–Hurwitz criteria; see Appendix) has suggested, in the study of the flying qualities, the use of some diagrams showing the stability boundary in the plane of a certain pair of parameters, in the hypothesis that the linear approximation is valid. Routh's or Hurwitz's criteria being, as will be seen, expressed in the form of sets of inequalities, the stability boundary is defined by the points in the plane of the chosen parameters in which at least one of the inequalities of the respective set becomes an equality.† Hence, the Routh–Hurwitz criteria yield critical test functions to be used for plotting the stability diagram instead of the characteristic roots of the system itself. The same critical test functions can be used according to the theorems of stability by the first approximation stated in the preceding section for systems of the form (5.1).

Let

$$\frac{dx}{dt} = A(p)x + N(t, x; p) \tag{5.4}$$

be the system in which emphasis was laid on the dependency of the constant matrix A and of the nonlinear parts on the system parameters. Let $\lambda_i(p)$, $i = 1, 2, \ldots, n$, be the eigenvalues of the matrix $A(p)$ and $\nu_i = \operatorname{Re} \lambda_i$. According to the theorems in the preceding section, the simultaneous satisfaction of the set of inequalities $\nu_i < 0$ implies a uniform asymptotic stability and that of at least one of the inequalities $\nu_i > 0$ $(i = 1, 2, \ldots, n)$ the instability of the trivial solution of set (5.4). Hence the stability boundary in the space of the parameters p will be formed of the totality of points

† "It is sometimes convenient to plot what are known as *stability boundaries*. These are obtained by setting one of the critical test functions equal to zero and allowing two of the airplane configuration or flight variables to change. For each value of one of these variables the vanishing of the test function fixes the value of the other. Thus, a curve is defined in the plane of the two variables which separates a region of stability from one of instability" [7, Section **6.11**].

See also [5, Section **11**] and, as of historical interest, the early papers of Gates and Zimmermann devoted to this subject [**15–17**].

belonging to this space, where at least one $\nu_j = 0$ equality occurs (j being any integer whatever between 1 and n).

The use of the direct procedure, consisting in actually finding the characteristic roots, is always possible, up-to-date computation techniques allowing these to be obtained with any desired degree of accuracy. The procedure is advantageous since knowledge of the characteristic roots not only gives the stability but also yields some basic data required to evaluate the quantitative characteristics of disturbed motion (e.g., the margin of stability proper to the linear model). If adequate computers are not available and if only the influence of certain parameters upon stability is sought (hence, in the phase of preliminary study or even in the rough draft design stage) the actual finding of the roots for systems of an order larger than 4 is not pursued, and it is avoided in the case of systems of an order higher than 2. In cases corresponding to characteristic equations of higher degree with parametric coefficients, one of the indirect procedures mentioned is resorted to; that is, some criteria are applied, indicating the sign of the real parts of the roots without actually finding their expression. Such criteria, widely used in the literature devoted to flight stability, are also those of Routh and Hurwitz (see Appendix I) and, more recently, other equivalent criteria making use of the operational technique, such as, for instance, the Mikhaylov–Leonhard–Cremer graphic criterion (independently formulated by the three authors [18–20]) of stability, or Nyquist's criterion [21].

In the conventional theory the stability boundary is defined in a (two-dimensional) plane. The use of only two basic parameters for defining the stability region simplifies computation considerably and facilitates the analysis of the influence of the system parameters upon stability. To profit by both the advantages shown, however, we consider plane sections in the parameter hyperspace or sets of plane sections, corresponding to constant values or sets of values of all the system parameters except the chosen pair of parameters. Using this procedure, we assume that the imposition of a variation to the pair of parameters of interest in the plane defined by the fixed values of other system parameters does not afford any incompatibility. It should be noted that the fulfillment of such a hypothesis is not evident.

Let p_i, $i = 1, 2, \ldots, m$ be the parameters of the system and p_1, p_2 the pair of parameters for which the stability diagram is plotted. Let it be assumed that p_1 and p_2 are conclusively dependent on other magnitudes q_j, $j = 1, 2, \ldots, l$. If the parameters p_i, $i \geqslant 3$, are not essentially dependent on q_j (their dependency on q_j is negligible), plane sections in the hyperspace p may be considered. If the variation of the magnitude q bears a noticeable influence on the parameters p_i, such a procedure becomes inadequate.

Let $f_k(p_1, p_2, \ldots, p_m) < 0$, $k = 1, 2, \ldots, n$, be the system of inequalities defining the stability region. Let it be assumed, for simplicity, that $l = 2$. Then q_1 and q_2 can be determined as functions of p_1 and p_2 from $p_1(q_1, q_2)$ and $p_2 = p_2(q_1, q_2)$ and the stability diagrams in the plane (p_1, p_2) can be plotted on basis of the inequalities $g_k(p_1, p_2) < 0$, $k = 1, 2, \ldots, n$ where $g_k(p_1, p_2) = f_k(p_1, p_2, p_3(p_1, p_2), \ldots)$.

If p_1 and p_2 vary approximately independently of the remaining system parameters, sets of diagrams can be plotted according to the inequalities $h_k^{\times}(p_1, p_2) < 0$, where $h_k^{\times}(p_1, p_2) = f_k(p_1, p_2, p_3^{\times}, p_4^{\times}, \ldots)$, $p_i^{\times} = \text{const}$.

Consider, for instance, the pair of parameters (C_{m_α}, C_{m_q}), used in the conventional stability diagrams for the longitudinal motion. The longitudinal damping coefficient C_{m_q} is conclusively dependent on some geometrical characteristics of the tail, and the static stability coefficient C_{m_α} depends on the same magnitudes and on the location of the center of gravity. (Since C_{m_q} is not dependent on the center-of-gravity location, C_{m_α} may also be made to vary independently of C_{m_q}. The pair of parameters (C_{m_α}, C_{m_q}) are also advantageous since they can be made to vary effectively even in relatively advanced stages of construction (flight test or even service).) The remaining parameters of system (2.16), such as C_D, C_L, C_{L_α}, may reasonably be considered independent of the foregoing magnitudes (except the derivative $C_{m_{\dot\alpha}}$, which always accompanies the damping coefficient C_{m_q}; actually, the diagram should represent the sum $C_{m_q} + C_{m_{\dot\alpha}}$ as a function of C_{m_α}). Hence, the simplified procedure can be applied for plotting the stability diagram in plane sections, parallel to the plane of the axes C_{m_α} and C_{m_q} in the hyperspace of the parameters.

In the case of lateral motion this is not always possible. For the lateral motion the weathercock stability coefficient C_{n_β} and the lateral aerodynamic derivative C_{l_β}, sometimes referred to as the dihedral-effect coefficient, or their ratio are first chosen as coordinates of the stability diagram. Indeed, the first magnitude depends essentially on the geometrical characteristics and the location of the vertical tail (viz., of the fin), and the second on the effective dihedral of the wing. Since these parameters only influence the performance characteristics to a negligible extent, they can be determined from stability considerations assigned to the functions C_{n_β} and C_{l_β}. (In this respect the use of the rotary derivatives C_{l_p}, C_{l_r}, C_{n_p} as coordinates of the stability diagrams is less desirable since they depend on the geometrical (particularly on the plan form) and aerodynamic characteristics of the wing, which have to be determined on the basis of considerations of performance.) Moreover, C_{n_β} obviously also depends, though to a smaller extent, on the geometrical and aerodynamic characteristics of the fuselage and the engine cars, which implies certain reservations. Special caution is demanded by the following fact. The same design characteristics as the ones connected with the vertical tail surfaces also influence other aerodynamic and rotary derivatives, such as C_{y_β}, C_{y_r}, and particularly C_{n_r}. Hence it is not always possible to consider them fixed. It is recommended to endeavor to express, for example, C_{n_r} as a function of C_{n_β} by eliminating the parameters, such as $p_1, p_2, \ldots, p_k$, that characterize the vertical tail surface in the system of equations in p_i: $f_i(C_{n_\beta}, C_{n_r}, p_1, p_2, \ldots, p_k) = 0$, $i = 1, 2, \ldots, k$. Because of the decisive role of the vertical tail surface parameters in the determination of C_{n_β} and C_{n_r}, their being independently selected is obviously out of question.

5.4. SAFE AND DANGEROUS ZONES OF THE STABILITY BOUNDARY

Sometimes merely plotting the stability boundary proves to be insufficient, a closer investigation of its properties becoming necessary. The notions of safe and dangerous zones of the stability boundary will be discussed later according to Bautin [22]. The discussion may prove useful in elucidating the concept of neutral dynamic stability (in the sense in which it is used in current literature).

Generally, it is not the (rigorous) stability boundary that is of practical interest for the behavior of a nonlinear system, but an entire domain of a certain size ("thickness") in its neighborhood. In this context, what is essential for the idea of a safe (or dangerous) stability boundary is that for a system located in the vicinity of the stability boundary in the parameter space, the behavior of the solution be determined by the behavior of the solution of a neighboring system located on the very boundary. Particularly for autonomous systems, the basic idea amounts to the fact that the nonlinear terms play a decisive role with respect to the practical stability of the solution, not only in the case where there is some zero or pure imaginary characteristic root, the others having negative real parts, but also in the case of small roots of the characteristic equation, or of roots having small real parts (those that are not small or with small real parts, having real parts that are negative).

An additional characteristic, from another—rather formal—point of view, of the idea of safe or danger stability boundary consists in reducing the investigation of the effective behavior of some systems standing in the vicinity of the stability boundary to the investigation of stability in the sense of Liapunov.

As already mentioned, the effective behavior of the solution is not always satisfactorily defined by its stability or instability property in the sense of Liapunov; thus, when the aim is to know if a given system behaves *satisfactorily* from the practical standpoint, the Liapunov stability used directly as a criterion may not be conclusive in two senses: it may be either too strong, too restrictive—when the investigation of ε_0-stability of the system becomes more suitable—or, on the contrary, too weak: the attraction region resulting from the property of stability according to Liapunov being liable to prove too narrow with respect to the size of current disturbances. In this latter case, the disturbances (i.e., initial conditions) not being included in the attraction region, a disturbed motion differing from the basic motion, hence an entirely undesirable behavior, may result although the system is stable and even asymptotically stable according to Liapunov.

Such situations, where the Liapunov stability does not directly characterize the effective behavior of the solution in the very domains that are of practical interest, appear in the first place in the vicinity of the stability boundary, corresponding in the case of autonomous systems to the existence of small characteristic roots or to roots with small real parts, in comparison to the value of the real parts of the other roots, when the latter are negative. The notions of safe and danger points of the stability boundary define the effective behavior of the solution for a system located in the vicinity of the stability boundary, precisely for cases of practical interest.

The significance of this statement will be more apparent from the definition of those two notions that will be given hereafter. Consider the system (5.2) and let $p^{(0)}$ be a point on the boundary of the stability region of the trivial solution of system (5.2). The distance between a point whatever p and the point $p^{(0)}$ will be denoted $\delta(p, p^{(0)})$.

The point $p^{(0)}$ will be called the *safe point* of the stability boundary if for each positive number ε there exist two functions $\eta_1(\varepsilon)$ and $\eta_2(\varepsilon)$ for which the simultaneous fulfillment of the inequalities $|x_0| < \eta_1$ and $\delta < \eta_2$ implies $|x(t, p)| < \varepsilon$, over the whole semiaxis $t \geqslant t$. The solution of the system (5.2) corresponding to the initial condition $x(t_0, p) = x_0$ was denoted $x(t, p)$. It should be noted that the position of point p is defined only by the distance δ, whether it is located in the stability or instability region.

Point $p^{(0)}$ will be a danger point of the stability boundary if it does not fulfill the conditions required by the foregoing definition of the safe point. In other words, in order for $p^{(0)}$ to be a danger point it is sufficient for a certain fixed number ε_1 to exist, for which, however small the initial conditions, for a sufficiently small distance δ, the corresponding solution of the system (5.2) becomes equal to the number ε_1.

For quasi-autonomous systems (and autonomous systems, exclusively considered in this chapter), (5.2) can be written in the form (5.4). (In the case of autonomous systems the nonlinear terms N are not, either, explicitly dependent on t.) It is recalled that the stability boundary defined in the preceding section will be formed of all the points in p-space where at least one equality $\nu_j = 0$ occurs, j being any integer whatever between 1 and n, ν_j representing the real parts of the eigenvalues of matrix A.

Now, on basis of the continuity of the characteristic roots with respect to the system parameters, to a small distance $\delta(p^{(0)}, p)$, hence to a system located in the neighborhood of the stability boundary, there will correspond at least one characteristic root having a small real part.

Finding out whether a given point $p^{(0)}$ of the stability boundary is or is not safe amounts to investigating the stability (in the sense of Liapunov) of the zero solution of the system

$$\frac{dx}{dt} = X(t, x; p^{(0)}) \tag{5.5}$$

corresponding to this point. If this solution is uniformly asymptotically stable (in the case of system (5.4), if $\nu_j(p^{(0)}) < 0$ for all the j subscripts from 1 to n), $p^{(0)}$ is a safe point of the boundary; hence, according to the definition above, any solution of the system (5.2) corresponding to a sufficiently small initial condition remains bounded, by bounds however small, if the distance $\delta(p^{(0)}, p)$ is sufficiently small, even if point p is in the instability region. The theorem above is implied by Malkin's first theorem of stability under persistent disturbances stated in Section 3.4. Asymptotic stability creates—if an intuitive although not rigorous analogy is used—a kind of stability margin that permits the maintenance of a behavior that is acceptable from the standpoint of stability even in the case of a slight change of the system. Continuing along the line of nonrigorous formulations and abusing terms, we could say that asymptotic stability constitutes a stable stability, in contrast with the (nonasymptotic) stability according to Liapunov, generally "unstable" for perturbations of the system parameters: in most cases a change, however small, of the latter, can turn it from stable into unstable. In this sense, $p^{(0)}$ appears as a danger point of the stability boundary if the zero solution of system (5.5) is stable without being asymptotically stable. We will have to deal with a danger point of the boundary, the moreso if the trivial solution of the system (5.5) corresponding to this point is unstable in the sense of Liapunov.† The statement will be illustrated in Appendix II by means of a simple system.

5.5. EMPIRICAL CONCEPT OF NEUTRAL STABILITY AND ITS MATHEMATICAL DEFINITION IN THE LITERATURE

In the literature of flight dynamics, neutral dynamic stability means (nonasymptotic) stability with respect to the variables considered. Since in the framework of the linear theory of autonomous systems, nonasymptotic stability situations can only appear in the case of at least one zero or purely imaginary characteristic root (the others being pseudo-negative; for brevity, pseudo-negative roots will mean roots having negative real parts) a direct connection was made between the existence of such roots and the (in fact empirical) property of neutral stability: in the language of the mathematical model it was agreed that by neutral stability we would understand the behavior corresponding to the existence of one or several zero

† More or less intuitive reasonings such as these obviously cannot replace rigorous demonstrations. In contrast with the theorem linking the notion of safe point with asymptotic stability on the boundary, rigorously deduced from Malkin's theorem, the *statement* above, based on the character, generally "unstable" with respect to the system changes, of the (nonasymptotic) stability property according to Liapunov, only points to a probable connection between this stability and the "danger" nature of the respective point on the boundary, a probability that is emphasized by the dissipation character of the processes connected with actual flight.

or imaginary characteristic roots.† Nevertheless, examining the meanings in which the term neutral dynamic stability is used in the literature, we can perceive the empirical content of the notion, connected with the low variation speed of the induced deviations. Thus, neutral stability is mentioned both in the case of an indifferent behavior of the yaw angle (see [8, page **417**]: "neutral stability in yaw, i.e. in azimuth") corresponding to a (nonasymptotic) uniform stability situation in the sense of Liapunov, as well as of some slightly damped or slowly diverging—phugoid [8, page **130**] or spiral‡—situations of asymptotic stability. Actual indifferent behavior corresponding to a zero characteristic root of the first-approximation system constitutes an exception of, generally, reduced probability. (For such an exceptional case see Section **6.3**.) Indeed, the domain where zero or imaginary roots exist, under conditions where the remaining characteristic roots are pseudo-negative, on the one hand, and the (nonasymptotic) stability domain according to Liapunov, on the other hand, are only superimposed in the case of systems of differential equations without nonlinear terms. If however the linear set is a first approximation system—as generally used in the theory of flight stability—nonasymptotic stability of the complete nonlinear set only constitutes one of the three possible situations and, it should be added, not the most probable: in most cases the nonlinear additions (for the case of pseudo-negative characteristic roots and of at least one zero or imaginary root) determine either asymptotic stability or instability. (See study of critical cases in [**13**, **23**, **24**].)

A case having interesting applications in the domain of aircraft stability, where (nonasymptotic) stability of the solution of the nonlinear complete system corresponds to a zero characteristic root of the linearized system is the one represented by systems of the type (**3.26**).

In this book systems of the form

$$\frac{dy}{dt} = Ay + Y(y, z), \qquad \frac{dz}{dt} = B(z)y$$

with $||B(z)||$ bounded or $B(z) = \text{const.}$, or of the form

† See [8, page **130**]: "When $E_1 = 0$ [E_1 denotes the free term of the characteristic equation], the characteristic equation has a zero root, representing a state of neutral stability;" and [5; 6, Section **4.10**; **11**, page **448**], etc.

‡ [8, page **421**]: "When $E_2 = 0$ the characteristic equation has a zero root, representing a state of neutral stability."

$$\frac{dy}{dt} = Ay + Y(y), \qquad \frac{dz}{dt} = B(z)y$$

are considered. In both, the variables y represent angular velocities, and z angular coordinates, the forces and moments induced by a disturbance being—at least in linear approximation—independent of the latter. The second equation of the two foregoing systems represents the kinematic relationship between the angles z and angular velocities y. They are the cases that will be considered of neutral behavior in azimuth ([7, pages 234–235] or [8, page 417]) for nonasymptotic stability with respect to the angle of bank when the variables β and ψ or β, $\hat{r}$, and ψ, respectively, are simultaneously constrained, or of the neutral behavior of the three Euler angles at quasi-zero speed of VTOL aircraft in the situations to be considered in Section 7.3.

However, with respect to the cases of a phugoid root or a spiral root, equal to zero, the behavior is determined by the nonlinear terms. In most cases only slightly stressed situations of instability in free flight (gentle divergence), easy to correct by control, are to be expected. Rigorously speaking, stability, even if neutral, is, however, out of question since in free flight, the deviations increase to any extent, given a sufficiently long time interval. For these cases it is recommended to investigate the behavior of disturbed motion for a finite time interval and to establish by estimates some practical criteria meant to assure quantitative indices (rate of increase, frequency) that are acceptable in relation to the dynamic characteristics of the controls. Obviously, suitable flight behavior here results not from the property of stability connected with an infinite time interval, but based on the continuity of the solution with respect to the initial conditions.

It should be noted that, from the standpoint of flight practice, there may be no essential difference between the situation of neutral stability (non-asymptotic stability in the sense of Liapunov or practical stability) on the one hand, and some instability situations characterized by acceptable quantitative indices on the other, both situations requiring, but also allowing, the action of the controls for stabilization and damping.

Moreover, the same category of situations requiring the action of the controls can also include those of asymptotic stability according to Liapunov having a weak rate of decay, the moreso as a local asymptotic stability may, in fact, mean, as was seen in Sections 3.5 and 5.4, practical instability.

Therefore, in these situations also, it seems more proper that, instead of investigating stability by qualitative processes, we resort to response calculations consisting, here, in estimating the rate of the increase (or of damping) and the frequency of oscillation of the induced deviations, as well as to other estimations, such as that of the size of the attraction region, in the case of stable asymptotic local systems (asymptotically stable by the first approximation).

To model mathematically the empirical concept of neutral stability connected with the low rate of variation of the disturbances, the mathematical notion of safe stability boundary in the space of the parameters is resorted to, which also allows a relation to be established with the traditional model of linear systems with zero characteristic roots.

5.6. MATHEMATICAL MODELS OF NEUTRAL DYNAMIC STABILITY; COMMENTS

The property that, to the pilot, seems to be quasi-dynamic indifference, currently called neutral stability, corresponds, from the mathematical standpoint, to either the local behavior of solutions of systems of the type (3.26) or the behavior of solutions of the systems located in the neighborhood of safe points of the stability boundary. Neutral stability being a property characterizing uncontrolled aircraft, these systems of equations will, obviously, correspond to free flight. From the standpoint of actual flight, neutral stability constitutes on the one hand an efficiency criterion of the stabilizing intervention (which must, however, be completed by the criterion linked with the oscillation frequency) and on the other hand, essentially, an indication of the necessity to reach stability by controls: thus, neutral stability warns of the necessity to improve the stability characteristics artificially, by means of the controls.

Moreover, as already mentioned, neutral dynamic stability is not the only situation requiring and partly allowing, in actual flight, stabilization by controls. The pilot draws the same conclusions and reacts about the same also in some cases of practical or theoretical slight instability, corresponding, for instance, to the neighborhood of dangerous boundary points. However, piloting reactions may be similar even in cases of free-flight asymptotic stability if the damping of the disturbance is not sufficiently rapid (e.g., in the case of a slight phugoid or spiral stability). This fact results in a decrease of the practical importance of the investigation of free-flight stability in situations insufficiently clear to be self-defining, hence, among others, in the case of neutral stability as well.

In other words, if the instability of the free basic motion is not too severe, or if this motion is not asymptotically stable with a sufficiently high rate of damping, the necessity to inquire into the stability of the controlled system stands out as a rule.

5.7. SOME REMARKS ON THE SELECTION OF CONTROLS: QUALITATIVE ASPECTS OF THE PROBLEM OF CONTROL EFFICIENCY

5.7.1. Introduction: Efficiency of Controls

This section is devoted to a short discussion on the validity of the model of partially controlled motions, along the lines suggested in Section 4.3. The question whether this model is adequate or not for investigating the stability of the respective systems amounts to verifying the fulfillment, for every case studied, of the conditions required by the general theory.

Roughly, it may be stated that for efficient action it is necessary that the rate of increase (e.g., the time to double amplitude) and that the oscillation frequency of the induced deviations not be too large. It was stated in the preceding section that neutral stability and quasi-neutral behavior of the aircraft, ensuring a slow variation in the amplitude of induced deviations, constitute criteria of efficiency in the stabilizing intervention of the controls. Within this section, by introducing the concept of time of response, it will be possible to specify the meaning of *too high* a rate of increase or *too high* a frequency. Further, another basic element of the efficiency of artificial stabilization will be considered: the secondary effect of control, modeled as persistent disturbances of the system considered.

Checking the fulfillment of the conditions required by theory is achieved by estimating the upper bound of constrained deviations starting from the instant controls are applied and estimating the time interval after which these deviations may be considered practically damped. From these estimates there also follows the degree of approximation of the results obtained on the basis of the auxiliary systems for problems such as the nature of the variation in time of the free variables or the estimation of their size and damping velocity. In this section, only a few essential qualitative aspects will be considered.

In Chapters 6 and 7 various cases of partial control will be analyzed, with a view to selecting the constrained variables: assuming that the conditions imposed by the general theory are fulfilled, we will seek those variables whose constraint permits attainment of the uniform asymptotic stability of the partially controlled basic motion with respect to the free variables. But it is first necessary to investigate the validity of the model. This amounts to the problem of the existence and—in the affirmative—the selection of the controls meant to assure the fulfillment of the conditions required by the theory. From a theoretical standpoint, therefore, investigation of the stability of a partially controlled system consists of the selection of the constrained variables and of the controls by which their constraint is achieved, with the aim of obtaining optimum stability characteristics.

The main factors influencing the fulfillment of the conditions of boundedness and damping required by the general theory will be reviewed here. For short, these conditions will be called conditions (C), or, taken separately, condition (CI) and condition (CII).

Let $\zeta(\varepsilon)$ be the upper bound, deduced by some estimating method, of the value of function $z(t)$, which ensures fulfillment of the inequality $|y| < \varepsilon$ for $t \geqslant t_0$. As in Chapter 4, z represents the constrained or controlled variables, and y the free ones; ε is a quantity deduced from empirical considerations. Further, let $z_{\text{ef}}(t)$ be the effective value, at every instant, of the vector z, reached as a result of the operation of controls, and $O(z_{\text{ef}})$ the order of magnitude of its upper bound on the semiaxis $t \geqslant t_0$, deducted empirically or by estimative computation. Then, conditions (C) are expressed in the form

$$\zeta(\varepsilon) > O(z_{\text{ef}}), \tag{CI}$$

$$\lim_{t\to\infty} z_{\text{ef}}(t) = 0. \tag{CII}$$

The potency of the controls selected for constraining the variables z to achieve fulfillment of conditions (C) will be called, conventionally, efficiency of controls. Among the factors influencing efficiency of controls, two will be considered hereafter, as being of decisive importance: (i) the factor linked with the time of response; and (ii) the one corresponding to the secondary effects of the controls.

5.7.2. Time of Response

By time of response will here be understood the interval having elapsed from the appearance of the signal of command (control input signal) in the form of a perturbation of the flying conditions in comparison to the reference conditions until the new corrected value of the input magnitude has been established. In other words, it is a matter of dealing with the duration of the entire complex process of the control system. In this time interval there is a first stage in which flight proceeds in the conditions of free motion with locked controls, namely, the duration of the process taking place between the appearance of the disturbance and the initial moment when the respective control organ is actuated (e.g., this moment indicates the beginning of elevator, rudder, etc., motion). In this first stage processes occur such as: recording of disturbance as a command signal by the sensor of the flight control system: the computing operation by which the signal received is compared with the possibility of the control system and an adequate decision is reached; the transmission of this decision of the computer, by means of the actuator, to the control element (e.g., control surface). In the case of a human pilot the stage includes the reaction time (i.e., "the time which elapses between the impact of a physical stimulus on some sense organ ... and the beginning of the responsive movement of hand or foot" [6, page 3]) and the time required for the motion to be transmitted to the aerodynamic control surface or throttle, or whatever, mechanically or through servomotors. In the final stage of the response time a transitory process occurs, characterized by a displacement of the control surface to a new setting and by the aerodynamic phenomena (or, in the case of thrust control, of combustion and mechanical), following which the magnitude of the control force or moment corresponding to the new position of controls is established. Obviously, in this last stage, flight conditions will vary between those corresponding to the initial and to the new setting of the controls, hence, usually, so as to restore the basic motion. Nevertheless, it will, for the time being, be assumed that flight continues in free conditions with locked controls during the entire response time interval, which will be denoted $[t_0, t_0 + \tau]$, t_0 being the instant when the disturbance appears.

Intuitively, it is obvious that, to obtain as rapid, and hence as efficient, as possible a stabilization, it is proper for the response time τ to be as short as possible. From the standpoint of the control efficiency defined by conditions (C), it is useful to compare the time of response with the two essential response characteristics of the aircraft in free flight: the variation speed (rate of change) of the amplitude of the induced deviation (increase or damping speed) and the period of oscillation.† Thus, in the case of the control of a variable having an unstable component, it is necessary to verify if, in the interval $[t_0, t_0 + \tau]$, the amplitude of the induced deviation does not exceed the upper bound $\zeta(\varepsilon)$, for which the validity of the results obtained on the basis of the respective auxiliary system is assured. A rough estimate of the allowable time lag as a function of $\zeta(\varepsilon)$ and of the rate of increase of amplitude is obtained, for instance, as follows. Let $z_{ef}(t) = f(t) + \gamma e^{\lambda t}$ with $\lim_{t\to\infty} f(t) = 0$ and with $\lambda > 0$ and $\gamma > 0$ ($\gamma e^{\lambda t}$ is the unstable component) be the variation of the controlled variable z in free flight. It is assumed that at the initial moment $t_0 = 0$, $\bar{f}(0) + \gamma < \zeta(\varepsilon)$ takes place, where $\bar{f}(t)$ represents the amplitude of the function $f(t)$. Then, from continuity considerations only, there results the existence of $\tau \neq 0$ (obviously $\tau > 0$) for which the inequality $\bar{f}(\tau) + \gamma e^{\lambda\tau} \leqslant \zeta(\varepsilon)$ occurs. The estimation of the allowable value of the time lag τ can be obtained from this inequality or, since $\bar{f}(\tau) \leqslant \bar{f}(0)$ ($f(t)$ representing the asymptotically stable mode of the solution element $z(t)$), from the inequality

$$\tau < \frac{1}{\lambda} \ln \frac{\zeta(\varepsilon) - \bar{f}(0)}{\gamma}$$

where λ, which is dependent on the system parameters, measures the rate of increase of the unstable mode; γ, besides the system parameters, is also dependent on the initial conditions (hence on the nature and magnitude of the disturbance).

The time of response $[t_0, t_0 + \tau]$ should also be compared with the oscillation period of the significant modes of the respective controlled variable. It seems obvious that a time of response of an order of magnitude equal to that of the monotone variation interval of the respective mode (equal to

† In the case of unsteady basic motion the variation of the induced deviations is not necessarily either monotonic or periodic. In this case, as will be seen in Chapter 9, it will be possible, instead of the period of oscillation, to use as criterion of comparison for the value of the time lag τ the lower bound of the time interval between two successive cancellations of the respective deviation or, to have a margin of safety, of the time interval in which the deviation varies monotonously. Since, generally, continuous functions having continuous derivatives are dealt with, the latter can be estimated on the basis of the smallest interval between two successive cancellations of the derivative.

the half-period of oscillation) may reverse the effect of the controls and hence lead to a stressing of the deviation. For acceptable efficiency it is necessary that the time of response be shorter than the quarter of a period ($\tau < T/4$).

Attention was drawn above to the fact that it is necessary to keep the condition expressed by the relation between the time of response and the period of oscillation only in the case of the significant modes. This statement will be made clear hereafter.

A mode of the controlled variable may be insignificant, from the point of view considered here, in two manners: (i) either as a result of heavy damping, or (ii) because of the negligible value of the respective mode, in comparison with the other modes of the controlled variable, during a certain time interval following the disturbance.

(i) Rapid damping of an oscillatory mode makes actuation of any correction control unnecessary and stabilizing controls are only applied in order to counter, among the remaining modes, those that are divergent or too slowly damped. For instance, let us consider the longitudinal motion of an aircraft having phugoid instability in free flight. As will be seen in the following chapter, instability cannot always be eliminated by constraining the longitudinal attitude. Let it be assumed that the solution of the system representing free flight, hence also its component $\theta(\hat{t})$, is composed of two aperiodic modes, of which one or both are divergent, and the rapid longitudinal oscillation (of high frequency and heavily damped). In a first short stage, the value of the term corresponding to the oscillatory mode is of the same order of magnitude as that of the aperiodic modes. If damping were not sufficiently rapid, the time of response of the control system should satisfy the inequality $\tau < T/4$ for a very small value of the period T. Obviously, such a condition would be difficult to fulfill in the case of manual piloting, or it would require an involved transmission and adjustment installation in the case of automatic control.

However, since the damping of the oscillatory mode is rather rapid (e.g., $t_{1/2} < \tau$) it may seem advantageous not to decrease the time of response τ, but on the contrary, to increase it by increasing the reaction time. Indeed, if the reaction time exceeds the time interval required to reduce the amplitude of oscillation below a certain value (at any rate below the size level of the aperiodic terms), very efficient constraint of the variable $\Delta\theta$ may be obtained. Such an increase of the reaction time is achieved naturally in the case of manual piloting, as the pilot only acts with the intention of canceling the phugoid motion after the rapid oscillation is practically completely damped, or to be more accurate, after the amplitude of this oscillation—in free flight—has decreased below a certain value. Quantitatively this value can be expressed either in fractions of the initial (which is also the maximum) value of amplitude, or independently of it. Thus, if $\gamma e^{\nu t}$ is the amplitude of the rapid oscillatory mode, this may be considered practically damped if $\gamma e^{\nu t} < \kappa\gamma$, $0 < \kappa < 1$, or if $\gamma e^{\nu t} < \varepsilon$, the quantities κ and ε being determined empirically. Then, the time interval after which the oscillation may be considered as damped will be $t_\kappa = (1/\nu) \ln \kappa$ or $t_\varepsilon = (1/\nu) \ln (\varepsilon/\gamma)$.

(ii) Another aspect that should be taken into account is the weight ratio of the controlled variable modes in a first short time interval that follows the appearance of a disturbance, particularly, the relative magnitude of the high-frequency oscillatory mode in comparison with the aperiodic or low-frequency ones. Let z_1 be a controlled variable, the function $z_1(t)$ having an expression of the form

$$z_1(t) = \gamma e^{\nu t} \sin(\omega t + \varphi) + \sum_{j=3}^{n} \gamma_j e^{\lambda_j t}$$

where the constant coefficients γ and γ_j depend on the system parameters and on the initial conditions, λ_j are the roots of the characteristic (nth-degree) equation, $\lambda_{1,2} = \nu \pm i\omega$, λ_j for $j = 3, 4, \ldots, n$ are real or complex. It is assumed that the frequency ω is relatively high: $\omega \gg \operatorname{Im} \lambda_j$, $j = 3, 4, \ldots, n$.

Obviously, in a sufficiently short time interval $[0, t_1]$ the constants γ and γ_j are conclusive with regard to the relative magnitudes of the various terms, whatever the sense and value of the real parts of the characteristic roots $\lambda_{1,2}$ and λ_j. The sign and value of the latter only become decisive after a certain time interval, longer than t_1, has elapsed. Now, if $\gamma \ll \sum_{j=3}^{n} \gamma_j$ and the time of response $\tau < t_1$, the high-frequency oscillatory term does not play a conclusive part in estimating control efficiency; hence the foregoing criteria, linked to the characteristics of this term (e.g., $\tau < T/4$ or $\tau > t_\varepsilon$) can be neglected.

As an implication of this aspect it will now be shown that an efficient constraint of the angular attitude variables is easier to achieve than that of the respective angular velocities.

Let

$$\dot{\eta}(t) = \gamma e^{\nu t} \sin(\omega t + \varphi) + \Gamma e^{Nt} \sin(\Omega t + \Phi) \tag{5.6}$$

be the variation of a component of the aircraft angular speed. It is assumed that γ and Γ are of the same order of magnitude, while $\omega \gg \Omega$ and $(\nu^2 + \omega^2)^{1/2} \gg (N^2 + \Omega^2)^{1/2}$. Since the frequency ω is high, if a very rapid damping of the first term—by means of a negative ν of high value—is not secured, it is most difficult to ensure efficient control of the angular speed. At the same time, an efficient control of the angular variable η may be easily achievable. Indeed, since

$$\eta(t) = K + \frac{\gamma}{(\nu^2 + \omega^2)^{1/2}} e^{\nu t} \sin(\omega t + \varphi') + \frac{\Gamma}{(N^2 + \Omega^2)^{1/2}} e^{Nt} \sin(\Omega t + \Phi') \tag{5.7}$$

from $O(\gamma) = O(\Gamma)$ (O denotes the order of magnitude of the quantity in parentheses) and $(\nu^2 + \omega^2)^{1/2} \gg (N^2 + \Omega^2)^{1/2}$, it readily results that in a certain—sufficiently short—time interval $[0, t_1]$, the prevailing term is the low-frequency term. In this case the total time response τ should be compared with t_1, which can considerably exceed the value of the rapid oscillation quarter period $1/(4\omega)$.

Taking the above into account, we can clearly see why it is easier to obtain an efficient constraint of the angle of bank ϕ and of the azimuth ψ than of the angular velocities p or r. If, for instance, the roots of the characteristic equation for the system of the first four equations (2.18) are λ_1, λ_2 (real), and $\lambda_{3,4} = \nu \pm i\omega$, the relations $|\lambda_1| > (\nu^2 + \omega^2)^{1/2} \gg |\lambda_2|$ with $\lambda_1 < 0$, $\lambda_2 \gtrless 0$, $\nu < 0$ may be taken as typical. (See discussion concerning the case of yaw angle constraint in Section 7.2.) Then the functions $\hat{p}(\hat{t})$ and $\hat{r}(\hat{t})$ will be of the form

$$\begin{aligned} \hat{p}(\hat{t}) &= \gamma_{21} e^{-|\lambda_1|\hat{t}} + \gamma_{22} e^{\lambda_2 \hat{t}} + \Gamma_2 e^{\nu \hat{t}} \sin(\omega \hat{t} + \varphi_2), \\ \hat{r}(\hat{t}) &= \gamma_{31} e^{-|\lambda_1|\hat{t}} + \gamma_{32} e^{\lambda_2 \hat{t}} + \Gamma_3 e^{\nu \hat{t}} \sin(\omega \hat{t} + \varphi_3). \end{aligned} \tag{5.8}$$

Generally, in a first short time interval after the disturbance none of the right-hand side terms appears a priori to be dominant. (In the expression of angular velocity $\hat{r}$, γ_{31} has a much lower value than γ_{32} or Γ_3, the latter two being usually of the same order of magnitude.) Following the relatively rapid convergence of the terms in $e^{-|\lambda_1|\hat{t}}$, the controls must act particularly on the other aperiodic mode as well as on the oscillations represented by the last terms of the two expressions. The higher the frequency ω, the more difficult it will be to act upon the direction and lateral oscillations by means of controls, particularly if the damping velocity, expressed by the quantity ν, is insufficient.

NOTE. The suitable relationship between damping and frequency is even specified in official *flying qualities requirements.* Thus, in Section 1.3 of [7, page 9] a diagram is reproduced from the requirements of the United States Air Force (*Flying Qualities of Piloted Airplanes,* USAF Spec. 1815-B, 1948) in which the damping of the lateral oscillation, expressed by the time required to reduce the amplitude to one half, $t_{1/2}$, is demanded as a function of the oscillation period T. The graph shows a linear decrease of the maximum admitted limit of the duration $t_{1/2}$, with the decrease of T, according to the equation $t_{1/2} = 2.5T - 3.5$, $t_{1/2}$ and T being in seconds, until the period equal to 2 sec below which a constant upper bound of the duration $t_{1/2}$, equal to 1.5 sec, is shown (Figure 5.1).

Hence, according to these requirements, if the amplitude of the lateral oscillation drops to half in less than 1.5 sec, the oscillation period may decrease any amount, and even tend toward zero. Obviously, for such low

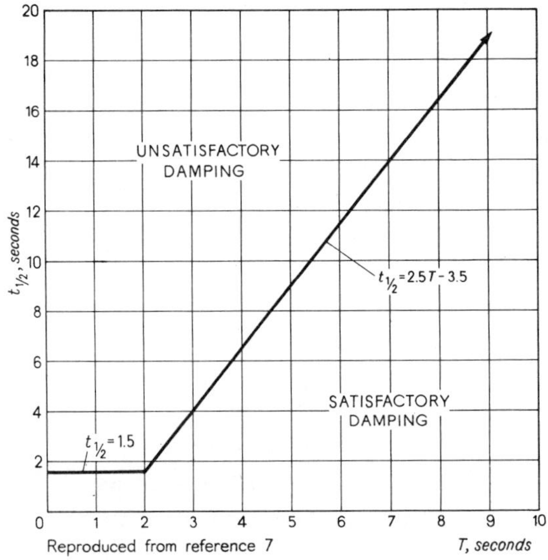

Figure 5.1. Reproduced from [7].

values of the period, the efficiency of controls is at best zero (controls eventually even contributing to stress the deviation). Practically, this conclusion means that, if the amplitude drops to half in less than 1.5 sec, the action of the controls to accelerate damping becomes useless, damping being considered satisfactory. The slow aperiodic mode, however, must still be cancelled by controls λ_2 being eventually positive (spiral divergence). Thus, a situation similar to the one described at point (i) is reached. Since the damping of the other two modes is sufficiently rapid, the slow aperiodic mode can be controlled (the spiral divergence can be eliminated) if the control forces appear after the heavily damped modes have been exhausted, provided, however, that the slow mode does not amplify beyond a certain limit during the time required for their practical damping (κ-damping or ε-damping). Obviously, the comparison is made with the longer of the damping times of the rapid modes, corresponding to the lateral oscillations ($|\nu| < |\lambda_1|$). Let κ be the ratio in which the amplitude of the lateral oscillations should be reduced in order for the latter to be considered practically damped and let k be the allowable upper bound of the increment factor for the slow aperiodic mode. The duration of the κ-damping of the oscillations being $t_\kappa = (1/\nu) \ln \kappa = (1/|\nu|) \ln(1/\kappa)$, the allowable upper bound of the rate of increase of the spiral mode expressed by the value of the positive real root λ_2 results immediately:

$$\lambda_2 \leqslant |\nu| \frac{\lg k}{\lg(1/\kappa)} \tag{5.9}$$

Figure 5.2 shows the allowable upper bound of the rate of increase of the spiral mode amplitude corresponding to the setting of an upper bound for this mode, equal to 1.5 times the initial value, as a function of the damping of the lateral oscillations expressed by means of the time to halve their amplitude, $t_{1/2}$. Up to $t_{1/2} = 0.5$ sec a very rapid decrease of the allowable λ_2 is found, with the increase of $t_{1/2}$ (hence with the weakening of damping). Now let us suppose that $k = 1.5$ and that the lateral oscillations may be considered practically damped for $\kappa = 0.1$. In this case when the time to damp to half the amplitude of the lateral oscillation is 1.5 sec the period may be arbitrarily small (constraint of the oscillatory mode may become useless) only if $\lambda_2 \leqslant 0.081$ sec^{-1}. It should be noted that generally acceptable damping of the lateral oscillation in free flight is obtained at the expense of spiral behavior; hence, if, because of design changes, λ_2 increases faster than $|\nu|$, the condition expressed by the foregoing inequality might not be satisfied. This situation can be avoided in two ways: either by the artificial increase of $|\nu|$ by including a yaw damper (having a small time lag for a large ω) in the control loop; or by eliminating the spiral mode indirectly, by constraint of the lateral attitude (see Section 7.2).

It will be shown hereafter that a constraint of the variable ϕ is possible even in the case of a high value of the frequency ω, which makes inefficient the control of the variables $\hat{p}$ and $\hat{r}$. This readily results from the relations (5.6) and (5.7). Indeed, $\dot{\phi} = \hat{p} + b_{43}\hat{r}$ (see (2.18)) or, taking relations (5.8) into account:

$$\frac{d\phi}{d\hat{t}} = \gamma_{41}e^{-|\lambda_1|\hat{t}} + \gamma_{42}e^{\lambda_2\hat{t}} + \Gamma_4 e^{\nu\hat{t}} \sin(\omega\hat{t} + \varphi_4)$$

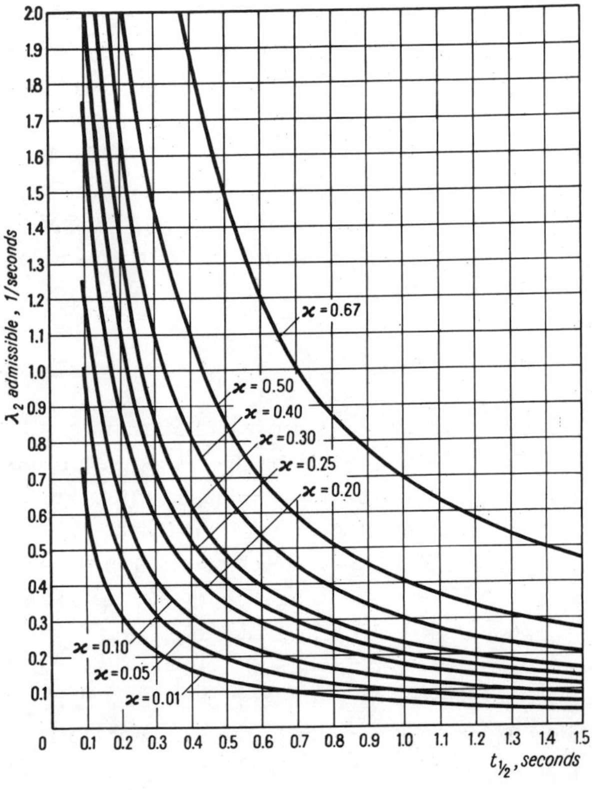

Figure 5.2.

where $\gamma_{41} = \gamma_{21} + b_{43}\gamma_{31}$, $\gamma_{42} = \gamma_{22} + b_{43}\gamma_{32}$, $\Gamma_4 = [\Gamma_2^2 + b_{43}^2\Gamma_3^2 + 2b_{43}\Gamma_2\Gamma_3 \cdot \cos(\varphi_2 - \varphi_3)]^{1/2}$, $\varphi_4 = \cos^{-1}[(\Gamma_2 \cos\varphi_2 + b_{43}\Gamma_3 \cos\varphi_3)/\Gamma_4]$. The coefficients γ_{41}, γ_{42}, and Γ_4 are approximately of the same order of magnitude, the three modes entering approximately with the same weight in the componency of the angular velocity in a first, sufficiently short stage after the disturbance. However, in the same stage, the contribution of the three modes becomes different in the case of the angle ϕ on account of the discrepancy between the values of the moduli of the characteristic roots shown above. Indeed, since

$$\phi(\hat{t}) = K_\phi + \frac{\gamma_{41}}{\lambda_1} e^{\lambda_1 t} + \frac{\gamma_{42}}{\lambda_2} e^{\lambda_2 t} + \frac{\Gamma_4}{(\nu^2 + \omega^2)^{1/2}} e^{\nu\hat{t}} \sin(\omega\hat{t} + \varphi_4')$$

and $|\lambda_1| > (\nu^2 + \omega^2)^{1/2} \gg \lambda_2$, the dominant variable term in the expression of function $\phi(\hat{t})$ is $(\gamma_{42}/\lambda_2)e^{\lambda_2\hat{t}}$. The dominant character of this term is not only restricted to a first short time interval, but, on the contrary, it grows stronger with time, according to the relations $|\lambda_2| < \nu \ll |\lambda_1|$ and, a fortiori, for $\lambda_2 > 0$. Thus, since the dominant mode is aperiodic, with a relatively slow variation (λ_2 small), the efficiency of the controls applied for the constraint of the lateral attitude is assured. This leads to the practical elimination of the (divergent or weakly convergent) spiral mode.

Hence, by a constraint of the lateral attitude (besides that of the azimuth), condition (5.9) is eliminated *ipso facto* and therefore sufficient damping implies satisfactory behavior of the aircraft in free flight. As it results from the diagram reproduced in Figure 5.1, the condition imposed on damping is not too severe, the requirement being the halving of the amplitude in 1.5 sec.

5.7.3. Secondary Effects of Controls

It was seen above that the efficiency of a control generally decreases as a result of the time lag in the appearance of the control force with respect to that of the disturbance force. However, action of the controls is usually also accompanied by parasite forces that may act adversely on efficiency. When a control is applied to constrain a variable, the entire dynamic picture is generally changed. This change is determined, on the one hand, by the change imposed on the constraint variable—and this actually represents the scope of the control, and on the other hand, by the uncalled-for consequences resulting from the application of controls, which will be called parasite or secondary effects. A secondary effect may diminish the strength of the control (ratio of the variation of constraint variables, obtained by applying the control, to that of the control variable) or it may change the motion undesirably by acting (directly or through the intermediate of the constrained variable) on the other state variables of the system. It should be noted that not all indirect effects are noxious. This will be illustrated by a simple example where the effects of the main longitudinal controls will be sketched: those of elevator deflection and change in thrust by throttle control.

The main effect of elevator deflection is the induction of a longitudinal control moment, hence, as regards the mathematical model, the change of the equation of moments about the lateral axis. Elevator deflection, however, also affects lift and drag, their modification appearing as secondary effects. (In most cases these secondary effects are small in the case of conventional aircraft; for tailless aircraft, however, the change in lift may be quite important.)

According to the common convention of signs, the elevator setting angle δ_e is considered positive if it increases the effective incidence of the horizontal tail surface, determining a nose-down pitching moment ($C_{m_\delta} < 0$), and an increase of the aircraft lift ($C_{L_\delta} > 0$). Whatever the sign of the angle δ_e, the passive drag of the aircraft may increase, and hence the drag coefficient will increase for $\delta_e > 0$, and for $\delta_e < 0$ it may either increase or decrease. Generally, the variation of drag, induced by elevator deflection, is sufficiently small to be negligible. This variation will, however, be taken into account here in order to better illustrate the secondary effects of elevator control. That is, it will be assumed that C_D increases whatever the sign of the deflection angle ($C_{D_\delta} > 0$ for $\delta_e > 0$ and $C_{D_\delta} < 0$ for $\delta_e < 0$).

We consider first the case of the constraint of the inclination angle γ of the flight path to the horizontal by control of the angle δ_e. If the aircraft has a nose-down pitching tendency ($\Delta\gamma < 0$), a nose-up control moment C_{m_c} is created by means of the elevator, through a negative setting angle leading to an increment of the aircraft incidence and through this to an increase of lift. Since the incidence varies faster than the velocity of the center of mass, the main resulting effect will be a correction of the flight path angle, with a tendency to bring it back to zero. However, concomitantly, as a result of the appearance of a negative setting angle, the slope of the lift curve $C_L = f(\alpha)$ decreases and hence there is also a decrease of the main effect due to the nose-up control moment C_{m_c}, equal, in the hypothesis of linear variation, to $C_{m_\delta}\delta_e$. Let $\Delta\alpha_c$ be the increment in angle of attack determined by the moment C_{m_c}. Then the resultant increase of the lift coefficient will be, in the hypothesis of linear variation, equal to $(C_{L_\alpha} - |\delta_e| \partial C_{L_\alpha}/\partial\delta_e) \cdot \Delta\alpha_c$. It can similarly be shown that the effect of the secondary variation of lift is also detrimental in the case of a positive deviation of the flight path inclination angle ($\Delta\gamma > 0$). The increase of the drag coefficient C_D leading to a change in the equation of the forces tangential to the flight path, has a stabilizing effect for $\Delta\gamma < 0$, countering the acceleration corresponding to the increase of the downward inclination of the flight path, and destabilizing in the case $\Delta\gamma > 0$, by favoring the decrease of the speed. For the same reasons, if constraint of the flight velocity is sought by elevator control, the secondary effect on the lift coefficient is determined for both $\Delta V > 0$ and $\Delta V < 0$, while direct influencing of the drag coefficient (in the hypothesis $C_{D_\delta}\delta_e > 0$) by the deflection δ_e is only detrimental for $\Delta V < 0$.

We now consider the case of flight velocity constraint using as control variable the magnitude of the thrust force T. The main effect of the variation of thrust consists in the modification of the forces along the center-of-gravity velocity line. However, since thrust generally also has a component normal on the flight path and it does not pass through the aircraft center of mass, its variation will influence directly the equilibrium of the normal forces and of the moments with respect to the lateral axis Oy. Generally both these secondary effects are detrimental.

Finally, the nonrigorous satisfaction of condition (4.3) for other reasons than the result of the time lag may also be included in the category of secondary effects. Thus, the condition of canceling the control forces at the instant the controlled variable vanishes may not be fulfilled, for instance, if this variable passes through the zero value during the transitory process at an instant when the canceling of the resultant of external forces and moments is not achieved. It is recalled that for such a situation the theory expounded in Chapter 4 remains valid if the deviation from equilibrium is sufficiently small (in accordance with the theorem of stability under persistant disturbances).

It should be noted that from the standpoint of the efficiency of controls, the secondary effects of the type described above are generally less dangerous than the effects linked to the time of response. However, a verification is essential whenever adopting some nonconventional design type. (Thus, for example, mention has been made of the relatively strong effect of elevator deflection on the lift in the case of tailless aircraft.) According to Section 4.3, such a verification implies (i) an estimate of the quantity $\delta(\varepsilon)$ for a certain ε determined by empirical considerations, (ii) an estimate of the additional term $R(t, y, z)$ in the right-hand side of equation (4.4) (i.e., of the actual magnitude of the additional force of disturbance), and (iii) a comparison of the two values found, $R(t, y, z)$ having to be smaller than $\delta(\varepsilon)$. If a linearized auxiliary system is used, then condition $|N_y(t, y, z) + R(t, y, z)| < \delta(\varepsilon)$ should be checked; N_y represents the nonlinear terms of the function $Y(t, y, z)$ in the right-hand side of equation (4.4). It may readily be seen that the following expression may be used to estimate the left-hand side of the last inequality.

$$|R(t,0,0)| + z\frac{\partial(Y+R)}{\partial z}(t, \theta y, \theta z) + yz\frac{\partial^2(Y+R)}{\partial y\,\partial z}(t, \theta y, \theta z)$$

with $0 \leqslant \theta y \leqslant y$, $0 \leqslant \theta z \leqslant z$ or, further,

$$|R(t,0,0)| + |z| \operatorname*{lub}_{y,z}\left|\frac{\partial(Y+R)}{\partial z}\right| + |yz| \operatorname{lub}\left|\frac{\partial^2(Y+R)}{\partial y\,\partial z}\right|.$$

APPENDIX I

CRITERIA OF ROUTH AND HURWITZ

The investigation of the solution of a system of linear ordinary differential equations with constant coefficients amounts, as is known, to the examination of an algebraic equation—called characteristic—of a degree equal to the order of the system. The sign of the real parts of the roots of this algebraic equation is decisive for the stability of the solution of the linear system; the necessary and sufficient condition of asymptotic stability is that all the roots should have their real parts negative. Routh's and Hurwitz's theorems offer, in different formulation, the necessary and sufficient condition for the roots of an algebraic equation of the nth degree, in λ,

$$a_0\lambda^n + a_1\lambda^{n-1} + \cdots + a_{n-1}\lambda + a_n = 0 \tag{1}$$

to only admit of roots with negative real parts.

Routh's criterion [4]. The table

$$\begin{array}{l} A_1^1 \quad A_1^2 \cdots A_1^k \cdots \\ B_1^1 \quad B_1^2 \cdots B_1^k \cdots \\ A_2^1 \quad A_2^2 \cdots A_2^k \cdots \\ B_2^1 \quad B_2^2 \cdots B_2^k \cdots \end{array}$$

$$
\begin{array}{cccc}
\cdot & \cdot\;\cdots & \cdot & \cdots \\
A_j^1 & A_j^2 \cdots & A_j^k & \cdots \\
B_j^1 & B_j^2 \cdots & B_j^k & \cdots \\
\cdot & \cdot\;\cdots & \cdot & \cdots
\end{array}
\tag{2}
$$

is drawn up according to the rule

$$A_1^j = a_{2j-2}, \qquad A_1^j = 0 \qquad \text{for} \quad j > \frac{n+2}{2},$$

$$B_1^j = a_{2j-1}, \qquad B_1^j = 0 \qquad \text{for} \quad j > \frac{n+1}{2},$$

$$A_j^k = -\begin{vmatrix} A_{j-1}^1 & A_{j-1}^{k+1} \\ B_{j-1} & B_{j-1}^{k+1} \end{vmatrix}, \qquad B_j^k = -\begin{vmatrix} B_{j-1}^1 & B_{j-1}^{k+1} \\ A_j^1 & A_j^{k+1} \end{vmatrix},$$

$$j = 2, 3, \ldots; \qquad\qquad k = 1, 2, \ldots .$$

It results that $A_j^k = 0$ and $B_j^k = 0$ for $j > (n+1)/2$ or for $k > (n+3)/2 - j$. Thus, the table contains $n+1$ rows and is triangular, the last two rows corresponding to $j = (n+1)/2$ having only one element.

THEOREM 1. *The necessary and sufficient condition for all the roots of equation* (1) *to have their real part negative is that all the elements of the first column of the table in* (2) *be positive.*

Hence, Routh's criterion amounts to verifying the sign of $n+1$ functions T_i, $(i = 1, 2, \ldots, n+1)$. It can be shown that the function T_{n+1} is equal to $a_n T_n$ multiplied by a positive factor. Therefore, the last condition $T_{n+1} > 0$ can be replaced by $a_n > 0$. If $a_n = 0$, $T_i > 0$, $i = 1, \ldots, n$, equation (1) admits a zero root and the situation $T_n = 0$ with $T_i > 0, i = 1, \ldots, n-1$ and $a_n > 0$, corresponds to a pure imaginary pair of roots.

Hurwitz's criterion [25]. The theorem above can also be formulated in the following manner. Equation (1) with $a_0 > 0$ is considered, and the following determinants are drawn up.

$$\Delta_1 = a_1, \qquad \Delta_2 = \begin{vmatrix} a_1 & a_0 \\ a_3 & a_2 \end{vmatrix}, \qquad \Delta_3 = \begin{vmatrix} a_1 & a_0 & 0 \\ a_3 & a_2 & a_1 \\ a_5 & a_4 & a_3 \end{vmatrix}, \ldots,$$

$$\Delta_n = \begin{vmatrix} a_1 & a_2 & 0 & 0 & 0 & 0 & 0 & \cdots & 0 \\ a_3 & a_2 & a_1 & a_0 & 0 & 0 & 0 & \cdots & 0 \\ a_5 & a_4 & a_3 & a_2 & a_1 & a_0 & 0 & \cdots & 0 \\ \cdot & \cdot & \cdot & \cdot & \cdot & \cdot & \cdot & \cdot & \cdot \\ a_{2n-1} & a_{2n-2} & a_{2n-3} & a_{2n-4} & & & & \cdots & a_n \end{vmatrix} \equiv a_n \Delta_{n-1},$$

where $a_k = 0$ for $k > n$.

THEOREM 2. *The necessary and sufficient condition for all the roots of equation* (1) *to have their real parts negative is that all the determinants* Δ_i, $i = 1, 2, \ldots, n$, *be positive.*

The Routh–Hurwitz conditions for the third-degree equation

$$a_0\lambda^3 + a_1\lambda^2 + a_2\lambda + a_3 = 0$$

with $a_0 > 0$ are

$$a_1 > 0, \quad a_1a_2 - a_3a_0 > 0, \quad a_3 > 0;$$

for the fourth-degree equation

$$a_0\lambda^4 + a_1\lambda^3 + a_2\lambda^2 + a_3\lambda + a_4 = 0$$

with $a_0 > 0$ the conditions are

$$a_1 > 0, \quad a_1a_2 - a_3a_0 > 0, \quad a_3(a_1a_2 - a_3a_0) - a_4a_1^2 > 0, \quad a_4 > 0.$$

APPENDIX II

EXAMPLE TO ILLUSTRATE THE NOTIONS OF SAFE AND DANGEROUS BOUNDARY

To illustrate some of the statements made in Section 5.4 we consider the equation

$$\frac{dx}{dt} = \mu x + ax^{2m+1} \tag{1}$$

where a is a constant, m any positive integer whatsoever, and μ the only parameter of the system, its sign being determined by that of the constant a in compliance with the relation $a\mu \leqslant 0$. The stability boundary will correspond to the zero value of the parameter μ, for small values of μ the equation (1) being situated in the vicinity of the boundary.

MARCOS G. ORTIZ M.

Let the Liapunov function $\mathscr{V} = (1/2a)x^2$ be considered, having its time derivative along the integral curves of equation (1) corresponding to $\mu = 0$, $d\mathscr{V}/dt = x^{2m+2}$, positive definite. It is immediately clear that the zero solution of the equation on the boundary will, according to some well-known Liapunov theorems, be asymptotically stable or unstable according as the constant a will be negative or positive: hence for $a < 0$, $\mu = 0$ will be the safe boundary, and for $a > 0$ the dangerous one.

Let parameter μ be different from zero and $a < 0$. For $a < 0$, $\mu = 0$ being the safe boundary, according to the definition given in Section 5.4, whatever the positive number ε, there exist $\eta_1(\varepsilon)$ and $\eta_2(\varepsilon)$ such that $|x(t)_0| < \eta_1$ and $\delta(p^{(0)}, p) = |\mu| < \eta_2$ should imply $|x(t)| < \varepsilon$ for any $t \geqslant t_0$. Indeed, integration of equation (1) yields

$$\left| a + \frac{\mu}{x^{2m}} \right| = \left| a + \frac{\mu}{x^{2m}(t_0)} \right| e^{-2\mu m t}.$$

There readily results $\lim_{t\to\infty} x^{2m}(t) = \mu/- a$ or $\lim_{t\to\infty}|x(t)| = (\mu/- a)^{1/2m}$. Besides $x = 0$ the equation admits two additional singular points: $x = \pm(\mu/- a)^{1/2m}$. For initial conditions satisfying the relation $|x_0| < (\mu/- a)^{1/2m}$ there correspond real solutions for which $x^{2m}(t) = \mu/(- a + a_0 e^{-2m\mu t})$, $a_0 = |a + \mu/x_0^{2m}|$, $t_0 = 0$; $|x(t)|$ will then be a monotone increasing time function $(dx^{2m}/dt) = 2m\mu a_0 e^{-2m\mu t}/(- a + a_0 e^{-2m\mu t})^2 > 0$ tending asymptotically toward $(\mu/- a)^{1/2m}$. If $x_0 > (\mu/- a)^{1/2m}$, the real solution will be obtained from $x^{2m}(t) = \mu/(- a - a_0 e^{-2m\mu t})$; since in this case $dx^{2m}/dt = 2ma_0\mu e^{-2m\mu t}/a + a_0 e^{-2m\mu t} < 0$, $|x(t)|$ will be monotone decreasing, tending asymptotically toward the same limit $(\mu/- a)^{1/2m}$.

It results that for equation (1) with $a < 0$, $\eta_1(\varepsilon) = \varepsilon$, $\eta_2(\varepsilon) = - a\varepsilon^{2m}$. Indeed, let $\delta = |\mu| < - a\varepsilon^{2m}$. Then, if $(\mu/- a)^{1/2m} < |x_0| < \varepsilon$, the solution in modulo being monotone increasing, tending asymptotically toward $(\mu/- a)^{1/2m} < \varepsilon$ will, obviously, always be smaller than ε; and if $|x_0| < (\mu/- a)^{1/2m} < \varepsilon$, the solution in modulo is monotone decreasing; hence $|x(t)| < |x(0)| < \varepsilon$ for $t \geqslant 0$.

Now let be $a > 0$ and $\mu \leqslant 0$. Since the solution of equation (1) is unstable for $\mu = 0$, $\mu = 0$ is a dangerous boundary. This means that in order to actually reach stability it is not sufficient for the equation to be situated in the stability region ($\mu < 0$), but it should also be far enough from the boundary (i.e., $|\mu|$ sufficiently large). Indeed

$$x\frac{dx}{dt} = x^2(\mu + ax^{2m-2}).$$

If $x_0 > (-\mu/a)^{1/(2m-2)} > 0$, dx/dt remains always positive and $x(t)$ is thus a strictly increasing function that overtakes, after a sufficiently large time interval, any given number, that is $\lim_{t\to\infty} x(t) = \infty$. If $x_0 < -(-\mu/a)^{1/(2m-2)} < 0$, the derivative dx/dt is negative along the entire semiaxis $t \geqslant t_0$, hence $x(t)$ is a strictly decreasing function, $\lim_{t\to\infty} x(t) = -\infty$, and consequently $x(t) < -N$, with $N > 0$ arbitrary, given a sufficiently large t, $(t \geqslant T(N))$.

Accordingly, although the trivial solution of equation (1) with $a > 0$, $\mu < 0$ is uniformly asymptotically stable in the sense of Liapunov, the equation admits divergent solutions for however small initial conditions in a sufficiently close vicinity of the boundary, $\mu = 0$, characterized by the inequality $|\mu| < a|x_0^{2m-2}|$.

REFERENCES

1. G. H. Bryan and W. E. Williams, "The Longitudinal Stability of Aerial Gliders," *Proc. Roy. Soc.* (*London*) **A73**(1904), 100–116.
2. G. H. Bryan, *Stability in Aviation*. Macmillan, London, 1911.
3. E. J. Routh, *A Treatise on the Stability of a Given State of Motion*. Macmillan, London, 1877.
4. E. J. Routh, *Dynamics of a System of Rigid Bodies* (6th ed.). Macmillan, London, 1905.
5. V. S. Vedrov, *Dynamic Stability of Aircraft* (in Russian). Oborongiz, Moscow, 1938.
6. W. J. Duncan, *The Principles of the Control and Stability of Aircraft*. Cambridge Univ. Press, London and New York, 1952.
7. B. Etkin, *Dynamics of Flight: Stability and Control*. Wiley, New York, 1959.
8. A. W. Babister, *Aircraft Stability and Control*. Macmillan (Pergamon), New York, 1961.
9. I. V. Ostoslavskiy, *Aircraft Aerodynamics* (in Russian). Oborongiz, Moscow, 1957.
10. C. D. Perkins and R. E. Hage, *Airplane Performance, Stability and Control*. Wiley, New York, 1949.
11. A. A. Lebedev and L. S. Chernobrovkin, *Flight Dynamics of Unmanned Flying Machines* (in Russian). Oborongiz, Moscow, 1962.
12. J. Renaudie, *Essais en vol*; *Performances et qualités de vol*, Tome 2. Dunod, Paris, 1960.
13. A. M. Liapunov, *Stability of Motion*, with a contribution by V. A. Pliss. Academic Press, New York, 1966.
14. R. Bellman, *Stability Theory of Differential Equations*. McGraw-Hill, New York, 1959.
15. S. B. Gates "A Survey of Longitudinal Stability below the Stall with an Abstract for Designers' Use," *Aeronaut. Res. Council Rept. & Memo.* **1118**(1928).
16. C. H. Zimmerman, *An Analysis of Longitudinal Stability in Power-off Flight with Charts for Use in Design*. NACA Rept. 521 (1935).

17. C. H. Zimmerman, *An Analysis of Lateral Stability in Power-off Flight with Charts for Use in Design.* NACA Rept. 589 (1937).
18. A. V. Mikhaylov, "Method of harmonic analysis in control theory" (in Russian), *Avtomat. Telemekh.* **1938**, 27.
19. A. Leonhard, "Neues Verfahren zur Stabilitätsuntersuchung," *Arch. Elektrotech.* **28**(1944), 17–28.
20. L. Cremer, "Ein neues Verfahren zur Beurteilung der Stabilität linearer Regelungs-Systeme," *ZAMM* **25–27**(1947).
21. H. Nyquist, "Regeneration theory," *Bell Syst. Tech. J.* **11**(1932).
22. N. N. Bautin, *Behavior of Dynamic Systems in the Neighborhood of the Stability Region Boundaries* (in Russian). Gostekhizdat, Leningrad, Moscow, 1949.
23. G. V. Kamenkov, "Investigation of a critical case in the problem of motion stability, I—II" (in Russian), *Kazan' Aviation Inst. Works* **4**(1935), 3–18; **5**(1936), 19–27.
24. G. V. Kamenkov, "On the stability of motion (in Russian), *Kazan' Aviation Inst. Works* **9**(1939).
25. A. Hurwitz, "Über die Bedingungen, unter welchen eine Gleichung nur Wurzeln mit negativen realen Teilen besitzt," *Math. Annal.* **46**(1895), 237–284.

Chapter 6

LONGITUDINAL STABILITY OF STEADY STRAIGHT FLIGHT

6.1. LONGITUDINAL STABILITY WITH LOCKED CONTROLS

Some general statements from the preceding chapter will be illustrated as we consider the stability of steady motion of an aircraft with locked controls and in a partially controlled condition as defined in Chapter 4. Our starting point in the study of longitudinal stability will be the linear system (2.16), corresponding to straight basic motion.

One of the dynamic variables that is best suited for control by the pilot, both on account of the nature of its variation and because it is possible to record this variation visually, is the longitudinal attitude of the aircraft. From here on, to enable us to consider the constraint of this variable, the following change of variables will be made in system (2.16):

$$\Delta\gamma = \Delta\theta - \Delta\alpha.$$

Then, for the case of locked controls ($\hat{M}_c \equiv 0$), this system becomes

$$\frac{d\Delta\hat{V}}{d\hat{t}} = a_{11}\Delta\hat{V} + a_{12}\Delta\alpha + a_{13}\Delta\theta,$$

$$\frac{d\Delta\alpha}{d\hat{t}} = a_{21}\Delta\hat{V} + a_{22}\Delta\alpha + a_{23}\Delta\theta + \Delta\hat{q},$$

$$\frac{d\Delta\theta}{d\hat{t}} = \Delta\hat{q},$$

$$\frac{d\Delta\hat{q}}{d\hat{t}} = a_{41}\Delta\hat{V} + a_{42}\Delta\alpha + a_{43}\Delta\theta + a_{44}\Delta\hat{q}, \tag{6.1}$$

where

$$a_{11} = -\frac{1}{\mu}\left[C_D + \tfrac{1}{2}\mathrm{M}C_{D_\mathrm{M}} - C_{T_V}\cos(\alpha - \varphi)\right],$$

$$a_{12} = -\frac{1}{\mu}\,[\tfrac{1}{2}C_{D_\alpha} + C_T\sin(\alpha-\varphi)] + \hat{g}\cos\gamma,$$

$$a_{13} = -\hat{g}\cos\gamma,$$

$$a_{21} = -\frac{1}{\mu}\,[C_L + \tfrac{1}{2}\mathbf{M}C_{L_\mathbf{M}} + C_{T_V}\sin(\alpha-\varphi)],$$

$$a_{22} = -\frac{1}{\mu}\,[C_{L_\alpha} + C_T\cos(\alpha-\varphi)] + \hat{g}\sin\gamma,$$

$$a_{23} = -\hat{g}\sin\gamma,$$

$$a_{41} = \frac{1}{i_B}\left\{C_m + \tfrac{1}{2}\mathbf{M}C_{m_\mathbf{M}} - C_{T_V}\hat{z}_T - \frac{1}{2\mu}C_{m_{\dot\alpha}}[C_L + \tfrac{1}{2}\mathbf{M}C_{L_\mathbf{M}} + C_{T_V}\sin(\alpha-\varphi)]\right\},$$

$$a_{42} = \frac{1}{i_B}\left\{\tfrac{1}{2}C_{m_\alpha} - \frac{1}{2\mu}C_{m_{\dot\alpha}}[C_{L_\alpha} + C_T\cos(\alpha-\varphi) - \mu\hat{g}\sin\gamma]\right\},$$

$$a_{43} = -\frac{1}{2i_B}\,\hat{g}\sin\gamma\, C_{m_{\dot\alpha}},$$

$$a_{44} = \frac{1}{2i_B}\,(C_{m_q} + C_{m_{\dot\alpha}}).$$

Hereunder some known statements and results of the conventional theory will briefly be recalled (see, e.g., [1, Chapter 6; 2, Pt. I, Chapter 9]). They concern certain typical solutions of the system (6.1) modeling the longitudinal disturbed motion of an aircraft with locked controls.

In conventional cases the characteristic equation

$$\lambda^4 + a_1\lambda^3 + a_2\lambda^2 + a_3\lambda + a_4 = 0 \tag{6.2}$$

with

$$a_1 = -(a_{11} + a_{22} + a_{44}),$$

$$a_2 = -a_{42} + a_{22}a_{44} + a_{11}(a_{22} + a_{44}) - a_{12}a_{21}$$

$$a_3 = -a_{11}(-a_{42} + a_{22}a_{44}) + a_{12}(-a_{41} + a_{21}a_{44}) - a_{13}a_{41}$$

$$a_4 = -a_{13}(a_{22}a_{41} - a_{21}a_{42})$$

of system (6.1) has either two pairs of complex-conjugate roots, or two complex-conjugate and two real roots. The classical theory also points to certain quantitative relations among these roots, namely, the two characteristic roots that in both cases mentioned are complex conjugate have both the absolute value of the real part and that of the imaginary part large in comparison with the moduli of the other two roots. For simplicity and in agreement with the generally adopted terminology, the first will be called "large roots," and the latter "small" or "phugoid roots."†

The large complex-conjugate roots have their real part negative, the mode corresponding to the disturbed motion being a strongly damped oscillation (it is practically completely exhausted in a few seconds) and of relatively high frequency, both as compared to the homologous characteristics of the remaining modes. (The absolute characterization of quantities such as damping velocity, and frequency is obviously meaningless. Nevertheless, for brevity, the respective terms of comparison will sometimes not be specified if no misunderstanding is involved. Thus, e.g., "high frequency" will mean a frequency of a greater order of magnitude as compared with that of either the remaining modes or the respective oscillations in other régimes.) This mode will be called hereafter the mode of short-period oscillations.

The rapid mode affects mainly the incidence angle and the rate of pitch: as a result of aircraft inertia, the rapid mode is practically exhausted before the deviation induced by a disturbance of the flight speed and flight path angle or that of the attitude becomes noticeable. Thus the discrepancy between the values of the two pairs of roots results not only in the existence of two physical components of the longitudinal induced motion, but also in their shift in time. In a first stage it practically only amounts to short-period oscillations of incidence and rate of pitch, which are damped in the first few seconds, followed by a stage essentially characterized by the relatively slow variation of the center-of-mass speed and path inclination. This allowed B. Melvill Jones (apparently for the first time, see [4]) to achieve separate simplified modeling of the two consecutive stages of longitudinal induced motion. He called the first stage, that of the short-period oscillations, *the rapid incidence adjustment*, the term that has remained in the literature.

† By extending this term, attributed by F. W. Lanchester [3, Chapter 2] to the theoretical paths of flying machines imagined by him ("aerodones"). The machines were conceived without thrust, while their paths were deduced on the basis of the assumption that air reaction is permanently proportional to the square of the speed of the center of mass relative to motionless atmosphere, and directed normally on the flight path (i.e., zero drag). Lanchester's hypotheses are adaptable to the induced motion in its phase where the mode corresponding to the "small" roots becomes dominant, which justifies the extension of the conventional term phugoid to this phase.

The characteristic roots corresponding to the relatively slow stage may have positive real parts. Generally, a slight phugoid instability is admitted in free flight (see, e.g., [5, Section 10–5, pages 401, 407]†) if this allows improved performance or maneuverability characteristics to be obtained. The study of the disturbed motion is, however, useful at this stage for considerations of handling qualities, since instability must be eliminated by controls. This is why the classical stability diagrams refer to phugoid roots. Hence, two Routh–Hurwitz relations will be sufficient instead of four (see Appendix I, Chapter 4), namely,

$$a_4 < 0 \tag{6.3}$$

and

$$a_1 a_2 a_3 - a_3^2 - a_1^2 a_4 > 0. \tag{6.4}$$

It can readily be shown that (6.3) maintains the real phugoid roots negative, while (6.4) maintains negative the real parts of the complex-conjugate phugoid roots.

Indeed, let the phugoid roots be real and let the system parameters vary until one of the two phugoid roots vanishes. The boundary of the asymptotic stability region has thus been reached; hence the parameters may be made to vary further, so that one root becomes positive. The vanishing of one of the roots obviously implies the vanishing of the free term; hence inequality (6.3) maintains the real phugoid roots negative.

If, initially, the phugoid roots are complex conjugate, the boundary of the asymptotic stability region will be reached, by continuous variation of the system parameters, when the real parts of these roots vanish. Hence, on the boundary there will be a pure imaginary pair of roots, that is, $\pm i\omega$. Putting $\lambda = \pm i\omega$, equation (6.2) becomes

$$\omega^4 \mp i a_1 \omega^3 - a_2 \omega^2 \pm i a_3 \omega + a_4 = 0$$

or, further

$$\omega^4 - a_2 \omega^2 + a_4 = 0, \qquad a_1 \omega^2 - a_3 = 0,$$

wherefrom, by eliminating ω, relation

† Kalatchev [6], relying on the opinion of a number of test pilots, recommends avoiding aperiodic divergence, while he finds oscillatory instability perfectly acceptable (see also [7, Section VI-3, page 401]).

$$a_1a_2a_3 - a_3^2 - a_1^2a_4 = 0$$

is obtained. Therefore the fulfillment of condition (6.4) implies that the negative sign of the real part of the complex phugoid root is maintained.

The classical diagrams for longitudinal stability are plotted on the basis of the relations (6.3) and (6.4), the coefficients a_1, a_2, a_3, and a_4 being considered functions of two parameters each, as, for example, aerodynamic damping and static stability ($C_{m_q} = f(C_{m_\alpha})$ diagrams), profile drag and lift ($C_{D_p} = f(C_L)$ diagrams), and so forth.

It has been stated above that most aircraft designs have phugoid instability in certain ranges of the flight régime. Indeed, let the expression of the free term a_4 of the characteristic equation (6.2) be written explicitly, using the following approximate† expressions of the a_{ik} coefficients (see (6.1))

$$a_{13} = -\hat{g},$$

$$a_{21} \cong -\frac{1}{\mu}(C_L + \tfrac{1}{2}\mathbf{M}C_{L_\mathbf{M}}),$$

$$a_{22} \cong -\frac{1}{2\mu}C_{L_\alpha},$$

$$a_{41} \cong \frac{1}{2i_B}(\mathbf{M}C_{m_\mathbf{M}} + a_{21}C_{m_{\dot\alpha}}),$$

$$a_{42} \cong \frac{1}{2i_B}(C_{m_\alpha} + a_{22}C_{m_{\dot\alpha}}).$$

It is readily seen that a_4 does not contain terms in $C_{m_{\dot\alpha}}$ and, with the accepted approximations, it has the expression

$$a_4 \approx \left(\frac{\hat{g}}{2i_B}\right)(a_{21}C_{m_\alpha} - a_{22}\mathbf{M}C_{m_\mathbf{M}})$$

and hence the condition $a_4 > 0$ may be replaced by

$$a_{21}C_{m_\alpha} - a_{22}\mathbf{M}C_{m_\mathbf{M}} > 0. \tag{6.5}$$

For a speed of the basic motion, for which the effect of the compressibility of the air on the aerodynamic characteristics is negligible, the latter inequality reduces to $-C_LC_{m_\alpha}/\mu > 0$ or even to $C_{m_\alpha} < 0$. (Inequality $a_4 > 0$ is some-

† Neglecting the terms in C_T and C_{T_V} in comparison with the remaining terms; in trimmed steady flight $C_m = 0$.

times referred to in the literature as the condition for static stability; see, e.g., [8, page 120], for this reason.) Hence, in an incompressible régime, static stability becomes a necessary and sufficient condition to avoid aperiodic phugoid instability. However, at high speed, in certain ranges, the situation may alter. Thus, for $C_L + \mathbf{M}C_{L_\mathbf{M}}/2 < 0$, a relation that is possible within a certain range beyond the critical Mach number, the role of static stability is reversed, sufficiently great static stability being liable to lead to non-fulfillment of condition (6.5) and hence to phugoid divergence. In this interval of the transonic region of flight speed the situation is made still worse by the presence of the second term in the left-hand side of inequality (6.5), which becomes negative together with $C_{m_\mathbf{M}}$.

Phugoid oscillations may also be unstable, mainly because of the change in sign of the coefficient a_{11} of system (6.1). Indeed, under the common assumption of the invariance of incidence during the phugoid phase, (6.1) reduces to a second-order system:

$$\frac{d\Delta\hat{V}}{d\hat{t}} = a_{11}\Delta\hat{V} + a_{13}\Delta\theta, \qquad \frac{d\Delta\theta}{d\hat{t}} = -\,a_{21}\Delta\hat{V}.$$

In the case of oscillatory phugoid motion the characteristic roots of the system are obviously complex conjugate, their real part reducing to $a_{11}/2$. It may usually (see the expression of coefficient a_{11} in equations (6.1)) become negative in the supersonic region of flight speed or at relatively low subsonic speeds.

Phugoid instability is eliminated by controls in order to constrain flight speed or flight path inclination, or both. The latter implies in the main a constraint of the longitudinal attitude, the incidence being less affected by the phugoid disturbed motion.† Direct constraint of incidence would, moreover, be rendered more difficult precisely by the dominant character of the rapid mode in the variation of the disturbance of this angle (requiring an elaborate adjustment installation with a very small time constant). To this would also be added the difficulties in detecting the incidence variation that cannot be measured by means of the instruments currently used for

† Nevertheless, the variation of incidence contains, obviously, slow modes which, however, are secondary in the general picture of this variation. In addition, as will be seen in Section 6.2, in the case of simultaneous constraint of speed and longitudinal attitude the slow modes of function $\Delta\alpha(t)$ are indirectly eliminated by the constraint of these two variables, leaving the rapid oscillation, which is dominant in the variation of incidence, to damp due to the inherent qualities of the aircraft.

determining the angular elements of motion (based on the principle of the gyroscope).

It should also be noted that direct constraint of the flight speed by throttle control does not result in satisfactory conditions either, on account of the relatively great delay with which thrust responds to throttle variation; nevertheless the slow variation in the speed of the center of mass justifies the discussion of this case of partial control. Hence, the following cases of partial control will be considered: constraint of pitch angle, constraint of flying speed, simultaneous constraint of these two variables, as well as the case of simultaneous constraint of the angle and the rate of pitch.

6.2. STABILITY OF PARTIALLY CONTROLLED LONGITUDINAL MOTION†

6.2.1. Partial Control Through Longitudinal Attitude Constraint

It is assumed that, by controls, the deviation of the pitch angle is maintained sufficiently small, in the sense of the theory expounded in Chapter 4. In this case, the linearized auxiliary system will be

$$\frac{d\Delta\hat{V}}{d\hat{t}} = a_{11}\Delta\hat{V} + a_{12}\Delta\alpha,$$

$$\frac{d\Delta\alpha}{d\hat{t}} = a_{21}\Delta\hat{V} + a_{22}\Delta\alpha + \Delta\hat{q},$$

$$\frac{d\Delta\hat{q}}{d\hat{t}} = a_{41}\Delta\hat{V} + a_{42}\Delta\alpha + a_{44}\Delta\hat{q}. \tag{6.6}$$

From the behavior of the solution of this system conclusions may be drawn in accordance with the theory, presented in Chapter 4, on the stability of the partially controlled steady motion considered. To this end the characteristic equation of system (6.6),

$$\lambda^3 + b_1\lambda^2 + b_2\lambda + b_3 = 0, \tag{6.7}$$

is studied. Coefficients b_i have the expressions

$$b_1 = -(a_{11} + a_{22} + a_{44}),$$

$$b_2 = -a_{42} + a_{22}a_{44} + a_{11}(a_{22} + a_{44}) - a_{12}a_{21},$$

† See [9, Section 4].

$$b_3 = -a_{11}(a_{22}a_{44} - a_{42}) + a_{12}(a_{21}a_{44} - a_{41}).$$

Equation (6.7) has, in typical situations, a pair of complex roots (which do not considerably differ from the pair of roots, conventionally called large, of equation (6.2) corresponding to free flight) and a real root. The sign of the latter depends, according to the Routh–Hurwitz criterion, on the sign of the free term b_3 and hence on that of the expression $a_{11}C_{m_\alpha} - (a_{11}a_{22} - a_{12}a_{21})C_{m_q} - a_{12}\mathbf{M}C_{m_\mathbf{M}}$. In many cases attitude control may be sufficient. Such cases are the ones in which the real root is negative and of a sufficiently large value to enable ensuring, besides rapid incidence adjustment, heavy damping of the rate of pitch and satisfactory convergence of flight speed deviation as well. The appearance of a positive real root implying dangerous divergence is unlikely. Situations may arise, however, where the real root is slightly positive or negative, failing to ensure sufficiently rapid convergence of the aperiodic mode. Stabilizing control will then be completed by the constraint, simultaneous with that of the pitch angle, of another variable: the derivative control of the longitudinal attitude (additional constraint of the pitching velocity) or flight speed control. These two cases of simultaneous control will be considered here.

6.2.2. Case of Simultaneous Constraint of the Angle and Angular Velocity of Pitch

The addition of derivative control generally has the effect, as will be shown here, of eliminating the oscillatory mode of the solution of the auxiliary system and, in some cases, of accelerating the convergence of this solution, which implies, for satisfactory response characteristics, analogous properties of the solution of the disturbed motion system in the conditions of the partial controls considered.

The auxiliary system corresponding to the simultaneous constraint of the variables $\Delta\theta$ and $\Delta\hat{q}$ will be

$$\frac{d\Delta\hat{V}}{d\hat{t}} = a_{11}\Delta\hat{V} + a_{12}\Delta\alpha, \qquad \frac{d\Delta\alpha}{d\hat{t}} = a_{21}\Delta\hat{V} + a_{22}\Delta\alpha, \tag{6.8}$$

having the characteristic equation

$$\lambda^2 - (a_{11} + a_{22})\lambda + a_{11}a_{22} - a_{12}a_{21} = 0. \tag{6.9}$$

In typical situations the value of the product $a_{12}a_{21}$ is lower than that of the product $a_{11}a_{22}$ and inferior to the quantity $(a_{11} + a_{22})/2$ in order of magnitude.

In agreement with this, the roots of equation (6.9) will be real and negative. Let it be assumed that the product $a_{12}a_{21}$ is negligible in comparison with $(a_{11} + a_{22})/2$. Then the roots of equation (6.9) will be approximately equal to a_{11} and a_{22}. To justify the supplementary use of derivative control for the longitudinal attitude of the aircraft it is necessary that $\min(|a_{11}|, |a_{22}|) \gg |\lambda_3|$ (by min is meant the smallest of the two magnitudes in parentheses), λ_3 being the real root of the characteristic equation (6.7) corresponding to the constraint of only the pitch angle.

6.2.3. Simultaneous Control of Longitudinal Attitude and Flight Speed

In this case of partial control the auxiliary system takes the form of the well-known system of equations for pure short-period oscillations

$$\frac{d\Delta\alpha}{d\hat{t}} = a_{22}\,\Delta\alpha + \Delta\hat{q}, \qquad \frac{d\Delta\hat{q}}{d\hat{t}} = a_{42}\,\Delta\alpha + a_{44}\,\Delta\hat{q}. \tag{6.10}$$

In the classical theory the system is obtained according to Jones's simplified assumptions, that is, that incidence adjustment is assumed a priori sufficiently rapid, and inertia and the inertia moment of the aircraft sufficiently large to allow us to neglect, in a first time interval, the variation in speed of the center of gravity and of the pitch angle induced by the disturbance (see [4, page 96]). In the given case, system (6.10), as an auxiliary system corresponding to the constraint of the variables $\Delta\hat{V}$ and $\Delta\theta$ was obtained on basis of the general hypotheses stated in Section 4.2 concerning the characteristics of the control system and the characteristics of the aircraft response to the controls applied with the intention of constraining $\Delta\hat{V}$ and $\Delta\theta$. It is recalled that these hypotheses were that: actuation of stabilizing controls induced by the appearance of a disturbance ceases at the instant the constrained deviations ($\Delta\hat{V}$ and $\Delta\theta$) vanish; and $\lim_{\hat{t}\to\infty} \Delta\hat{V}(\hat{t}) = 0$ and $\lim_{\hat{t}\to\infty} \theta(\hat{t}) = 0$. The realization of the hypotheses regarding the asymptotic behavior of the controlled variables (quality of the controls applied) permits us to obtain from the solution of the auxiliary system some valid information regarding the partially controlled motion as a whole. Thus, according to Proposition 2, Section 4.2, uniform asymptotic stability of the trivial solution of system (6.10) implies the uniform asymptotic stability of the basic motion of the aircraft in the conditions of the partial control set. Moreover, due to the damping of the controlled variables (providing the damping is sufficiently rapid) the behavior of the solution of system (6.10) also satisfactorily models the actual behavior of the free variables $\Delta\alpha$ and $\Delta\hat{q}$ as regards the character of the variation, the damping velocity, and so on.

As known from the simplified theory of rapid longitudinal disturbed motion, the characteristic equation of system (6.10),

$$\lambda^2 - (a_{22} + a_{44})\lambda + a_{22}a_{44} - a_{42} = 0, \tag{6.11}$$

has a pair of complex roots with values of both the real and imaginary parts sufficiently large to ensure heavily damped short period oscillations

$$\lambda_{1,2} = \frac{a_{22} + a_{44}}{2} \pm i \left[- a_{42} - \left(\frac{a_{22} - a_{44}}{2} \right)^2 \right]^{1/2}.$$

The rate of damping measured by $(a_{22} + a_{44})/2$ is generally much higher in value than that obtained by the partial control so far considered. However, certain reservations are required regarding the given case of a simultaneous constraint of the variables $\Delta\hat{V}$ and $\Delta\theta$ because of the difficulty in obtaining satisfactory direct control of the flight speed. Direct constraint of speed is achieved by throttle control implying considerable time lag of the response; this leads to lower accuracy in making the foregoing hypotheses satisfactory. However, on account of the inertia of the aircraft a certain compensation occurs as a result of the slow variation of speed.

6.2.4. Constraint of Flight Speed

It will be shown that constraint of flight speed *alone* is insufficient to stabilize some current steady-state motions of the aircraft, even if the reservations previously mentioned regarding the quality of the control by throttle control are not taken into account. Accordingly, the assumption will be made that the use of the controls allows us to impose a sufficiently small upper bound to the variable $\Delta\hat{V}$ and to cause its damping out ($\lim_{\hat{t}\to\infty} \Delta\hat{V}(\hat{t}) = 0$). The corresponding auxiliary system will be

$$\frac{d\Delta\alpha}{d\hat{t}} = a_{22}\Delta\alpha + a_{23}\Delta\theta + \Delta\hat{q},$$

$$\frac{d\Delta\theta}{d\hat{t}} = \Delta\hat{q},$$

$$\frac{d\Delta\hat{q}}{d\hat{t}} = a_{42}\Delta\alpha + a_{43}\Delta\theta + a_{44}\Delta\hat{q}, \tag{6.12}$$

having the characteristic equation

$$\lambda^3 + c_1\lambda^2 + c_2\lambda + c_3 = 0 \tag{6.13}$$

where

$$c_1 = -(a_{22} + a_{44}),$$

$$c_2 = -(a_{22}a_{44} - a_{42}) + a_{43},$$

$$c_3 = a_{22}a_{43} - a_{23}a_{42}.$$

Consequently c_1 is identical to the coefficient of λ in equation (6.11) and c_2 differs from the free term of this equation by the quantity a_{43}. The coefficients $a_{23} = -\hat{g}\sin\gamma$ and $a_{43} = -(C_{m_{\dot{\alpha}}}/2i_B)\hat{g}\sin\gamma$ are, generally, small (in comparison with the coefficients a_{22}, a_{42}, and a_{44}) both on account of the factor $\hat{g} = C_L/2\mu$ and, for not too abrupt inclinations of the path, of the factor $\sin\gamma$. Then c_3 will also be smaller in comparison with c_1 and c_2 and hence the characteristic equation (6.13) will have one real root λ_1 and a pair of complex-conjugate roots very close to those of equation (6.11) representing heavily damped short-period oscillations. The real root λ_1 corresponds to an aperiodic mode that may be divergent. Indeed, it may be seen that the sign of the free term c_3 of equation (6.13)

$$c_3 = -\hat{g}\sin\gamma\left(\frac{a_{22}C_{m_{\dot{\alpha}}}}{2i_B} - a_{42}\right)$$

is generally given by the factor $\sin\gamma$, the terms in the parentheses being positive. Hence the condition $c_3 > 0$ imposed by the Routh–Hurwitz criteria is only fulfilled for flight along a descending path ($\gamma < 0$). In the case of a climbing path there appears a divergent mode of the solution of auxiliary system (6.12) showing the insufficiency of partial control by constraint of only flight speed.

For the horizontal basic motion a zero characteristic root appears, and hence the linearized auxiliary system (6.12) ceases to be conclusive. Considering, according to Liapunov, the nonlinear system, the method of the critical cases generally shows that instability of the auxiliary system is also maintained for $\gamma = 0$.

6.3. A NEUTRALLY STABLE MODEL FOR CERTAIN LONGITUDINAL MOTIONS

In Section 5.5 we mentioned that if the linear first-approximation system has at least one zero characteristic root or a pair of imaginary roots, it ceases to be conclusive and examination of the nonlinear system becomes compulsory in order to investigate stability. The trivial solution of the latter may be, in the situation shown, either asymptotically stable or unstable, or it may enjoy neutral (i.e., nonasymptotic) stability. The remark was also made that from the standpoint of handling characteristics as determined from the pilot's opinion, the three situations have the common feature that they do not assure quick enough damping of the disturbance, and therefore require the application of corrective controls.

It was finally mentioned that in current cases the situation of theoretical neutral stability for the zero solution of the nonlinear system seldom appears. Such a situation, will, however, be considered hereunder for certain longitudinal motions of an aircraft, rather in order to illustrate the singularity of a neutrally stable behavior.

To this end, a system of equations of the longitudinal disturbed motion will be set up.

$$\frac{dx}{dt} = Ax + X(x). \tag{6.14}$$

Here, A is an $n \times n$ matrix having one of its eigenvalues zero with simple elementary divisors, while the remaining ones have negative real parts. Function X, denoting the nonlinear terms, is such that, by linear transformation, the system can be brought to the form

$$\frac{dy}{dt} = By + Y(y, z_0) \tag{6.15}$$

where z_0 is a parameter, B a Hurwitzian $(n-1) \times (n-1)$ matrix, and function Y containing the nonlinear terms in y satisfies the condition

$$\frac{\partial Y}{\partial y}(0, z_0) = 0$$

and is therefore arbitrarily small for y sufficiently small.

From the fact that, in relation to y, the function Y contains only terms of at least the second degree, it results that for a sufficiently small value of parameter z_0 there is a unique $y_0(z_0)$, so that $By_0(z_0) + Y(y_0(z_0), z_0) = 0$ and $\lim_{z_0 \to 0} y_0(z_0) = 0$. This means that system (6.15) has, in the neighborhood of the origin, the unique stationary point $y_0(z_0)$ that, as can readily be seen, is asymptotically stable.

Therefore, any solution of system (6.14) will tend asymptotically toward a constant, depending on the initial conditions, and small together with them. Hence, any motion satisfactorily modeled by system (6.6) will tend toward a steady-state motion that, for a sufficiently small initial disturbance, is arbitrarily close to the undisturbed motion. This statement is not true for every choice of the state variables.†

To build system (6.14) we first use the well-known equations of the fixed-control longitudinal disturbed motion referred to wind axes for the steady basic motion, assumed, for simplicity, straight and level:

$$\frac{d\Delta\hat{V}}{d\hat{t}} = a_{11}\Delta\hat{V} + a_{12}\Delta\alpha + a_{13}\Delta\theta + f_1(\Delta\hat{V}, \Delta\alpha, \Delta\theta, \Delta\hat{q}),$$

$$\frac{d\Delta\alpha}{d\hat{t}} = a_{21}\Delta\hat{V} + a_{22}\Delta\alpha + \quad \Delta\hat{q} + f_2(\Delta\hat{V}, \Delta\alpha, \Delta\theta, \Delta\hat{q}),$$

$$\frac{d\Delta\theta}{d\hat{t}} = \quad \Delta\hat{q},$$

$$\frac{d\Delta\hat{q}}{d\hat{t}} = a_{41}\Delta\hat{V} + a_{42}\Delta\alpha + a_{44}\Delta\hat{q} + f_4(\Delta\hat{V}, \Delta\alpha, \Delta\theta, \Delta\hat{q}); \qquad (6.16)$$

f_1, f_2, and f_4 stand for the respective nonlinear terms; coefficients a_{ij} are the same as in system (6.1). The following additional assumptions are made.

(i) The free term of characteristic equation (6.2) of the linear system is zero (i.e., $a_{41} = a_{21}a_{42}/a_{22}$); it will therefore have a zero root; the remaining characteristic roots are one of them real and negative and two complex conjugate with negative real part.

(ii) $f_1(\Delta\hat{V}, \Delta\alpha, \Delta\theta, \Delta\hat{q}) = f(\Delta\hat{V}, \Delta\alpha, \Delta\theta, \Delta\hat{q}) \neq 0$ and $f_i(\Delta\hat{V}, \Delta\alpha, \Delta\theta, \Delta\hat{q}) \approx 0$ for $i \neq 1$.

Hence, system (6.14) will be specified in the form

$$\frac{dx_1}{dt} = a_{11}x_1 + a_{12}x_2 + a_{13}x_3 + f(x_1, x_2, x_3, x_4),$$

$$\frac{dx_2}{dt} = a_{21}x_1 + a_{22}x_2 + x_4,$$

† This proviso, obviously involved in the very definition of stability as a mathematical concept, will be illustrated by an example at the end of this section.

$$\frac{dx_3}{dt} = x_4,$$

$$\frac{dx_4}{dt} = \frac{a_{21}a_{42}}{a_{22}} x_1 + a_{42}x_2 + a_{44}x_4, \tag{6.17}$$

with $t = \hat{t}$, $x_1 = \Delta\hat{V}$, $x_2 = \Delta\alpha$, $x_3 = \Delta\theta$, $x_4 = \Delta\hat{q}$. To bring this system to the form (6.15) a linear transformation with the matrix

$$\begin{pmatrix} 1 & 0 & 0 & 0 \\ 0 & 1 & 0 & 0 \\ 0 & 0 & 1 & 0 \\ 0 & \dfrac{a_{42}}{a_{22}} & \dfrac{a_{42}}{a_{22}} + a_{44} & \dfrac{1}{a_{22}} \end{pmatrix},$$

is first made,† yielding

$$\frac{dx_1}{dt} = a_{11}x_1 + a_{12}x_2 + a_{13}x_3 + g(x_1, x_2, x_3, z),$$

$$\frac{dx_2}{dt} = a_{21}x_1 + \left(\frac{a_{42}}{a_{22}} + a_{22}\right) x_2 + \left(-\frac{a_{42}}{a_{22}} + a_{44}\right) x_3 + \frac{1}{a_{22}} z,$$

$$\frac{dx_3}{dt} = \frac{a_{42}}{a_{22}} x_2 + \left(-\frac{a_{42}}{a_{22}} + a_{44}\right) x_3 + \frac{1}{a_{22}} z,$$

$$\frac{dz}{dt} = 0,$$

where g denotes the aggregate of nonlinear terms in x_1, x_2, x_3 of the first equation. Since $z = z_0 = \text{const.}$, the system sought (6.15) will be

$$\frac{dy_1}{dt} = a_{11}y_1 + a_{12}y_2 + a_{13}y_3 + g(y_1, y_2, y_3, z_0),$$

$$\frac{dy_2}{dt} = a_{21}y_1 + \left(\frac{a_{42}}{a_{22}} + a_{22}\right) y_2 + \left(-\frac{a_{42}}{a_{22}} + a_{44}\right) y_3 + \frac{z_0}{a_{22}},$$

$$\frac{dy_3}{dt} = \frac{a_{42}}{a_{22}} y_2 + \left(-\frac{a_{42}}{a_{22}} + a_{44}\right) y_3 + \frac{z_0}{a_{22}}, \tag{6.18}$$

† The transformation is obtained by replacing one of the system variables, namely, x_4, by a new variable $z = \mu_1x_1 + \mu_2x_2 + \mu_3x_3 + \mu_4x_4$ so that, by virtue of the linear part in (6.17), the derivative dz/dt is identically zero (i.e., z is a first integral of the linearized system).

where $y_1 = x_1 = \Delta\hat{V}$, $y_2 = x_2 = \Delta\alpha$, $y_3 = x_3 = \Delta\theta$, $z_0 = -a_{42}x_2 + (a_{42} - a_{22}a_{44})x_3 + a_{22}x_4 = \text{const.}$, if $\partial g/\partial y_i\,(0, 0, 0, z_0) = 0$, $i = 1, 2, 3$.

It can readily be seen that the eigenvalues of the matrix of the linearized system (6.18) are identical with the nonzero roots of equation (6.2), where $a_{21}a_{42}/a_{22}$ was substituted for a_{41}. Since these roots have been assumed with negative real parts, in compliance with the reasoning above, system (6.18) only has, in the neighborhood of the origin, one stationary point that is asymptotically stable. This stationary point is arbitrarily close to the origin if the initial disturbance, and hence z_0, is sufficiently small. That is, any solution (x_1, x_2, x_3) of this system will, for $t \to \infty$, tend toward the steady-state solution $x_1 = \nu_1$, $x_2 = \nu_2$, $x_3 = \nu_3$ where the constants ν_1, ν_2, and ν_3, dependent on the parameter z_0, are the roots of the following system of algebraic equations.†

$$a_{11}\nu_1 + a_{12}\nu_2 + a_{13}\nu_3 + g(\nu_1, \nu_2, \nu_3, z_0) = 0,$$

$$a_{21}\nu_1 + \left(\frac{a_{42}}{a_{22}} + a_{22}\right)\nu_2 + \left(-\frac{a_{42}}{a_{22}} + a_{44}\right)\nu_3 + \frac{z_0}{a_{22}} = 0,$$

$$\frac{a_{42}}{a_{22}}\nu_2 + \left(-\frac{a_{42}}{a_{22}} + a_{44}\right)\nu_3 + \frac{z_0}{a_{22}} = 0.$$

Indeed, after the change of variables

$$x_1 = \xi_1 + \nu_1, \quad x_2 = \xi_2 + \nu_2, \quad x_3 = \xi_3 + \nu_3,$$

system (6.18) becomes

† Since $\partial g/\partial y_i\,(0, 0, 0, z_0) = 0$, $i = 1, 2, 3$, the Jacobian reduces, at the origin, to the determinant

$$\begin{vmatrix} a_{11} & a_{12} & a_{13} \\ a_{21} & \dfrac{a_{42}}{a_{22}} + a_{22} & -\dfrac{a_{42}}{a_{22}} + a_{44} \\ 0 & \dfrac{a_{42}}{a_{22}} & -\dfrac{a_{42}}{a_{22}} + a_{44} \end{vmatrix}.$$

The matrix of system (6.18) being Hurwitzian (therefore having no zero eigenvalues) this determinant is different from zero at the origin. Then, according to the existence theorem of implicit functions, the system of algebraic equations in ν_i will have one and only one set of roots, holomorphic functions of z_0 that vanish for $z_0 = 0$.

$$\frac{d\xi_1}{dt} = \left(a_{11} + \frac{\partial g}{\partial x_1}(\nu_1, \nu_2, \nu_3, z_0)\right)\xi_1 + \left(a_{12} + \frac{\partial g}{\partial x_2}(\nu_1, \nu_2, \nu_3, z_0)\right)\xi_2$$
$$+ \left(a_{13} + \frac{\partial g}{\partial x_3}(\nu_1, \nu_2, \nu_3, z_0)\right)\xi_3 + \Xi_1(\xi_1, \xi_2, \xi_3, z_0),$$
$$\frac{d\xi_2}{dt} = a_{21}\xi_1 + \left(\frac{a_{42}}{a_{22}} + a_{22}\right)\xi_2 + \left(-\frac{a_{42}}{a_{22}} + a_{44}\right)\xi_3,$$
$$\frac{d\xi_3}{dt} = \frac{a_{42}}{a_{22}}\xi_2 + \left(-\frac{a_{42}}{a_{22}} + a_{44}\right)\xi_3, \tag{6.19}$$

Ξ_1 containing the nonlinear terms in ξ_1, ξ_2, and ξ_3, and obviously vanishing at the origin. Since $\partial g/\partial x_i\ (0, 0, 0, z_0) = 0$ by hypothesis, $\partial g/\partial x_i$ are small together with ν_j and hence with z_0 (see note to page 138), which is equal to $\sum_i \mu_i x_i(t_0)$. Consequently, for sufficiently small initial conditions, $\partial g/\partial x_i\ (\nu_1, \nu_2, \nu_3, z_0)$, $i = 1, 2, 3$ are arbitrarily small, the matrix of system (6.19) is therefore Hurwitzian if that of system (6.18) is so and the zero solution of (6.19) is uniformly asymptotically stable. It results that $x_i \to \nu_i$ for $t \to \infty$.

Hence, the disturbed motion acceptably described by system (6.17) tends toward a steady-state motion that (in the state variables considered) for a sufficiently small initial disturbance is arbitrarily close to the basic motion.

For more definiteness a particular case of function f and the initial conditions will be dealt with. That is, it will be assumed that the only higher-degree term is the one in $\Delta\alpha^2$. In other words, if $C_D = C_{D_p} + kC_L^2(\alpha)$, where the parasite drag C_{D_p} and the coefficient k characterizing induced drag are supposed constant, then $f = a\,\Delta\alpha^2$, with $a = k(C_{L_\alpha}^2(\alpha_0) + C_L(\alpha_0)C_{L_{\alpha^2}}(\alpha_0)) \approx kC_{L_\alpha}^2(\alpha_0)$. In these formulas α_0 denotes the undisturbed incidence angle, $C_{L_\alpha} = \partial C_L/\partial\alpha$, $C_{L_{\alpha^2}} = \partial^2 C_L/\partial\alpha^2$. As initial instant, the instant t_0 is considered when a vertical gust of short duration causes a sudden change of incidence angle (i.e., $\Delta\alpha(t_0) = \Delta\alpha_0 \neq 0$), while $\Delta\hat{V}(t_0) = \Delta\theta(t_0) = \Delta\hat{q}(t_0) = 0$. Then $z_0 = -\,a_{42}\,\Delta\alpha_0$ (see notations after system (6.18)) and the induced motion will tend asymptotically toward the following steady-state motion.

$$\hat{V} = \hat{V}_1, \quad \alpha = \alpha_1, \quad \theta = \theta_1, \quad \hat{q} = 0, \tag{6.20}$$

with

$$\hat{V}_1 = \hat{V}_0 - \frac{a_{22}}{a_{21}}[-\,\delta + (\delta^2 + a_{42}\eta\,\Delta\alpha_0)^{1/2}],$$

$$\alpha_1 = \alpha_0 - \delta + (\delta^2 + a_{42}\eta\,\Delta\alpha_0)^{1/2},$$

$$\theta_1 = \alpha_0 + \frac{a\eta}{-a_{13}}[-\Delta\alpha_0 - \delta + (\delta^2 + a_{42}\eta\,\Delta\alpha_0)^{1/2}],$$

where $\eta = a_{13}a_{42}/[a(-a_{42} + a_{22}a_{44})]$ and $\delta = (-a_{12} + a_{11}a_{22}/a_{21})/2a + \eta/2$. For $\Delta\alpha_0 = 0$, V_1, α_1, and θ_1 are obviously equal to V_0, α_0, and $\theta_0 = \alpha_0$, respectively; hence the steady-state motion (6.20) is arbitrarily close, with respect to $\hat{V}$, α, θ, and $\hat{q}$, to the basic motion $\hat{V} = \hat{V}_0$, $\alpha = \alpha_0$, $\theta = \alpha_0$, $\hat{q} = 0$.

Accordingly, the uniform straight level flight of an aircraft, described—allowing for the assumptions stated—by the zero solution of system (6.17) whose linear part has a zero characteristic root and the remainder with negative real parts, is neutrally (i.e., nonasymptotically) stable under symmetric disturbances. Induced motion tends, for $t \to \infty$, toward a new steady-state motion arbitrarily close to the basic motion, as regards the state variables used, for sufficiently small disturbances. The limit steady-state motion is described, under the same assumptions, by the solution (6.20). In other words, the aircraft has neutral phugoid stability.

It is readily seen that this is an exceptional case, linked with the special form of the system. Systems of the form (6.17) model the longitudinal disturbed motion sufficiently correctly, for example, for basic motion in the incompressible régime at speeds close to the maximum flight speed. As soon as the coefficients of the other nonlinear terms, namely, the second- or higher-order derivatives with respect to $\hat{V}$, α, θ, or $\hat{q}$ of the lift or longitudinal moment function are no longer negligible (e.g., in certain ranges of flight speed in the compressible régime or close to stalling speed), the system containing these terms, although it has the same linear part as (6.17), may become unstable. Moreover, the very stability of the trivial solution of system (6.16) in the adopted hypotheses is a precarious one that only obtains in the case where coefficients a_{jk} are rigorously constant.

Finally, we note that the neutral stability deduced here is in a way fallacious because of the model resorted to. Indeed, the stability of the considered basic motion is ensured only with respect to the variables chosen (i.e., with respect to speed, incidence, longitudinal attitude, or flight path angle and rate of pitch). It is, however, no longer so with respect to some other state variables, left aside when making the model (e.g., with respect to height). Indeed, according to the results above, the flight path angle will asymptotically tend toward the constant magnitude $\theta_1 - \alpha_1$, which is generally different from zero (although small for small values of $\Delta\alpha_0$). Therefore, the induced flight attitude differs from the basic level to any extent, given a sufficiently long time interval elapsed from the instant the disturbance occurred; that is, the basic motion considered is divergent with respect to height. Hence the control of height by, for example, pitch angle constraint, is compulsory if phugoid stability with respect to the state variables used is only neutral.

All these facts emphasize the necessity of considering the stability of controlled flight whenever the free basic motion is neutrally stable, and that of aiming at an asymptotic stability with high enough rate of decay.

6.4. EFFECT OF PERTURBATION OF HEIGHT ON THE LONGITUDINAL STABILITY CHARACTERISTICS IN STEADY LEVEL FLIGHT

In the preceding sections of this chapter the forces induced by the variation of flight height were assumed negligible. In compliance with such a hypothesis it has been possible (i) to renounce considering an additional dynamic variable, namely, the induced deviation of flight altitude Δh and (ii) to consider undisturbed flight along a path differing from level flight as a steady-state motion, and hence to consider the disturbed motion pertaining to it as autonomous, consequently treating the stability problem by means of the conventional theory. In this section the hypothesis above will be eliminated, hence, to be able to resort further to the model of linear systems with constant coefficients, the undisturbed flight path will have to be horizontal. The more general case of motions differing from level flight will be considered in Chapter 10.

The necessity of considering the effect of variation in altitude on some characteristics of disturbed motion in the case of a horizontal basic motion—where, prima facie, such an effect would appear entirely negligible—was mentioned for the first time, according to [11], by Scheubel in [12]. This study was later developed and broadened by Neumark in [11]. (See also [2, Chapter 14].) As shown in these studies, in the case of horizontal basic motion the variation of external forces, induced by height disturbance, is comparable to the variation of these forces, determined by deviations of other kinematic parameters, such as flight speed and incidence, from their undisturbed values. Hence, an additional dynamic variable Δh has to be brought into the system of equations of disturbed motion. Obviously, this necessitates estimating the partial derivatives of the external forces and moments with respect to this new variable. As regards the constant undisturbed value of flight altitude, it becomes a parameter of the system, characterizing the basic motion, similar to the undisturbed flight speed, incidence, and so on.

Renouncing the hypothesis of neglecting height disturbance and taking into account that the aerodynamic forces **D** and **L**, the longitudinal aerodynamic moment M, and the thrust magnitude T are dependent on the altitude of flight, we thus modify system (2.3) until it becomes, for horizontal and straight ($\gamma \equiv 0$) free flight ($M_c \equiv 0$):

$$m\frac{d\Delta V}{dt} = \left[-\frac{\partial \mathbf{D}}{\partial V} + \frac{\partial T}{\partial V}\cos(\alpha - \varphi)\right]\Delta V + \left[-\frac{\partial \mathbf{D}}{\partial \alpha} - T\sin(\alpha - \varphi)\right]\Delta\alpha$$

$$- mg\,\Delta\gamma + \left[-\frac{\partial \mathbf{D}}{\partial h} + \frac{\partial T}{\partial h}\cos(\alpha - \varphi)\right]\Delta h,$$

$$\frac{d\Delta\alpha}{dt} = \Delta q - \frac{d\Delta\gamma}{dt},$$

$$mV\frac{d\Delta\gamma}{dt} = \left[\frac{\partial \mathbf{L}}{\partial V} + \frac{\partial T}{\partial V}\sin(\alpha - \varphi)\right]\Delta V + \left[\frac{\partial \mathbf{L}}{\partial \alpha} + T\cos(\alpha - \varphi)\right]\Delta\alpha$$

$$+ \left[\frac{\partial \mathbf{L}}{\partial h} - \frac{\partial T}{\partial h}\sin(\alpha - \varphi)\right]\Delta h,$$

$$B\frac{d\Delta q}{dt} = \left(\frac{\partial M}{\partial V} - \frac{\partial T}{\partial V}z_T\right)\Delta V + \frac{\partial M}{\partial \alpha}\Delta\alpha + \frac{\partial M}{\partial q}\Delta q + \frac{\partial M}{\partial \dot{\alpha}}\frac{d\Delta\alpha}{dt}$$

$$+ \left(\frac{\partial M}{\partial h} - \frac{\partial T}{\partial h}z_T\right)\Delta h,$$

$$\frac{d\Delta h}{dt} = V\Delta\gamma.$$

To bring this system to nondimensional form, the units introduced in Section 2.3 will generally be used. As the unit of air density ρ, the one corresponding to a fixed altitude, for example, to sea level or to the flight altitude and denoted ρ_0, will be adopted. The symbol σ, instead of $\hat{\rho}$, will denote the nondimensional magnitude ρ/ρ_0. To simplify the notation further, the dimensionless magnitude $\Delta h/(\bar{c}/2)$, proportional to the induced deviation of flight altitude, will be denoted ζ.

Further, the following dimensionless parameters are introduced.

$$\sigma_\zeta = \frac{\bar{c}}{2\rho_0}\frac{d\rho}{dh}, \qquad a_\zeta = \frac{\bar{c}}{2V_0}\frac{da}{dh}, \qquad C_{T_\sigma} = \frac{1}{V_0^2 S}\frac{\partial T}{\partial \rho}.$$

Then

$$\frac{\partial}{\partial h}(\mathbf{D},\mathbf{L}) = \frac{S}{\bar{c}}\rho_0 V_0^2(\hat{V}^2 C_{(D,L)}\sigma_\zeta - \sigma\hat{V}\mathbf{M}^2 C_{(D,L)_\mathbf{M}} a_\zeta),$$

$$\frac{\partial M}{\partial h} = \tfrac{1}{2}S\rho_0 V_0^2(\hat{V}^2 C_m\sigma_\zeta - \sigma\hat{V}\mathbf{M}^2 C_{m_\mathbf{M}} a_\zeta),$$

$$\frac{\partial T}{\partial h} = 2\frac{S}{\bar{c}}\rho_0 V_0^2 C_{T_\sigma}\sigma_\zeta,$$

or, taking into account that $\mathbf{M} = \hat{V}/\hat{a}$, $\hat{a} = a/V_0$, $\hat{V} = V/V_0$, we have

$$\frac{\partial}{h}(\mathbf{D},\mathbf{L}) = \frac{S}{\bar{c}}\rho_0 V_0^2\hat{V}^2\left(C_{(D,L)}\sigma_\zeta - \sigma\frac{\hat{V}}{\hat{a}^2}C_{(D,L)_\mathbf{M}} a_\zeta\right),$$

$$\frac{\partial M}{\partial h} = \tfrac{1}{2}S\rho_0 V_0^2\hat{V}^2\left(C_m\sigma_\zeta - \sigma\frac{\hat{V}}{\hat{a}^2}C_{m_\mathbf{M}} a_\zeta\right).$$

The nondimensional system will then be

$$\begin{aligned}\mu_0\frac{d\Delta\hat{V}}{d\hat{t}} &= [-\sigma\hat{V}(C_\mathbf{D} + \tfrac{1}{2}\mathbf{M}C_{D_\mathbf{M}}) + C_{T_V}\cos(\alpha-\varphi)]\,\Delta\hat{V}\\ &\quad + [-\tfrac{1}{2}\sigma\hat{V}^2 C_{D_\alpha} - C_T\sin(\alpha-\varphi)]\,\Delta\alpha - \mu_0\hat{g}\,\Delta\gamma\\ &\quad + [-\tfrac{1}{2}(\hat{V}^2 C_D\sigma_\zeta - \hat{V}\mathbf{M}^2 C_{D_\mathbf{M}} a_\zeta) + C_{T_\sigma}\sigma_\zeta\cos(\alpha-\varphi)]\,\zeta,\end{aligned}$$

$$\frac{d\Delta\alpha}{d\hat{t}} = \Delta\hat{q} - \frac{d\Delta\gamma}{d\hat{t}},$$

$$\begin{aligned}\mu_0\hat{V}\frac{d\Delta\gamma}{d\hat{t}} &= \left[\sigma\hat{V}(C_L + \tfrac{1}{2}\mathbf{M}C_{L_\mathbf{M}}) + C_{T_V}\sin(\alpha-\varphi)\right]\Delta\hat{V}\\ &\quad + [\tfrac{1}{2}\sigma\hat{V}^2 C_{L_\alpha} + C_T\cos(\alpha-\varphi)]\,\Delta\alpha\\ &\quad + [\tfrac{1}{2}(\hat{V}^2 C_L\sigma_\zeta - \sigma\hat{V}\mathbf{M}^2 C_{L_\mathbf{M}} a_\zeta) + C_{T_\sigma}\sigma_\zeta\sin(\alpha-\varphi)]\,\zeta,\end{aligned}$$

$$\begin{aligned}i_B\frac{d\Delta\hat{q}}{d\hat{t}} &= [\sigma\hat{V}(C_m + \tfrac{1}{2}\mathbf{M}C_{m_\mathbf{M}}) - C_{T_V}\hat{z}_T]\,\Delta\hat{V} + \tfrac{1}{2}\sigma\hat{V}^2 C_{m_\alpha}\,\Delta\alpha\\ &\quad + \tfrac{1}{2}\sigma\hat{V}^2 C_{m_q}\,\Delta\hat{q} + \sigma\hat{V}^2 C_{m_{\dot{\alpha}}}\frac{d\Delta\alpha}{d\hat{t}}\\ &\quad + [\tfrac{1}{2}(\hat{V}^2 C_m\sigma_\zeta - \sigma\hat{V}\mathbf{M}^2 C_{m_\mathbf{M}} a_\zeta) - C_{T_\sigma}\sigma_\zeta\hat{z}_T]\,\zeta,\end{aligned}$$

$$\frac{d\zeta}{d\hat{t}} = \sin\gamma\,\Delta\hat{V} + \hat{V}\cos\gamma\,\Delta\gamma,$$

or an equivalent one if the ratio V/a is substituted for the Mach number. Further, the products $C_T \cos(\alpha - \varphi)$ and $C_T \sin(\alpha - \varphi)$ can be replaced, according to the conditions of the equilibrium of forces along the directions of the tangent and of the normal on the path, by $\sigma C_D/2$ and $-\sigma C_L/2 + \mu_0 \hat{g}$, respectively. Since the basic motion is uniform, $\hat{V} = 1$. Then, making the change of variables $\Delta\gamma = \Delta\theta - \Delta\alpha$ and denoting the independent variable (nondimensional time) by unaccented t, we obtain, for the last system,

$$\frac{d\Delta\hat{V}}{dt} = a_{11}\Delta\hat{V} + a_{12}\Delta\alpha + a_{13}\Delta\theta + a_{15}\zeta,$$

$$\frac{d\Delta\alpha}{dt} = a_{21}\Delta\hat{V} + a_{22}\Delta\alpha + \Delta\hat{q} + a_{25}\zeta,$$

$$\frac{d\Delta\theta}{dt} = \Delta\hat{q},$$

$$\frac{d\Delta\hat{q}}{dt} = a_{41}\Delta\hat{V} + a_{42}\Delta\alpha + a_{44}\Delta\hat{q} + a_{45}\zeta,$$

$$\frac{d\zeta}{dt} = -\Delta\alpha + \Delta\theta, \tag{6.21}$$

where

$$a_{11} = \frac{1}{\mu_0}\left[-\sigma(C_D + \tfrac{1}{2}\mathbf{M}C_{D_\mathbf{M}}) + C_{T_V}\cos(\alpha - \varphi)\right],$$

$$a_{12} = -\frac{1}{2\mu_0}\sigma(C_{D_\alpha} - C_L) - \hat{g},$$

$$a_{13} = -\hat{g},$$

$$a_{15} = \frac{1}{\mu_0}\left[-\tfrac{1}{2}(C_D\sigma_\zeta - \sigma\mathbf{M}^2 C_{D_\mathbf{M}} a_\zeta) + C_{T_\sigma}\sigma_\zeta\cos(\alpha - \varphi)\right]$$

$$a_{21} = -\frac{1}{\mu_0}\left[\sigma(C_L + \tfrac{1}{2}\mathbf{M}C_{L_\mathbf{M}}) + C_{T_V}\sin(\alpha - \varphi)\right],$$

$$a_{22} = -\frac{1}{2\mu_0}\sigma(C_{L_\alpha} + C_D),$$

$$a_{25} = -\frac{1}{\mu_0}\left[\tfrac{1}{2}(C_L\sigma_\zeta - \mathbf{M}^2 C_{L_\mathbf{M}} a_\zeta) + C_{T_\sigma}\sigma_\zeta\sin(\alpha - \varphi)\right],$$

$$a_{41} = \frac{1}{i_B}\left\{\tfrac{1}{2}\sigma \mathbf{M} C_{m_\mathbf{M}} - C_{T_V}\hat{z}_T - \frac{1}{2\mu_0}\sigma C_{m_{\dot\alpha}}[\sigma(C_L + \tfrac{1}{2}\mathbf{M} C_{L_\mathbf{M}}) + C_{T_V}\sin(\alpha - \varphi)]\right\},$$

$$a_{42} = \frac{1}{2i_B}\sigma\left[C_{m_\alpha} - \frac{1}{2\mu_0}\sigma^2 C_{m_{\dot\alpha}}(C_{L_\alpha} + C_D)\right],$$

$$a_{44} = \frac{1}{2i_B}\sigma(C_{m_q} + C_{m_{\dot\alpha}}),$$

$$a_{45} = \frac{1}{i_B}\left\{-\tfrac{1}{2}\sigma \mathbf{M}^2 C_{m_\mathbf{M}} a_\zeta - C_{T_\sigma}\sigma_\zeta \hat{z}_T - \frac{1}{2\mu_0}\sigma C_{m_{\dot\alpha}}[\tfrac{1}{2}(C_L\sigma_\zeta - \sigma\mathbf{M}^2 C_{L_\mathbf{M}} a_\zeta)\right.$$

$$\left. + C_{T_\sigma}\sigma_\zeta \sin(\alpha - \varphi)]\right\}.$$

To simplify, it will hereafter be assumed that $z_T = 0$, $\alpha - \varphi = 0$, and C_{T_V} will be neglected. Then, the coefficients a_{jk} of the system above will be

$$a_{11} = -\frac{\sigma}{\mu_0}(C_D + \tfrac{1}{2}\mathbf{M} C_{D_\mathbf{M}}),$$

$$a_{12} = -\frac{\sigma}{2\mu_0}(C_{D_\alpha} - C_L) - \hat{g},$$

$$a_{13} = -\hat{g},$$

$$a_{15} = -\frac{1}{\mu_0}[\tfrac{1}{2}(C_D\sigma_\zeta - \sigma\mathbf{M}^2 C_{D_\mathbf{M}} a_\zeta) - C_{T_\sigma}\sigma_\zeta],$$

$$a_{21} = -\frac{\sigma}{\mu_0}(C_L + \tfrac{1}{2}\mathbf{M} C_{L_\mathbf{M}}),$$

$$a_{22} = -\frac{\sigma}{2\mu_0}(C_{L_\alpha} + C_D),$$

$$a_{25} = -\frac{1}{2\mu_0}(C_L\sigma_\zeta - \sigma\mathbf{M}^2 C_{L_\mathbf{M}} a_\zeta),$$

$$a_{41} = \frac{\sigma}{i_B}\left[\tfrac{1}{2}\mathbf{M} C_{m_\mathbf{M}} - \frac{\sigma}{2\mu_0}C_{m_{\dot\alpha}}(C_L + \tfrac{1}{2}\mathbf{M} C_{L_\mathbf{M}})\right],$$

$$a_{42} = \frac{\sigma}{2i_B}\left[C_{m_\alpha} - \frac{\sigma}{2\mu_0}C_{m_{\dot\alpha}}(C_{L_\alpha} + C_D)\right],$$

$$a_{44} = \frac{\sigma}{2i_B}(C_{m_q} + C_{m_{\dot\alpha}}),$$

$$a_{45} = -\frac{\sigma}{2i_B}\left[\mathbf{M}^2 C_{m_\mathbf{M}} a_\zeta + \frac{1}{2\mu_0} C_{m_{\dot\alpha}}(C_L \sigma_\zeta - \sigma \mathbf{M}^2 C_{L_\mathbf{M}} a_\zeta)\right], \tag{6.22}$$

σ and $\mathbf{M}$ (or, in place of the Mach number, the nondimensional speed of sound in the undisturbed motion, $\hat{a}$) being constant parameters defined by the constant altitude of undisturbed flight. Since, hereafter, we will study the effect of the variation of flight altitude during the transitory process induced by a disturbance and not the static influence of these parameters of the basic motion, the parameter σ may be considered equal to 1, air density at the altitude of undisturbed flight (and not that at sea level) being taken as ρ_0. This unit can obviously be chosen arbitrarily, since its choice bears no influence whatever on the results of the reasoning. Hence, hereafter, σ will be assumed equal to 1. Then, to make the effect of the disturbance of flight altitude stand out, system (6.1), deduced in the hypothesis where this effect was neglected, may be used as a term of comparison.

It can easily be seen that the characteristic equation of system (6.21) where the coefficients a_{ik} are considered constant, having the expressions (6.22) with $\sigma = 1$, may be written in the form

$$\lambda^5 + a_1\lambda^4 + (a_2 + k_2)\lambda^3 + (a_3 + k_3)\lambda^2 + (a_4 + k_4)\lambda + k_5 = 0 \tag{6.23}$$

where a_1, a_2, a_3, and a_4 are the coefficients of the characteristic polynomial (6.2) of the system (6.1), and k_2, k_3, k_4, and k_5, which represent the influence of altitude variation, have the expressions

$$k_2 = a_{25},$$

$$k_3 = a_{15}a_{21} - a_{25}(a_{11} + a_{44}),$$

$$k_4 = -a_{15}a_{21}a_{44} + a_{25}(a_{11}a_{44} - a_{42}) + a_{45}a_{22},$$

$$k_5 = a_{15}(a_{22}a_{41} - a_{21}a_{42}) + a_{25}[a_{11}a_{42} - a_{41}(a_{12} + a_{13})]$$
$$+ a_{45}[-a_{11}a_{22} + a_{21}(a_{12} + a_{13})].$$

Generally, the additional quantities k_i are sufficiently small in comparison with a_i† not to influence the sign and relative order of magnitude of the

† In the numerical example computed in [11] for various values of the static stability coefficient, k_2 varies from 2.5×10^{-5} to 1.38×10^{-4} times a_2, k_3 from 3.5×10^{-3} to 1.79×10^{-2} times a_3, and k_4 from 2.55×10^{-2} up to 1.46×10^{-1} times a_4.

coefficients (compared with one another). The coefficient k_5 being small in comparison with the other coefficients,† equation (6.24) will have a real root relatively small in value. ($k_5 = 0$ corresponds to the boundary of the convergence region, hence for k_5 relatively small we are in the vicinity of this frontier, defined by low values for at least one of the characteristic roots.) There will correspond to this root, connected with the consideration of flight altitude variation, a slow aperiodic mode, convergent or divergent, of the disturbed motion.‡ The remaining roots—in typical cases—have a configuration similar to those of equation (6.2): a pair of complex-conjugate roots with large negative real parts and with the imaginary part likewise high in value, to which corresponds a strongly damped high-frequency oscillation (rapid incidence adjustment) and a pair of small in modulo real or complex roots, corresponding to the phugoid disturbed motion, formed of either two slow aperiodic modes, each of which might be convergent or divergent, or a weakly damped or slightly increasing in amplitude, low-frequency oscillation.

Generally, the small real root λ_5 corresponding to the effect of altitude variation can be determined with satisfactory approximation for the practical purpose in view from the equation obtained by setting to zero the last two terms of the characteristic polynomial of the system

$$(a_4 + k_4)\lambda + k_5 = 0;$$

hence

$$\lambda_5 \approx -\frac{k_5}{a_4 + k_4} \approx -\frac{k_5}{a_4}.$$

The large complex roots only differ with negligible quantities from those of equation (6.2): they can be obtained approximately from the equation resulting when the first three terms of the characteristic polynomial (6.24) are set to zero:

$$\lambda^2 + a_1\lambda + a_2 + k_2 = 0.$$

† In the same numerical example k_5 is less than $3.79 \times 10^{-5}a_1$, $3.29 \times 10^{-5}a_2$, $1.29 \times 10^{-3}a_3$, and $5.1 \times 10^{-3}a_4$.

‡ In the numerical example in [11] this root is positive, corresponding to a divergence; it is, however, of very low value, viz., 5.7–30.7 times smaller than the value of the real part of the corresponding phugoid roots and 1360–7650 times smaller than that of the real part of the respective short-period oscillation roots.

The real part of the roots of this equation, estimating the damping velocity of the rapid oscillations, is equal to $a_1/2$, and hence independent of altitude variation, and the imaginary part equal to $i(a_2 + k_2 - a_1^2/4)^{1/2}$ is generally slightly affected by the variation of the flight altitude. (In the numerical example in [11], the difference between the value of the imaginary parts in cases where altitude variation is neglected or taken into consideration does not exceed 0.004%.)

Considering the partially controlled motion as corresponding to speed and to altitude constraint, it is possible to obtain another estimate of damping speed and of the frequency of oscillation characterizing rapid incidence adjustment. The auxiliary system

$$\frac{d\Delta\alpha}{dt} = a_{22}\Delta\alpha + \Delta\hat{q} + a_{25}\zeta,$$

$$\frac{d\Delta\hat{q}}{dt} = a_{42}\Delta\alpha + a_{44}\Delta\hat{q} + a_{45}\zeta,$$

$$\frac{d\zeta}{dt} = -\Delta\alpha, \tag{6.24}$$

has the characteristic polynomial

$$P(\lambda) = \lambda^3 - (a_{22} + a_{44})\lambda^2 + (a_{22}a_{44} - a_{42} + a_{25})\lambda - a_{25}a_{44} + a_{45} \tag{6.25}$$

with its free term small in comparison with the coefficients of the terms of a degree higher than zero. Hence, the solution corresponding to any initial condition whatever of the system (6.24) will be composed of a weakly convergent or gently divergent aperiodic mode and a comparatively rapid oscillation having the same rate of decay and a frequency almost identical (the difference being practically negligible) to the one described by the solution corresponding to the same initial conditions of system (6.10). However, the presence of the slow aperiodic mode prevents the auxiliary system (6.24) from being entirely conclusive, suggesting the necessity for a simultaneous constraint of the flight altitude also.

The main effect of the variation in altitude induced by a disturbance in the case of level basic motion concerns the phugoid motion, that is, as will be seen hereunder, the frequency of the phugoid oscillations. The considerations above regarding the typical configuration of the roots of equation (6.23), hence that of the harmonics of the disturbed motion described by the solution of system (6.21), justify the separation of the slow mode on basis of the classical hypotheses stated in Section 6.1. Admitting, therefore, that after a rapid return of the incidence to its undisturbed value, its induced deviation remains negligible in the subsequent stages of the disturbed motion, for these stages, setting $\Delta\alpha \equiv 0$ and returning to the dynamic variable $\Delta\gamma$ instead of $\Delta\theta$ ($\Delta\theta = \Delta\gamma + \Delta\alpha = \Delta\gamma$), we separate from (6.21) the following system.

$$\frac{d\Delta\hat{V}}{dt} = a_{11}\Delta\hat{V} + a_{13}\Delta\gamma + a_{15}\zeta,$$

$$\frac{d\Delta\gamma}{dt} = -a_{21}\Delta\hat{V} - a_{25}\zeta,$$

$$\frac{d\zeta}{dt} = \Delta\gamma, \tag{6.26}$$

which only contains the variables $\Delta\hat{V}$, $\Delta\gamma$, and ζ and describes, with the approximation implied by the admitted hypotheses, the phugoid motion. In the hypothesis where the effect of variation in altitude is neglected, namely, $a_{15} = a_{25} = 0$, it becomes useless to take the last equation into consideration, the approximate phugoid equations then reducing to the known system, namely,

$$\frac{d\Delta\hat{V}}{dt} = a_{11}\Delta\hat{V} + a_{13}\Delta\gamma, \qquad \frac{d\Delta\gamma}{dt} = -a_{21}\Delta\hat{V}. \tag{6.27}$$

It is assumed, for definiteness, that the phugoid motion is periodic and hence that the roots $\lambda_{1,2}^{(0)}$ of the characteristic equation of system (6.27)

$$\lambda^2 - a_{11}\lambda + a_{13}a_{21} = 0 \tag{6.28}$$

are complex conjugate

$$\lambda_{1,2}^{(0)} = \frac{a_{11}}{2} \pm i\omega^{(0)}, \qquad \omega^{(0)} = \left(a_{13}a_{21} - \frac{a_{11}^2}{4}\right)^{1/2}. \tag{6.29}$$

The characteristic equation of system (6.26) being

$$\lambda^3 - a_{11}\lambda^2 + (a_{13}a_{21} + a_{25})\lambda + a_{15}a_{21} - a_{25}a_{11} = 0, \tag{6.30}$$

it is assumed—according to what happens in current cases—that the value of the free term $a_{15}a_{21} - a_{25}a_{11}$ is small in comparison with the other coefficients or, which amounts to the same thing, that the value of one of the roots is much lower than the modulus of the other two: $|\lambda_3| \ll |\lambda_{1,2}|$. A very slow aperiodic mode, without practical importance, will correspond to this small root. The other two roots can be obtained with satisfactory approximation by neglecting the free term; their expressions will then be

$$\lambda_{1,2} = \frac{a_{11}}{2} \pm i\omega, \qquad \omega = \left(a_{13}a_{21} - \frac{a_{11}^2}{4}\right)^{1/2}.$$

As can be seen, if ω is real, and hence the roots λ_1 and λ_2 complex conjugate, the real part of the latter two coincides with that of the roots of equation (6.28) corresponding to the disregard of the effect of height. The imaginary part $i\omega$ differs, however, from $i\omega^{(0)}$ and, since a_{25} is not, generally, negligible in comparison with the magnitude $a_{13}a_{21} - a_{11}^2/4$ (it is even possible for large Mach numbers in the supersonic range to exceed it), it results that the oscillation frequency estimated by means of the magnitude ω is strongly influenced by the variation in altitude induced by a disturbance, even in the case of a level basic motion.

To estimate the effect of variation in altitude, the magnitudes σ_ζ and a_ζ comprised in the expression of coefficient a_{25} (see (6.22)) will first be estimated. The characteristics of standard atmosphere will be resorted to in order to illustrate the phenomenon qualitatively. It should, however, be noted that the local variations of air density and sound velocity can differ considerably from the standard ones and this variation may influence the result in a decisive manner, particularly in the case of level basic motion being considered here. Hence, to study the actual or effective quantitative aspects, for computation, it is necessary to take into account all the available information regarding the properties of the atmosphere in the special conditions of the problem.† The great majority of this information is obviously of a statistical nature, the results being obtained in the form of probable quantities. For all practical purposes, however, a theory concerned with the rather qualitative aspect of the influence of altitude variation on the disturbed motion (for which the International Standard Atmosphere constitutes a satisfactory basis) seems to be sufficient at present.

† We must, i.e., consider geographical coordinates with the respective orographic characteristics, actual flight altitude as well as other characteristics influencing the structure and intensity of atmospheric turbulence, or other aerologic properties. Such data, obtained particularly from experimental studies and forming the object of special investigations in an area situated somewhere near the boundary between aerology and flight dynamics (see, e.g., [13]), are still far from being complete or even satisfactory. It is therefore premature to try to elaborate a theory of the influence of the variation of altitude on flight stability more subtle than the one presented in this section.

At first some matters regarding the definition of standard atmosphere will be recalled. As known, its data are computed by assuming the air as dry, considered to be a perfect gas and the gravity field as constant (see, e.g., [14, Chapter 1; 5, Appendix]); in other words, the validity of the equation of state

$$\frac{p}{\rho} = gRT, \tag{6.31}$$

R and g being constant, is admitted. Here, T represents the absolute temperature and p the air pressure. To define the standard atmosphere it is, in addition, assumed that the absolute temperature T is, within the troposphere, linearly dependent on altitude according to equation

$$T = T_0 - \alpha_T h, \qquad \alpha_t = \text{const.} \tag{6.32}$$

while it is constant in the stratosphere, and that p, ρ, and T vary as functions of altitude, being independent of the other coordinates of the point in the atmosphere with respect to a reference system fixed to the earth and not explicitely dependent on time.† The constants comprised in formulas (6.31) and (6.32)—for air density ρ in kilograms (force) $\times$ seconds squared per meter to the fourth power (kg sec^2/m^4), altitude h in meters (m), atmospheric pressure p in kg/m^2—have the approximate values: $g = 9.81$ m/sec^2; $R = 29.27$ m/(°C); $\alpha_T = 0.0065$(°C)/m; $T_0 = 288$°K; $T_{\text{str}} = 216.5$°K. (In R and α_T, (°C) means centigrade degrees; T_0 and T_{str} are given in degrees Kelvin, absolute temperature, denoted °K.)

Further, by writing the equilibrium condition of an element of air volume in the atmosphere, we obtain the equation

$$\frac{dp}{dh} = -\rho g. \tag{6.33}$$

† Actually, the state parameters p, ρ, and T are both point and time functions, the nonlinear functions $p(t, h)$, $\rho(t, h)$, and $T(t, h)$ varying with the longitude and latitude of the place through its position with respect to the sun, the relief, the weather characteristics, etc. For the standard atmosphere, using average values, p, ρ, and T finally only depend on altitude. Thus, e.g., formula (6.32) not only represents an approximation of temperature variation in the troposphere as a function of h (at a given geographical point and a given instant $T = T(h)$ being, nevertheless, nonlinear), but also an average of these approximations in time and on the entire surface of the earth. The actual procedure consisted in computing the average of measurements carried out along the parallel of 40° northern latitude (see, e.g., [5, Appendix]).

The speed of sound a, it is also recalled, may be expressed by means of the state parameters p and ρ by the simple relation $a = (dp/d\rho)^{1/2}$, which, in the case of an adiabatic process, becomes

$$a = \left(\kappa \frac{p}{\rho}\right)^{1/2}, \qquad \kappa = \text{const.},$$

or, taking the equation of state (6.31) into account,

$$a = (\kappa g R T)^{1/2}, \tag{6.34}$$

the adiabatic constant κ being taken the same as for a two-atom gas (according to the assumption that air is dry), namely, $\kappa = 1.4$.

With the foregoing, the approximate expressions by means of which the nondimensional parameters σ_ζ and a_ζ will be estimated are immediately obtained. Indeed, from (6.31) and (6.33) there results

$$\frac{dp}{dh} = -\frac{p}{RT} \tag{6.35}$$

and from (6.32)

$$\frac{dT}{dh} = -\alpha_T; \tag{6.36}$$

hence, differentiating (6.31) and taking (6.35) and (6.36) into account, we obtain

$$\frac{d\rho}{dh} = -g(1 - \alpha_T R)\frac{\rho^2}{p}$$

and, likewise, from (6.34) and (6.36)

$$\frac{da}{dh} = -\frac{\kappa g \alpha_T R}{2a}.$$

Since the basic motion considered is horizontal, the approximation

$$mg = \tfrac{1}{2}\rho V^2 S C_L \tag{6.37}$$

will be admitted. Then, taking into account that $\sigma_\zeta = (\bar{c}/2\rho_0)\, d\rho/dh$, $a_\zeta = (\bar{c}/2V_0)\, da/dh$, $m = (1/2)\rho_0 S\bar{c}\mu_0$ where ρ_0 represents constant density at the flight altitude (not necessarily at $h = 0$), there results

$$\begin{aligned} \sigma_\zeta &= -k_1 \frac{\mathbf{M}^2 C_L}{\mu_0}, \qquad & k_1 &= \frac{\kappa}{2}(1 - \alpha_T R), \\ a_\zeta &= -k_2 \frac{\mathbf{M} C_L}{2}, \qquad & k_2 &= \frac{\kappa}{2}\alpha_T R. \end{aligned} \tag{6.38}$$

With the standard data above the dimensionless constants k_1 and k_2 have the approximate values $k_1 = 0.567$, $k_2 = 0.133$. The oscillation frequency of the solution of system (6.27), obtained by neglecting the effect of altitude variation is given by (6.29), whose square, accounting for (6.37), is now expressed by

$$\omega^{(0)2} = \frac{C_L}{2\mu_0}(C_L + \tfrac{1}{2}\mathbf{M}C_{L_\mathbf{M}}) - \frac{1}{\mu_0^2}(C_D + \tfrac{1}{2}\mathbf{M}C_{D_\mathbf{M}})^2. \tag{6.39}$$

The effect of altitude variation for standard troposphere immediately stands out by writing the expression for ω^2 taking into account the formulas (6.22), (6.38) as well as the adopted approximation (6.37):

$$\omega^2 = \frac{C_L}{2\mu_0}[C_L(1 + k_1\mathbf{M}^2) + \tfrac{1}{2}\mathbf{M}C_{L_\mathbf{M}}(1 - k_2\mathbf{M}^2)] - \frac{1}{\mu_0^2}(C_D + \tfrac{1}{2}\mathbf{M}C_{D_\mathbf{M}})^2. \tag{6.40}$$

It can be seen that (6.40) is obtained from (6.39) by applying to the aerodynamic parameters C_L and $C_{L_\mathbf{M}}$ in the first parenthesis of the latter, the correction factors $(1 + k_1\mathbf{M}^2)$ and $(1 - k_2\mathbf{M}^2)$, respectively ($k_1 \approx 0.567$, $k_2 \approx 0.133$). For large Mach numbers in the supersonic range the value of these correction factors becomes considerable, ω differing considerably from $\omega^{(0)}$, the aspect of its dependence on the flight régime (on the actual Mach number) also changing.

For standard stratosphere relation (6.36) is replaced by $d\mathbf{T}/dh = 0$, whence

$$\sigma_\zeta = -k_3\bar{c}, \qquad a_\zeta = 0$$

where the constant k_3, having the dimension of the inverse of a length $1/[l]$, for example, 1/m, is

$$k_3 = \frac{1}{2RT_{\text{str}}} \approx 0.0007875\ \text{m}^{-1}$$

and ω^2 acquires the expression

$$\omega^2 = \frac{C_L}{2\mu_0}(C_L + \tfrac{1}{2}\mathbf{M}C_{L_\mathbf{M}} + k_3\bar{c}\mu_0) - \frac{1}{\mu_0^2}(C_D + \tfrac{1}{2}\mathbf{M}C_{D_\mathbf{M}})^2.$$

Hence, for the conditions of standard stratosphere the correction term $k_3\bar{c}\mu_0$, which can also be quite important, appears in the first parenthesis of (6.39).

For simplicity, everywhere so far, the dimensional unit for density has been considered to be the one corresponding to the constant altitude of undisturbed flight. To make the static influence of the variation of air density stand out quantitatively—in the case of various level flight altitudes—the dimensional unit initially adopted, ρ_0 (density at sea level), will again be resorted to. Since σ occurs as a factor at the first power in the expressions that yield a_{11}, $a_{22} + a_{44}$, $a_{13}a_{21}$, and a_{42}, it can be seen that the dependence is a simple one: the rate of damping of the rapid oscillations increases together with σ; the stable or, rather unexpectedly, the unstable character of the phugoid oscillations is stressed by the increase of σ (an increase of the rate of damping corresponding to it for $a_{11} < 0$ and the acceleration of instability for $a_{11} > 0$); the value of the real phugoid roots also seems to increase with σ. Therefore the divergence, if any, results accelerated by the increase of σ being, hence, the slighter the higher the flight altitude. The influence of parameter σ on the oscillation frequency is not as clear for the general case.

The static influence of sound velocity variation with flight altitude is less important and may usually be neglected.

REFERENCES

1. B. Etkin, *Dynamics of Flight: Stability and Control.* Wiley, New York, 1959.
2. A. W. Babister, *Aircraft Stability and Control.* Macmillan (Pergamon), New York, 1961.
3. F. W. Lanchester, *Aerial Flight, Part II: Aerodonetics.* Constable, London, 1908.
4. B. M. Jones, *Dynamics of the Airplane* (Volume 5 of the series *Aerodynamic Theory,* W. F. Durand, ed.). Springer, Berlin, 1935.
5. C. D. Perkins and R. E. Hage, *Airplane Performance, Stability and Control.* Wiley, New York, 1949.
6. G. S. Kalatchev, "On a measure of longitudinal dynamic stability of aircraft" (in Russian), *Central N. E. Joukovski Hydro-Aerodynamic Inst.* (ŤAGI) *Works* **365**(1938).
7. I. V. Ostoslavskiy, *Aircraft Aerodynamics* (in Russian). Oborongiz, Moscow, 1957.
8. W. J. Duncan, *The Principles of the Control and Stability of Aircraft.* Cambridge Univ. Press, London and New York, 1952.
9. T. Hacker, "Stability of Partially Controlled Motions of an Aircraft," *J. Aerospace Sci.* **28**(1961), 15–26.
10. A. M. Liapunov, *Stability of Motion.* Academic Press, New York, 1966.
11. S. Neumark, "Longitudinal stability, speed and height," *Aircraft Eng.* **22**(1950), 323–334.
12. F. N. Scheubel, "The Effect of Density Gradient on the Longitudinal Motion of an Aircraft" (R. T. P. transl. no. 1739), from *Luftfahrtforsch.* **19**(1942), 132–136.
13. N. Z. Pinus, S. M. Shmeter, and V. D. Reshetov, *Atmospheric Turbulence Generating Bump* (in Russian), (editor-in-chief: N. Z. Pinus). Gimiz, Moscow, 1962.
14. R. von Mises, *Theory of Flight.* McGraw-Hill, New York, 1945.

Chapter 7

LATERAL STABILITY OF STEADY REGIME FLIGHT

7.1. STABILITY WITH LOCKED CONTROLS

System (2.18), reproduced here for convenience, will be used as a starting point.

$$\frac{d\beta}{d\hat{t}} = b_{11}\beta + b_{12}\hat{p} + b_{13}\hat{r} + b_{14}\phi,$$

$$\frac{d\hat{p}}{d\hat{t}} = b_{21}\beta + b_{22}\hat{p} + b_{23}\hat{r},$$

$$\frac{d\hat{r}}{d\hat{t}} = b_{31}\beta + b_{32}\hat{p} + b_{33}\hat{r},$$

$$\frac{d\phi}{d\hat{t}} = \hat{p} + b_{43}\hat{r},$$

$$\frac{d\psi}{d\hat{t}} = b_{53}\hat{r}, \tag{7.1}$$

where

$$b_{11} = \frac{1}{2\mu} C_{y_\beta}, \qquad b_{12} = \frac{1}{2\mu} C_{y_p}, \qquad b_{13} = \frac{1}{2\mu} C_{y_r} - 1,$$

$$b_{14} = \frac{1}{2\mu} C_{L_0} \cos \theta_0,$$

$$b_{21} = \frac{1}{2i_A} C_{l_\beta}, \qquad b_{22} = \frac{1}{2i_A} C_{l_p}, \qquad b_{23} = \frac{1}{2i_A} C_{l_r},$$

$$b_{31} = \frac{1}{2i_C} C_{n_\beta}, \qquad b_{32} = \frac{1}{2i_C} C_{n_p}, \qquad b_{33} = \frac{1}{2i_C} C_{n_r},$$

$$b_{43} = -\tan \theta_0, \quad b_{53} = \sec \theta_0;$$

β, $\hat{p}$, $\hat{r}$, ϕ, ψ are the nondimensional deviations of the respective quantities induced by the disturbance; the symbols showing quantities corresponding to the basic motion are indicated by the subscript zero. System (7.1) represents the lateral asymmetric disturbed motion of the aircraft with controls fixed in neutral position, in the hypotheses stated in Section 2.3. The controls in the nondeflected neutral position are trimmed for the given steady straight level motion.

The characteristic equation of the system

$$\lambda^5 + A_1\lambda^4 + A_2\lambda^3 + A_3\lambda^2 + A_4\lambda = 0 \tag{7.2}$$

where

$$A_1 = -(b_{11} + b_{22} + b_{33}) = -\frac{1}{2\mu} C_{y_\beta} - \frac{1}{2i_A} C_{l_p} - \frac{1}{2i_C} C_{n_r},$$

$$A_2 = b_{11}b_{22} + b_{22}b_{33} + b_{33}b_{11} - b_{12}b_{21} - b_{13}b_{31} - b_{23}b_{32} - b_{14}$$

$$= \frac{1}{4\mu i_A}(C_{y_\beta}C_{l_p} - C_{l_\beta}C_{y_p}) + \frac{1}{4\mu i_C}(C_{y_\beta}C_{n_r} - C_{n_\beta}C_{y_r})$$

$$+ \frac{1}{4i_A i_C}(C_{l_p}C_{n_r} - C_{n_p}C_{l_r}) + \frac{1}{2i_C}C_{n_\beta} - \frac{1}{2\mu}C_{L_0}\cos\theta_0,$$

$$A_3 = -b_{11}(b_{22}b_{33} - b_{23}b_{32}) + b_{12}(b_{21}b_{33} - b_{23}b_{31}) - b_{13}(b_{21}b_{32} - b_{22}b_{31})$$

$$+ b_{14}(b_{22} + b_{23} + b_{31}b_{43}) = \frac{1}{8\mu i_A i_C}[C_{y_\beta}(C_{l_r}C_{n_p} - C_{n_r}C_{l_p}) + C_{y_p}(C_{l_p}C_{n_r}$$

$$- C_{n_\beta}C_{l_r}) + (-C_{y_r} + 2\mu)(C_{l_\beta}C_{n_p} - C_{n_\beta}C_{l_p})$$

$$+ 2C_{L_0}\cos\theta_0(-i_C C_{l_\beta} + i_A C_{n_\beta}\tan\theta_0),$$

$$A_4 = -b_{14}[-b_{21}b_{33} - b_{23}b_{31} + b_{43}(b_{21}b_{32} - b_{22}b_{31})]$$

$$= \frac{C_{L_0}\cos\theta_0}{8\mu i_A i_C}[C_{l_\beta}C_{n_r} - C_{n_\beta}C_{l_r} + \tan\theta_0(C_{l_\beta}C_{n_p} - C_{n_\beta}C_{l_p})],$$

has, as can be seen, a zero root; hence the condition imposed by the theorem of stability by the first approximation (see Section 3.4) is not satisfied, since the trivial solution of the system (7.1) is not uniformly asymptotically stable. Generally, the so-called critical cases of one or several zero roots of the characteristic equation are studied by considering the nonlinear system. In the present case, however, the problem is easier to solve. Indeed, since the first four equations of system (7.1) do not contain the variable ψ, they determine completely, as a function of the initial conditions, the behavior of the four variables they contain: β, $\hat{p}$, $\hat{r}$, and ϕ. Particularly, the trivial solution of the system of the first four equations may be uniformly asymptotically stable. That the zero solution of system (7.1) as a whole is not, in this case, also uniformly asymptotically stable is only due to the component ψ, with respect to which the system will only be (nonasymptotically) uniformly stable, the variable tending as $t \to \infty$ toward a constant limit value differing from zero, depending on the initial conditions. Indeed, by integrating the system of the first four equations of (7.1), $\hat{r} = \hat{r}(\hat{t})$ is determined, and by means of this function, from the last equation we obtain $\psi = \psi(\hat{t})$ in the form

$$\psi(\hat{t}) = \psi(\hat{t}_0) + b_{53}\int_{\hat{t}_0}^{\hat{t}} \hat{r}(s)\, ds.$$

But, since the system without the last equation is uniformly asymptotically stable even exponentially, there obviously results

$$\lim_{t\to\infty} \psi(\hat{t}) = \psi(\hat{t}_0) + b_{53}\left(\int \hat{r}(s)\, ds\right)_{\hat{t}_0} = \psi(\hat{t}_0) + b_{53}\mathbf{P}(\hat{t}_0), \tag{7.3}$$

$\mathbf{P}(\hat{t})$ representing the primitive of function $\hat{r}(\hat{t})$. System (7.1) is a particular case of systems of the form (3.26). As it results from the proposition stated in Section 3.8 concerning these systems, the result above regarding uniform asymptotic stability with respect to the components β, $\hat{p}$, $\hat{r}$, ϕ and the existence of an asymptote differing from zero and depending on the initial conditions for the component ψ (which implies (nonasymptotic) uniform stability with respect to the latter) remain valid even when the nonlinear terms neglected in the first four equations are dependent on the variable ψ.

The foregoing amount to the fact that no aircraft currently considered dynamically stable in the conditions of fixed controls keeps a constant heading in flight: after every disturbance and as a function of this, it will tend toward some other heading, characterized by the limit given in (7.3) of the value of angle ψ. Although small, these variations, when they add up during flight, may in the end induce a noticeable deviation of the aircraft flight path. Therefore, heading control is demanded by a constraint imposed on angle ψ, the required action being carried out, in the main, by means of the rudder (manually or automatically, the sensing element reacting to angular deviations of the aircraft symmetry plane with respect to a vertical plane fixed to the earth).

7.2. LATERAL STABILITY IN PARTIAL CONTROL CONDITIONS

7.2.1. Constraint of the Angle of Yaw

This case leads us to use as an auxiliary system the first four equations in (7.1),

$$\frac{d\beta}{d\hat{t}} = b_{11}\beta + b_{12}\hat{p} + b_{13}\hat{r} + b_{14}\phi, \qquad \frac{d\hat{p}}{d\hat{t}} = b_{21}\beta + b_{22}\hat{p} + b_{23}\hat{r},$$
$$\frac{d\hat{r}}{dt} = b_{31}\beta + b_{32}\hat{p} + b_{33}\hat{r}, \qquad \frac{d\phi}{dt} = \hat{p} + b_{43}\hat{r}, \tag{7.4}$$

usually considered in the literature for fixed-controls lateral stability.

A few of the known conclusions deriving from the investigation of this system will now be summed up. In typical situations the characteristic equation

$$\mathbf{A}(\lambda) \equiv \lambda^4 + A_1\lambda^3 + A_2\lambda^2 + A_3\lambda + A_4 = 0 \tag{7.5}$$

has one negative real root of relatively high absolute value; one real root, generally (relatively) close to zero, positive or negative; and, finally, a pair of complex roots whose imaginary part has, as compared to the latter, a relatively large value.

To the large real root λ_1 there corresponds a rapid convergence, contributing with the highest weight to the disturbed motion about the longitudinal axis Ox (rolling motion). The contribution of this rapid mode to the composition of the disturbed motion about axis Oz, as well as the lateral translation of the aircraft center of mass, is negligible as a result of the high value of the moment of inertia C about the axis Oz and of the mass m, namely, the inertia of the aircraft center of mass, which oppose the development of a rapid motion. Hence, the rapid aperiodic mode of asymmetric disturbed motion is reduced, practically, to pure rolling, given the relatively low value of the moment of inertia A (particularly for slender aircrafts). This mode is therefore sometimes called the rolling mode of lateral disturbed motion (or rapid convergence in roll). It can be modeled, with a good approximation, by the equation of moments about the longitudinal axis of the aircraft (second equation of system (7.4)) in which are neglected the deviation induced by the sideslip angle β and the angular velocity of yaw $\hat{r}$, that is, by equation $d\hat{p}/d\hat{t} = b_{22}\hat{p}$; the corresponding characteristic equation will have the root $\lambda_1' = b_{22} = C_{l_p}/2i_A$, generally very close in value to the large real root λ_1 of equation (7.5). Thus, in the numerical example given in [1, page 232], the difference hardly exceeds

1%. Hence, the rolling motion corresponding to the large characteristic real root is influenced by the aerodynamic damping in roll (defined by the rotary derivative C_{l_p}) and by the moment of inertia about the longitudinal axis Ox. Such a qualitative conclusion appears a priori evident. On account of the rapid convergence of this aperiodic motion, its control is beyond practical concern.

The slow aperiodic mode corresponding to the small real root λ_2 of the characteristic equation (7.5) contributes with the highest weight to the componency of the variation law of the angle of bank ϕ. The slow aperiodic components of the variables β, $\hat{p}$, and $\hat{r}$ being relatively small, there result weak aerodynamic disturbance moments and forces for this mode. Since for current aircraft the small real root may also be positive, the aperiodic mode of the corresponding disturbed motion may be divergent. In the case of fixed controls (including those acting on the azimuth of the axis Ox), as a result of the monotone increase of the variable $\hat{r}$ (implying a rather pronounced divergence of the angle ψ); the monotone increase of the angle of bank, and the fact that sideslip is negligible, there results, for $\lambda_2 > 0$, a disturbed motion consisting of a continuous quasicorrectly banked turn of decreasing radius [1, page 231]. Hence, by extension, the slow aperiodic mode of lateral disturbed motion is usually called the spiral mode. However, in the given case of the Ox axis azimuth constraint (divergence of variable ψ prevented) the slow aperiodic mode corresponding to the small real root λ_2 of the characteristic equation (7.5) is reduced, mainly, to the divergence of the variable ϕ, hence to a slow rotation about the longitudinal axis. Therefore, the constraint of the variable ψ is insufficient to eliminate spiral divergence and it must be accompanied by the constraint of the angle of bank (see the case of simultaneous constraint of the variables ψ and ϕ, discussed later).

The possibility of eliminating spiral divergence by design can be made to stand out by means of stability diagrams representing the curve $A_4(P) = 0$, P being a point in the space of the system parameters. There corresponds to the loss of aperiodic stability due to the variation of the system parameters the passing of the free term A_4 of the characteristic equation, from the domain $A_4 > 0$ to the one where $A_4 < 0$. For $A_4 = 0$ there obviously corresponds the boundary situation $\lambda_2 = 0$.

Condition $A_4 > 0$ amounts to

$$C_{l_\beta}C_{n_r} - C_{n_\beta}C_{l_r} + \tan\theta_0(C_{l_\beta}C_{n_p} - C_{n_\beta}C_{l_p}) < 0. \tag{7.6}$$

As mentioned in Section 5.3, to ensure inherent design qualities meant to favor lateral dynamic stability of the aircraft without noticeably influencing the characteristics of performance and those of longitudinal stability, it is recommended to resort to the modification of the wing dihedral γ, of the rudder-fin unit area, and—if possible—of the distance between the aerodynamic center of fin and rudder aft of the aircraft center of mass. The variation of the wing dihedral amounts to the variation of the lateral static stability coefficient C_{l_β}; the other magnitudes included in the right-hand side of inequality (7.6) may be considered as independent of γ. Among the parameters included in (7.6), the variation of the vertical tail area will influence the weathercock stability C_{n_β} and the rotary derivative C_{n_r}. Then (7.6) can be written in the form

$$C_{l_\beta}C_{n_r} + k_1 C_{l_\beta} + k_2 C_{n_\beta} < 0,$$
$$k_1 = C_{n_p} \tan\theta_0 \approx \text{const.}, \quad k_2 = -(C_{l_r} + C_{l_p}\tan\theta_0) \approx \text{const.} \tag{7.7}$$

Generally, the dependence of the aerodynamic and rotary derivatives on the design parameters, flight régime, and so forth, is determined experimentally, due to the theoretical difficulty in obtaining the analytic evaluation, particularly of the influence of aerodynamic interference among the various component elements of the aircraft (wing, fuselage, etc.). Thus, by means of wind tunnel tests, the curves C_{n_β} and C_{n_r} are obtained as functions, for instance, of the parameter $\hat{S}_F = S_F/S$, where S_F indicates the rudder-fin area, and, eliminating the parameter $\hat{S}_F$ by some graphic procedure, the curve representing the function $C_{n_r} = f(C_{n_\beta})$ is obtained.

Then the curve

$$C_{l_\beta}(f(C_{n_\beta}) + k_1) + k_2 C_{n_\beta} = 0 \tag{7.8}$$

will represent the spiral stability boundary in the plane of the parameters C_{l_β}, C_{n_β}.

Sometimes, in order to obtain some rough information (on the order of magnitude), the dependency of the rotary derivative C_{n_r} on the parameter $\hat{S}_F$ (i.e., accordingly, also on C_{n_β}) is neglected. Then the boundary will be represented by the straight line

$$k_3 C_{l_\beta} + k_2 C_{n_\beta} = 0 \tag{7.9}$$

where $k_3 = k_1 + C_{n_r}$. Since in level flight, at flying speeds close to the maximum, $\tan\theta_0 = \tan\alpha_0 \approx \alpha_0$ is small, the spiral stability boundary is sometimes plotted on basis of the simplified equation

$$k_3' C_{l_\beta} + k_2' C_{n_\beta} = 0 \tag{7.10}$$

where $k_3' = C_{n_r}$, $k_2' = C_{l_\beta}$.

Finally, an oscillatory mode, usually having a short period (of the order of seconds or even fractions of a second), will correspond to the pair of complex roots $\lambda_{3,4} = \nu \pm i\omega$. The respective component of disturbed motion is referred to as lateral oscillation or "Dutch roll." It affects, to a greater or lesser extent, according to the case, depending on the system parameters, both the sideslip angle β and the rotations about the Ox and Oz axes.

Generally the real part ν of the complex roots is relatively small. It is advisable to try to avoid a positive ν by design, hence by the inherent stability qualities of the aircraft, the elimination of the oscillatory instability by controls being made more difficult by the high frequency of lateral oscillation.

Even if negative, however, too low a value of the real part is not desirable, since too weak damping of lateral oscillation is inconvenient in piloting, particularly at a low flying speed, in evolutions requiring a high degree of concentration, such as takeoff and landing. Consequently, particularly at a low speed, means are sought to improve lateral stability artificially (and primarily to improve the characteristics of the oscillatory mode), by the use of an automatic yaw damper (see [2]).

As in the case of longitudinal motion, it can be shown that the oscillatory stability boundary is determined by means of Routh's discriminant:

$$R = A_1 A_2 A_3 - A_3^2 - A_1^2 A_4 = 0, \tag{7.11}$$

the region of oscillatory stability being defined by the inequality $R > 0$.

In the plane of the parameters C_{l_β} and C_{n_β}, the curve $R = 0$ (representing the oscillatory stability boundary) forms, together with the curve $A_4 = 0$ given by (7.8), the asymptotic stability region boundary of the auxiliary system (7.4). The classical dilemma that faces the designer when considering this diagram is well known: too low a C_{n_β}/C_{l_β} ratio implies oscillatory instability; too high, spiral instability. The manner of solving the dilemma by renouncing spiral stability is also well known.

The variation of static weathercock stability and static lateral stability (i.e., the rolling moment derivative due to sideslip, sometimes referred to as the "dihedral effect") is achieved by modifying the design characteristics.

However, as the plan-form of the aircraft, its mass and implicitly, through S, b, and m, the relative density coefficient μ are determined by performance requirements, and the center-of-gravity location (hence also the nondimensional moment of inertia i_C) by requirements of longitudinal stability, there remain as main free parameters for the improvement of the asymmetric static stability (lateral and weathercock) the rudder-fin area S_F and the wing dihedral γ. Thus, the variation of parameter C_{l_β} is obtained by varying the dihedral angle of the wing (once the plan-form is fixed, the possibility of influencing this parameter by varying the wing-sweep angle is eliminated), while that of C_{n_β} is obtained by modifying the rudder-fin area. But the modifying of the latter implies not only the variation of C_{n_β}, but also of other parameters of the system, such as C_{y_β}, C_{y_r}, C_{n_r}. The increase of the sideslip-induced side force $(C_{y_\beta}\beta)$ and hence of the aircraft response to side gusts sets a limit to the increase of the vertical tail area.

Therefore, in order to obtain the curve $R = 0$ (as well as the curve $A_4 = 0$) in the (C_{l_p}, C_{n_r}) plane, the variation of these parameters through the intermediary of S_F should also be taken into account. The graphic solution of this dependency also requires, besides the plotting of C_{y_β} and C_{n_r} versus $\hat{S}_F$, and so on, the supplementary operation of eliminating the parameter $\hat{S}_F$ by a graphical procedure, which implies, besides an increase in the number of operations to be carried out, the introduction of some new errors inherent in the procedure used.

It therefore seems reasonable to consider the graphs $A_4 = 0$ and $R = 0$ directly in the plane of the parameters (γ, S_F) or (l_F, S_F) with $\gamma = \text{const}$. By using one of the latter two diagrams, the influence of some parameters usually considered separately, such as the ratios C_{n_β}/C_{l_β}, C_{n_r}/C_{n_β} (see [3, Section 6.5]), is accounted for simultaneously in the same graph.

7.2.2. Simultaneous Control of Both the Yaw and the Bank Angles

Since, as already mentioned, the ensurance of spiral stability by design is renounced in favor of oscillatory stability, the former must be obtained by means of the controls. Even for spiral-stable aircraft, interference of the controls remains necessary in order to accelerate spiral convergence. The elimination of spiral instability, the other two modes being maintained almost unchanged, is achieved by the constraint of the angle of bank ϕ, obviously in addition to correcting the heading.

Let

$$\mathbf{B}(\lambda) = \lambda^3 + B_1\lambda^2 + B_2\lambda + B_3 = 0 \tag{7.12}$$

be the characteristic equation of the auxiliary system

$$\frac{d\beta}{d\hat{t}} = b_{11}\beta + b_{12}\hat{p} + b_{13}\hat{r},$$

$$\frac{d\hat{p}}{d\hat{t}} = b_{21}\beta + b_{22}\hat{p} + b_{23}\hat{r},$$

$$\frac{d\hat{r}}{d\hat{t}} = b_{31}\beta + b_{32}\hat{p} + b_{33}\hat{r}, \tag{7.13}$$

corresponding to this case of partial control.

It can easily be seen that the characteristic polynomial $\mathbf{A}(\lambda)$ of the system (7.4) may be written in the form

$$\mathbf{A}(\lambda) \equiv \lambda\mathbf{B}(\lambda) - b_{14}(b_{21} + b_{31}b_{43})\lambda + A_4$$

and that

$$B_1 = A_1, \quad B_2 = A_2, \quad B_3 = A_3 + b_{14}(b_{21} + b_{31}b_{43}).$$

The root λ_2 of equation $\mathbf{A}(\lambda) = 0$ is usually very small and it can be determined, as already mentioned, with a negligible error, from the last two terms of the equation: $\lambda_2 \approx A_4/A_3$ with $|A_4| \ll |A_3|$. The coefficient $b_{14}(b_{21} + b_{31}b_{43})$ is also small in value in comparison with A_1, A_2, A_3, $|\lambda_1|$ and the moduli of the roots λ_3 and λ_4. Therefore the roots of the equation $\mathbf{B}(\lambda) = 0$ will be, respectively, close to the roots λ_1, λ_3, and λ_4 of the equation $\mathbf{A}(\lambda) = 0$ and, hence, the solution of the system (7.4) will be composed of a heavily damped aperiodic component and of an oscillatory component corresponding to the pair of complex roots close to λ_3 and λ_4. The slow aperiodic mode has been eliminated as a result of the elimination of the small real root. As in the preceding case and for the same reasons, the rapid aperiodic mode will practically affect only the rolling motion; the large root $\lambda_1' \approx \lambda_1$ is, in this case also, obtained from the same equation, that of moments about the longitudinal axis, simplified by neglecting the terms in β and $\hat{r}$.

The angle of bank ϕ being, generally, the most intensively affected by the slow aperiodic mode (hence also by the spiral divergence) and relatively slightly influenced by the oscillatory mode, it was to be expected that, by its constraint, the spiral mode should be removed while the oscillatory mode is little influenced. Hereafter, by means of the cases treated, an endeavor will be made to show that (i) in typical conditions, the constraint of the lateral attitude ϕ is necessary in order to prevent by controls the aperiodic instability of the asymmetric motion, and that (ii) in the same conditions, the simultaneous constraint of the angles ψ and ϕ constitutes the optimum case of partial control of asymmetric motion, any achievable supplementary control or any other combination of constraint variables generally leading, in comparison with this case, to less suitable flying qualities; (iii) there will implicitly result, this being obvious according to the general theory, that a subjection of the system to partial controls by constraining certain variables may lead to a deterioration of the stability characteristics and even to the loss of stability of a system that is stable in the conditions of free flight. The reservation made regarding the possibility of constraint for other variables, besides ψ and ϕ, is demanded by the fact that those variables (i.e., β, $\hat{p}$, and $\hat{r}$) are generally strongly influenced by the oscillatory mode. In the case of too high a frequency of the lateral oscillations, however, the achievement by controls of the conditions required by the general theory (see Section 4.2) becomes difficult or even impossible. This applies in the first place to the condition for the constraint variable to have at all times a sufficiently small upper bound. Therefore, for the theoretical cases considered hereafter, this condition will be assumed as being fulfilled.

We will now consider a few cases of partial control, some of which are only of theoretical interest; however, the statements above will be made clear with their help.

NOTE. The independent consideration of the constraint of the various variables (e.g., ψ, $\hat{r}$ and β, or ϕ and $\hat{p}$) for which essentially the same group of controls is applied, is justified by the different aspect of their variation. Thus, in the cases considered above, it was tacitly assumed that it was possible to constrain angle ψ by means of controls, leaving free the angle of sideslip β and the yawing velocity. The assumption was based on the fact that, because of the slow variation, by accumulation, of the deviation of the longitudinal axis azimuth, the necessity of controls appears at relatively long time intervals. Note that in these intervals between two consecutive actuations of the rudder, both the angular velocity r and the sideslip β induced by disturbance are directly and rather strongly affected by the generally rapid lateral oscillation. Analogously, the constraint of the angle of bank ϕ will be assumed to be independent of the variables $\hat{p}$ and $\hat{r}$ that compose the angular velocity $\dot{\phi}$.

7.2.3. Yaw and Sideslip Angles Under Constraint

It will be assumed that by adequate operation of the controls (in the first place of the rudder), both ψ and β are maintained sufficiently small in the sense of the theory expounded in Chapter 4. Then the corresponding auxiliary system will be

$$\frac{d\hat{p}}{d\hat{t}} = b_{22}\hat{p} + b_{23}\hat{r},$$

$$\frac{d\hat{r}}{d\hat{t}} = b_{32}\hat{p} + b_{33}\hat{r},$$

$$\frac{d\phi}{d\hat{t}} = b_{43}\hat{r}, \tag{7.14}$$

with the characteristic equation

$$\lambda(\lambda^2 + C_1\lambda + C_2) = 0 \tag{7.15}$$

where

$$C_1 = -(b_{22} + b_{33}), \qquad C_2 = b_{22}b_{33} - b_{23}b_{32}.$$

As it can be seen, like (7.1), system (7.14) is of the type (3.26). Therefore, according to the proposition in Section 3.8, the uniform asymptotic stability of the zero solution of the system of the first two equations implies, for sufficiently small initial conditions, a limit of the variable ϕ, however small, different from zero, when $\hat{t}$ tends toward infinity.

For current values of the parameters the system formed of the first two equations in (7.14) admits of two negative real characteristic roots: the one of a value close to the large real root of equation (7.5) to which corresponds a rapid convergence in almost pure roll (the coefficient of aerodynamic damping in roll C_{l_p} is predominant in the expression of this root), and the other noticeably smaller in value than the first (but which can nevertheless considerably exceed the value of the small root of equation (7.5) corresponding to the spiral mode). A more or less slow convergent nonperiodic mode corresponds to the latter root.

Although the motion as a whole is thus stable in the Liapunov sense, the situation is, nevertheless, unsatisfactory since after each disturbance the bank draws away aperiodically, little by little, from its undisturbed value ($\phi_0 = 0$), these repeated motions being liable finally to lead to a noticeable deviation of the aircraft lateral attitude, the necessity to control the angle ϕ now appearing. Through the supplementary constraint of this variable the auxiliary system will be reduced to the first two (7.14) equations considered above.

7.2.4. Constraint of the Rotation in Yaw

It is assumed that by operating the controls, success is achieved in imposing sufficiently small upper bounds to the values of the variables $\hat{r}$ and ψ in the sense required by the general theory, as well as their asymptotic canceling. The corresponding auxiliary system will be

$$\frac{d\beta}{d\hat{t}} = b_{11}\beta + b_{12}\hat{p} + b_{14}\phi,$$

$$\frac{d\hat{p}}{d\hat{t}} = b_{21}\beta + b_{22}\hat{p},$$

$$\frac{d\phi}{d\hat{t}} = \hat{p}, \tag{7.16}$$

having the characteristic equation

$$\lambda^3 + D_1\lambda_2 + D_2\lambda + D_3 = 0 \tag{7.17}$$

where

$$D_1 = -(b_{11} + b_{22}), \quad D_2 = b_{11}b_{22} - b_{12}b_{21}, \quad D_3 = -b_{14}b_{21}.$$

It can usually be assumed that $b_{12} = 0$ $(C_{y_p} = 0)$. From Hurwitz's criterion we have that this partially controlled system will be uniformly asymptotically stable if and only if there appears a simultaneous fulfillment of the inequalities $D_1 > 0$, $D_3 > 0$, $D_1D_2 - D_3 > 0$, equivalent to

$$i_A C_{y_\beta} + \mu C_{l_p} < 0, \tag{7.18a}$$

$$C_{l_\beta} C_{L_0} \cos\theta_0 < 0, \tag{7.18b}$$

$$-C_{y_\beta} C_{l_p}\left(\frac{1}{\mu} C_{y_\beta} + \frac{1}{i_A} C_{l_p}\right) + 2C_{l_\beta} C_{L_0} \cos\theta_0 > 0. \tag{7.18c}$$

Since in current aircraft design the aerodynamic derivatives C_{y_β} and C_{l_β}—in the case of positive wing dihedral γ or positive wingsweep, or of both simultaneously—and the damping-in-roll derivative C_{l_p} are negative, the first two inequalities are always satisfied. (Positive sideslip, corresponding to the lateral component of center-of-mass velocity directed in the positive sense of the axis Oy (see Figure 7.1) determines an aerodynamic force (lateral lift) opposed to this sense, hence, according to the adopted convention, negative. Wherefrom $C_{y_\beta} < 0$.) The condition expressed by the third

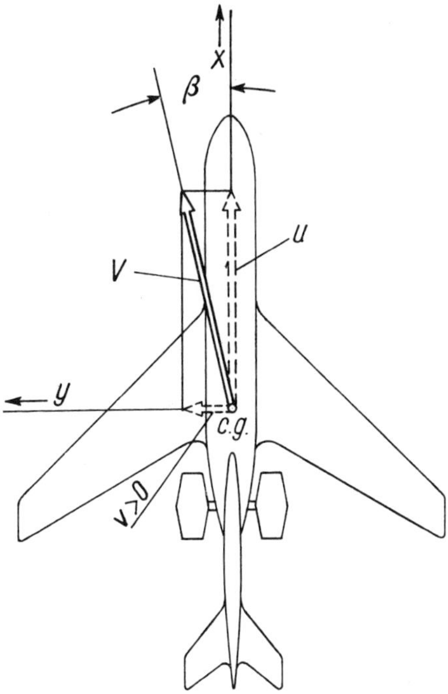

Figure 7.1.

inequality can, however, not be satisfied precisely in the case of aircraft with spiral stability in free flight with locked controls. Indeed, as a result of the present trend of using high wing loading (i.e., for a given mass, small wing area) and low aspect ratio for high speeds, aerodynamic damping in roll becomes weak and the side-force derivative C_{y_β} is likewise small, so that by increasing the dihedral effect† with the aim of assuring spiral stability

† The dihedral is positive if, for the aircraft in normal position, the tip chord is situated above the root chord. Positive wingsweep corresponds to a plan-form as in Figure 7.1 (sweep-back). By dihedral effect is understood the rolling moment due to sideslip, determined by either $\gamma > 0$ or sweep-back, or both. The measure of the dihedral effect is the lateral static stability coefficient C_{l_β}.

in free flight, the negative term of inequality (7.18c), namely, $2C_{l_\beta}C_{L_0} \cos \theta_0$, may become prevailing, particularly at low flight speeds, hence at high incidences, implying high values for C_{L_0}. In other words, the aircraft laterally stable in free flight becomes unstable by rotation constraint about the axis Oz. The explanation of this can probably be found in the fact that this partial control excludes aerodynamic damping in yaw, playing, as a result of the generally high value of the respective rotary derivative C_{n_r}, an important part in stabilizing the motion of an aircraft subjected to an asymmetric disturbance.

It can readily be seen that the instability resulting from the nonfulfillment of condition (7.18c) is eliminated by the constraint of ϕ. Indeed, in this case, the characteristic equation of the pertinent auxiliary system reduces to

$$\lambda^2 + D_1\lambda + D_2 = 0,$$

having two negative real roots. Neglecting, as above, the term in C_{y_p}, these roots will be $C_{l_p}/2i_A$ and $C_{y_\beta}/2\mu$. The first characterizes a rapid convergence, hence, essentially, the character of the solution will be given by the second root, of relatively small value. Therefore, the solution of the auxiliary system corresponding to the constraint of the variables $\hat{r}$, ϕ, and ψ will be uniformly asymptotically stable, but with a precarious rate of damping because of the relatively low value of the ratio $C_{y_\beta}/2\mu$.

It should be noted that the nonfulfillment of the condition given by Routh–Hurwitz's discriminant (7.18c) is related to the preponderance, in the left-hand side of the inequality, of the term containing as a factor the measure C_{l_β} of the dihedral effect. In other words, the discriminant is negative when the attainment of spiral stability in free flight is sought to the detriment of oscillatory stability. But, as already mentioned, in order to obtain acceptable flying qualities, the present trend is just the opposite: the attainment of lateral oscillatory stability is sought to the detriment of spiral stability. Sometimes the canceling of the dihedral effect or even a negative dihedral effect ($C_{l_\beta} > 0$) is reached, implying nonfulfillment of condition (7.18b) and hence the appearance of a gentle divergence of the considered partially controlled motion (corresponding to the spiral divergence of free flight). In this manner the oscillatory stability of this motion is, however, obtained. Nevertheless, for aircrafts with high wing loading and low aspect ratio, the damping of the oscillations proves too slow for the requirements of proper piloting. The acceleration of damping could, on the one hand, be obtained by the accentuation of the negative dihedral effect, a rather limited possibility in practice, and by the artificial increase of the lateral damping coefficient C_{l_p}.

The constraint of the variable $\hat{r}$ in the sense required by the general theory usually encounters great difficulties. All that can be obtained practically is an acceleration of damping in yaw by introducing an automatic damper that detects the angular velocity r, and not a sufficiently small upper bound for any $t \geqslant t_0$ of the amplitude value. In this manner the conditions required by the theory for the partial control considered (constraint of $\hat{r}$ and ψ) cannot generally be fulfilled. At the same time, however, the control of angular velocity r allows stability to be improved in other cases of partial control or in free flight.

7.2.5. Rapid Convergence in Pure Roll

To avoid the troublesome influence of the term in C_{l_β} in the left-hand side of inequality (7.18c) a supplementary control may be resorted to in order to constrain sideslip of the aircraft. It can, however, be seen that although the instability shown above is eliminated, the supplementary constraint of the angle ϕ still remains necessary.

As a result of the constraint of the variables β, $\hat{r}$, and ψ, the auxiliary system is reduced to the approximate equation of the disturbed motion in roll determining $\hat{p}(\hat{t}, \hat{t}_0)$ obtained by neglecting the deviations β and $\hat{r}$ in the moment equation about the Ox axis of the free disturbed motion system (7.1) to which the kinematic relationship between the angle ϕ and the angular velocity $\hat{p}$—simplified according to the rule for forming auxiliary equations—is added.

$$\frac{d\hat{p}}{d\hat{t}} = b_{22}\hat{p}, \tag{7.19a}$$

$$\frac{d\phi}{d\hat{t}} = \hat{p}. \tag{7.19b}$$

From equation (7.19a) there results the component $\hat{p}$ of the solution of the auxiliary system in the known form of rapid convergence in pure roll: $\hat{p}(\hat{t}) = \hat{p}(\hat{t}_0)\exp[(C_{l_p}/2i_A)(\hat{t} - \hat{t}_0)]$ and from (7.19b) the component ϕ having the expression $\phi(\hat{t}) = \phi(\hat{t}_0) - \hat{p}(\hat{t}_0)(C_{l_p}/2i_A)\{1 - \exp[(C_{l_p}/2i_A)(\hat{t} - \hat{t}_0)]\}$. Obviously, the angle of bank $\phi(\hat{t})$ will tend toward a magnitude generally differing from zero, depending on the initial conditions $\hat{p}(\hat{t}_0)$, $\phi(\hat{t}_0)$ and small together with the latter two, namely, toward $\phi(\hat{t}_0) - \hat{p}(\hat{t}_0)C_{l_p}/2i_A$. To avoid a troublesome value of the bank angle ϕ, liable to appear consequently to repeated disturbances, the correction of lateral attitude by controls (hence the additional constraint of angle ϕ) becomes necessary from time to time.

Such a partial control would be ideal as regards the result, but not the means of achieving it. Indeed, a very strong lateral stability is obtained by leaving only the rapid aperiodical mode free. This partial control, however, involves the imposition of efficient bounds on the variables β and $\hat{r}$ that are strongly affected by the short-period lateral oscillation; hence, such a partial control grows troublesome. Thus, of the two conditions imposed by the general theory (see Section (4.2)) with reference to a constraint of bounds on the value of the controlled induced deviations at any instant $t \geqslant t_0$ and to their damping after a certain time interval, generally only the second one can be satisfactorily fulfilled. For this reason the case studied above and, to a certain extent, the two preceding ones are only, strictly speaking, of theoretical interest.

7.2.6. Oscillation in the Oxy plane (Yaw and Sideslip)

Constraining the latteral attitude ϕ and rolling velocity p, we get an auxiliary system of a type similar to system (3.26)

$$\frac{d\beta}{d\hat{t}} = b_{11}\beta + b_{13}\hat{r},$$

$$\frac{d\hat{r}}{d\hat{t}} = b_{31}\beta + b_{33}\hat{r},$$

$$\frac{d\psi}{d\hat{t}} = b_{53}\hat{r}, \tag{7.20}$$

characterized by the behavior, qualitatively different, of the solution regarding the variable ψ from the one resulting from the first two linear equations in β and r. If the trivial solution of the system of the first two equations is uniformly asymptotically stable, the component ψ of the solution of system (7.20) will be (nonasymptotically) stable and $\lim_{\hat{t}\to\infty} \psi(\hat{t}, \hat{t}_0) = l(\beta(\hat{t}_0), \hat{r}(\hat{t}_0), \psi(\hat{t}_0)) \neq 0$ (see Section 3.8, the proposition referring to system (3.26)). It was shown that, in this situation, the necessity arises to correct the heading, the rudder being used at relatively long time intervals.

For current types of aircraft and in conventional flight régimes, the system of the first two equations is uniformly asymptotically stable. The roots of the respective characteristic equation of the second degree, even in the case of aircraft having not too great a weathercock stability coefficient, will be complex with their real part negative. Aerodynamic damping in roll being excluded by constraint of the angular velocity p, the damping of oscillations in the Oxy plane ($\nu = (1/4)(C_{y_\beta}/\mu + C_{n_r}/i_C)$), will be weaker than that of the lateral oscillations for free flight or for partial control by bank-angle constraint (without simultaneously constraining the rolling velocity). Therefore, in this respect, the partial control considered leads to a deterioration of the quality of flight, which is, however, offset to a certain extent by the decrease of the oscillation frequency. To accelerate damping, an automatic yaw damper may also be resorted to here (i.e., an automatic control system reacting to deviations of the aircraft angular velocity about the Oz axis: $\dot{\psi} = b_{53}\hat{r} = (\sec \theta_0)\hat{r}$, with θ_0 and hence also b_{53}, constant). It will be shown here how the effect of a yaw damper influences the equations.

An ideal automatic control system will be considered, assuring, without any time lag, rudder deflection by an angle proportional to the deviation r of angular velocity about the axis Oz,

$$\zeta = g_r \hat{r} \tag{7.21}$$

where g_r is constant. The angle ζ is assumed small together with the deviation r, hence, in linear approximation also resorted to when establishing the equations (7.20) the additional aerodynamic moment determined by rudder deflection can then be expressed in the form $\Delta N = (\partial N/\partial\zeta)\zeta$ and $\Delta C_n = C_{n_\zeta}\zeta$, with $C_{n_\zeta} = \partial C_n/\partial\zeta$. The expression of the additional aerodynamic side force coefficient is similar: $\Delta C_y = C_{y_\zeta}\zeta$, where $C_{y_\zeta} = \partial C_y/\partial\zeta$. Then, the auxiliary system corresponding to the constraint of rolling for an aircraft provided with a yaw damper is obtained from (7.20) by the adequate modification of the first two equations:

$$\begin{aligned}
\frac{d\beta}{d\hat{t}} &= b_{11}\beta + \left(b_{13} + \frac{g_r C_{y_\zeta}}{2\mu}\right)\hat{r} \equiv b_{11}\beta + b_{13}'\hat{r},\\
\frac{d\hat{r}}{d\hat{t}} &= b_{31}\beta + \left(b_{33} + \frac{g_r C_{n_\zeta}}{2i_C}\right)\hat{r} \equiv b_{31}\beta + b_{33}'\hat{r},\\
\frac{d\psi}{d\hat{t}} &= b_{53}\hat{r},
\end{aligned} \tag{7.22}$$

where $b_{13}' = -1 + (C_{y_r} + g_r C_{y_\zeta})/2\mu$ and $b_{33}' = (C_{n_r} + g_r C_{n_\zeta})/2i_C$. Usually $(C_{y_r} + g_r C_{y_\zeta})/2\mu$ is negligible in comparison with 1 and hence $b_{13}' \approx b_{13}$, the first equation remaining unchanged. Then the rate of damping of the oscillations described by the first two equations of the system will be

$$\nu' = \frac{1}{4}\left[\frac{C_{y_\beta}}{\mu} + \frac{C_{n_r} + g_r C_{n_\zeta}}{i_C}\right] = \nu + \frac{g_r C_{n_\zeta}}{4 i_C},$$

that is, larger than the one corresponding to system (7.20), by a quantity proportional to the constant g_r of the automatic damper.

An increase of the oscillation period T also corresponds to the increase of the damping coefficient C_{n_r} by the appearance of a control moment ΔC_n

$$T \approx 2 i_C^{1/2}\left[-2C_{n_\beta} - \frac{1}{4}\left(\frac{1}{\beta} C_{y_\beta} + C_{n_r} + g_r C_{n_\zeta}\right)^2\right]^{-1/2}.$$

It can be seen that starting from a certain value of the product $g_r C_{n_\zeta}$ the motion becomes nonperiodic.

7.3. COUPLED MOTIONS. A PARTICULAR CASE: HOVERING FLIGHT STABILITY OF VTOL AIRCRAFT

7.3.1. Coupling of Longitudinal and Lateral Motions

As was seen in Section 2.2, to be able to resolve the system of Euler equations (2.4) modeling the flight of an aircraft into two independent systems corresponding to the longitudinal and the asymmetric motions, some additional hypotheses were necessary, namely, hypotheses 1, 3, 4, and 5 of Section 2.2. In most cases encountered in current design and flight practice these assumptions are perfectly acceptable. There are, however, also cases where, on account of design considerations, the nature of the intended evolution (basic motion), or the intensity of the disturbance, or all three, the necessity appears of foregoing one or several of the hypotheses adopted. Then, coupling terms occur, making imperative the simultaneous consideration of system (2.4). In a certain period (between 1953 and 1958) important space was devoted in the literature to the case of the maneuvers that include a rolling motion with an angular velocity exceeding a certain limit (hence assumption 5 in Section 2.2 is nonvalid). The necessity of stabilization by controls has been pointed out (this stabilization being achieved by using automatic control of the elevator geared to the deviations of the pitching velocity $\dot{\theta}$, the gearing coefficients being proportional to the square of the rate of roll [4]). (A more complete analysis of this case, based on the existing literature, is given in [5, Pt. III, Chapter 17].)

The study of spin also requires that we take into consideration the complete system. The gyroscopic moments due to the spinning rotors of the engines constitute a source of coupling terms. Indeed, in the case of large turbojet engines, assumption 3, Section 2.2, is liable to lose its validity. In [6], this effect is analyzed in the hypothesis that linearization is admissible (assumption 4, Section 2.2). An improvement in lateral oscillation damping and a weakening of rapid longitudinal oscillation damping is found. (Both the lateral and longitudinal short-period oscillations change as a result of coupling of the symmetric and asymmetric motions, but their change is relatively slight, thus justifying the maintainance of the two denominations.) In conventional aircraft types and at current flight speeds the effect of the gyroscopic coupling terms proves to be weak. It becomes significant, however, in the case of a precarious margin of stability, corresponding to weak damping in yaw and pitch, that is, for small values of the rotary derivatives C_{n_r} and C_{m_q}, respectively, but particularly for low flight speeds, the aerodynamic moments varying with the square of the latter. For flying speeds tending toward zero, the gyroscopic terms become predominant in the moment equations. This case will be considered later. It corresponds to the hovering flight of VTOL aircraft.

7.3.2. Equations of Disturbed Motion

The starting point is system (2.4) and (2.5) where the variables u, v, w, p, q, r, ϕ, $\Delta\theta$, ψ represent disturbances, and the aerodynamic forces and moments only appear as induced by the disturbance; these are ΔX, ΔY, ΔZ, ΔL, ΔM, and ΔN. This system will be

$$m\left(\frac{du}{dt} + qw - rv\right) = T\cos\varphi - mg\sin(\theta_0 + \Delta\theta) - \Delta X,$$

$$m\left(\frac{dv}{dt} + ru - pw\right) = mg\sin\phi\cos(\theta_0 + \Delta\theta) + \Delta Y,$$

$$m\left(\frac{dw}{dt} + pv - qu\right) = T\sin\varphi - mg\cos\phi\cos(\theta_0 + \Delta\theta) + \Delta Z; \quad (7.23)$$

$$A\frac{dp}{dt} - F\frac{dq}{dt} - E\frac{dr}{dt} + \frac{dh_{r_x}}{dt} = Dq^2 - Dr^2 + Epq + (B - C)qr$$
$$- Frp - qh_{r_z} + rh_{r_y} + \varDelta L,$$
$$- F\frac{dp}{dt} + B\frac{dq}{dt} - D\frac{dr}{dt} + \frac{dh_{r_y}}{dt} = Er^2 - Ep^2 - Dpq + Fqr + (C - A)rp$$
$$- rh_{r_x} + ph_{r_z} + \varDelta M,$$
$$- E\frac{dp}{dt} - D\frac{dq}{dt} + C\frac{dr}{dt} + \frac{dh_{r_z}}{dt} = Fp^2 - Fq^2 + (A - B)pq - Eqr + Drp$$
$$- ph_{r_y} + qh_{r_x} + \varDelta N; \tag{7.24}$$

$$\frac{d\phi}{dt} = p - q\sin\phi\tan(\theta_0 + \varDelta\theta) - r\cos\phi\tan(\theta_0 + \varDelta\theta),$$
$$\frac{d\varDelta\theta}{dt} = q\cos\phi - r\sin\phi,$$
$$\frac{d\psi}{dt} = \sec(\theta_0 + \varDelta\theta)(q\sin\phi + r\cos\phi). \tag{7.25}$$

For the basic motion characterized by $u_0 = v_0 = w_0 = p_0 = q_0 = r_0 = \phi_0 = \psi_0 = 0$, $\theta_0 \neq 0$ and hence $X_0 = Y_0 = Z_0 = L_0 = M_0 = N_0 = 0$. The foregoing systems taken together represent the disturbed motion of the aircraft (which was in a hovering state before the disturbance). So far no supposition has been made regarding the size of the disturbances and hence regarding the possibility of linearizing equations (7.23)–(7.25) with respect to the dynamic variables. As will be seen, the uncontrolled system appears, in all probability, to be unstable with respect to part of the variables, hence any assumption of this kind requires rigorous justifications and precautions. For simplicity it will be admitted that the aircraft is perfectly symmetric and hence $D = F = 0$, and that $h_{r_x} = \text{const.}$, $h_{r_y} = 0$, $h_{r_z} = \text{const.}$, implying $dh_{r_x}/dt = dh_{r_y}/dt = dh_{r_z}/dt = 0$. Generally, $C > B > A \gg E$.

To make the analysis simpler, the theory of partially controlled motions will be resorted to. It is natural to think first of constraining the motion of the center of mass to which correspond, as an auxiliary system, the equations (7.25) and (7.24) with $\varDelta L = \varDelta M = \varDelta N = 0$. The collected system is formally identical with that of the equations about the center of mass, considered to be fixed, of the aircraft *cum* spinning rotors system. However, such an auxiliary equation set generally proves to be inconclusive Indeed, since it is of the type (3.26), the variables ϕ, $\Delta\theta$, and ψ will, for t tending toward infinity, admit an asymptote, generally differing from zero, depending on the initial conditions; hence the trivial solution of the auxiliary system does not fulfill the condition required by the propositions stated in Chapter 4 regarding partially controlled motions, that of being uniformly asymptotically stable. It is easy to show that the situation does not change if, in addition, the pair of variables $(q, \Delta\theta)$ characterizing rotation about the lateral axis is constrained, and if instead of this component of rotation one of the other two, (p, ϕ) or (r, ψ), is constrained, an unstable auxiliary system is reached. It seems proper to consider the case corresponding to the constraint of the motion of the mass center and of the aircraft attitude, equations (7.24) with $\Delta L = \Delta M = \Delta N = 0$ resulting as an auxiliary system. Hence, taking the assumptions above into account, the following system will be considered.

$$\frac{dp}{dt} = \frac{1}{AC - E^2}[-(C - E)h_{r_z}q + E(A - B + C)pq + C(B - C - E)qr],$$

$$\frac{dq}{dt} = \frac{1}{B}[h_{r_z}p - h_{r_x}r - E(p^2 + q^2) + (C - A)rp],$$

$$\frac{dr}{dt} = \frac{1}{AC - E^2}[(A - E)h_{r_x}q + A(A - B + E)pq - E(A - B + C)qr]. \tag{7.26}$$

7.3.3. Effect of Rotation Dampers

The characteristic equation of the linear part of system (7.26) has a zero root and a pair of pure imaginary roots $\pm i\omega$ with $\omega^2 = [(A - E)h_{r_x}^2 + (C - E)h_{r_z}^2]/[B(AC - E^2)]$; hence, the behavior of the solution will be determined by the higher-degree terms. It can be assumed that the zero solution of the complete system (7.26) will, in the best case, be uniformly stable (nonasymptotically). Uniform (but nonasymptotic) stability in the large is readily evidenced for the case where the body axes considered coincide with the principal inertia axes. Indeed, the time derivative of the positive definite function

$$T_0 = \tfrac{1}{2}(Ap^2 + Bq^2 + Cr^2)$$

(representing total kinetic energy for $E = 0$ and $h_{r_x} = h_{r_z} = 0$ for the mass center considered as a fixed point) along the integral curves of the system

$$\frac{dp}{dt} = -\frac{h_{r_z}}{A}q + \frac{B - C}{A}qr,$$

$$\frac{dq}{dt} = \frac{h_{r_z}}{B}p - \frac{h_{r_x}}{B}r + \frac{C - A}{B}rp,$$

$$\frac{dr}{dt} = \frac{h_{r_x}}{C} + \frac{A - B}{C}pq, \tag{7.27}$$

obtained from (7.26) by setting $E = 0$, vanishes identically, the stated property of the trivial solution deriving therefrom.

To obtain a (locally) uniformly asymptotically stable auxiliary system, linear damping terms will be introduced in system (7.26): $-c_p p$ in the first equation, $-c_q q$ in the second, and $-c_r r$ in the third (c_p, c_q, c_r being positive constants). These additional terms model the action of some dampers in roll, pitch, and yaw, respectively, by means of which artificial stability of rotation about the center of mass is obtained. Therefore, the following linear first-approximation system will further be considered.

$$\frac{dp}{dt} = -c_p p - \left(\frac{C - E}{AC - E^2}\right)h_{r_z}q,$$

$$\frac{dq}{dt} = \left(\frac{h_{r_z}}{B}\right)p - c_q q - \left(\frac{h_{r_x}}{B}\right)r,$$

$$\frac{dr}{dt} = \left(\frac{A - E}{AC - E^2}\right)h_{r_x}q - c_r r. \tag{7.28}$$

It has the characteristic equation

$$\lambda^3 + (c_p + c_q + c_r)\lambda^2 + \left[\frac{A - E}{B(AC - E^2)}h_{r_x}^2 + \frac{C - E}{B(AC - E^2)}h_{r_z}^2 + c_p c_q + c_q c_r + c_r c_p\right]\lambda + \frac{A - E}{B(AC - E^2)}h_{r_x}^2 c_p + \frac{C - E}{B(AC - E^2)}h_{r_z}^2 c_q + c_p c_q c_r = 0. \tag{7.29}$$

It can be seen that, in a certain sense, the gyroscopic terms support the stabilizing action of the rotation dampers. It is thus obvious that for $h_{r_x} = h_{r_z} = 0$, the roots of equation (7.29) are, respectively, $-c_p$, $-c_q$, and $-c_r$, hence the necessity to artificially dampen all the angular velocity components about the aircraft center of gravity ($c_p \neq 0$, $c_q \neq 0$, and $c_r \neq 0$). Now, let

$h_{r_x} \neq 0$, $h_{r_z} = 0$. The conditions given by the Hurwitz criterion (see Chapter 5, Appendix) for equation (7.29) being

$$c_p + c_q + c_r > 0, \qquad \text{(i)}$$

$$c_p c_q c_r + \frac{A-E}{B(AC-E^2)} h_{r_x}^2 c_p + \frac{C-E}{B(AC-E^2)} h_{r_z}^2 c_r > 0, \qquad \text{(ii)}$$

$$2c_p c_q c_r + c_p^2(c_q + c_r) + c_q^2(c_r + c_p) + c_r^2(c_p + c_q)$$
$$+ \frac{C-E}{B(AC-E^2)} h_{r_z}^2 (c_p + c_q) + \frac{A-E}{B(AC-E^2)} h_{r_x}^2 (c_q + c_r) > 0, \qquad \text{(iii)}$$

it stands out that uniform asymptotic stability is also ensured with a single damper, in either roll or yaw. Artificial damping of only the rate of pitch leads to the appearance of a zero root of the characteristic equation, hence to a case situated at the boundary of aperiodic instability. The boundary of oscillatory instability (corresponding to the vanishing of Routh's discriminant when relation (iii) becomes an equality) can only be reached by foregoing artificial damping ($c_p = c_q = c_r = 0$), or by damping only a single component of angular velocity, in conjunction with the vanishing of a certain component of the gyroscopic moment ($c_p = c_q = h_{r_x} = 0$ or $c_q = c_r = h_{r_z} = 0$).

It should be recalled that the partially controlled motion corresponding to the constraint—among others—of flight attitude is here considered. For better efficiency of the controls applied for this purpose it is advisable to ensure as rapid as possible a damping of all the components of angular velocity of the aircraft about its center of mass. From this standpoint the use of dampers for all these components seems advisable.

7.3.4. Simultaneous Use of Rotation Dampers and Static Stabilizers

The case studied in this section is not limited strictly to hovering flight; it also includes, practically, evolutions at quasi-zero speed (inducing negligible aerodynamic moments and forces), such as vertical landing and takeoff. To be correctly performed, these evolutions require great attention on the part of the pilot and it is proper to relieve him of the burden of stabilizing the flight and particularly of constraining the flight attitude, leaving in his care, at most, the stabilizing of the center-of-mass velocity components, which is easier to achieve due to the relatively large inertia of the aircraft. This suggests a new case of mixed control (automatic and manual) to be considered, by using the following physical model: the variables characterizing the center-of-mass motion (u, v, w) are constrained by the pilot, while the ones defining motion around the center of mass and aircraft angular attitude (p, q, r), (ϕ, θ, ψ) are countered by automatic control. A scheme of direct automatic control will be used, creating a control moment about each of the three body axes. The control system is geared both to angular deviations and to those of the respective angular velocities according to the scheme

$$M_{c_x} = \frac{AC - E^2}{C}(c_p p + c_\phi \phi), \tag{7.30a}$$

$$M_{c_y} = B(c_q q + c_\theta \Delta\theta), \tag{7.30b}$$

$$M_{c_z} = \frac{AC - E^2}{A}(c_r r + c_\psi \psi). \tag{7.30c}$$

The terms that contain the angular deviations ϕ, $\Delta\theta$, and ψ, respectively, represent the static stabilizing effect (creation of restoring moments) about the respective axes resulting from the appearance of an angular deviation; although their effect is not entirely analogous to that due to aerodynamic incidence, since they are geared to attitude deviations, these stabilizing terms in conventional flight régimes increase the effect of (aerodynamic) intrinsic static stability. Aerodynamic static stability with respect to the three axes: lateral static stability (dihedral effect), longitudinal static stability, and weathercock (directional) static stability (represented by the static derivatives C_{l_β}, C_{m_α}, and C_{n_β}, respectively) refer to the deviations of the relative position of the aircraft with respect to the airflow and are induced by either a deviation of the aircraft attitude with respect to the ground, or an atmospheric disturbance (lateral or vertical gusts), or both.

Response time (time lag) is neglected in equations (7.30). (The effect of time lag will be discussed in Section 11.4.) The following mathematical model corresponds to this physical model: a partially controlled system by constraint of the variables u, v, w, admitting the linearized auxiliary system

$$\frac{dp}{dt} = -c_p p - h_{12} q - c_\phi \phi,$$

$$\frac{dq}{dt} = h_{21} p - c_q q - h_{23} r - c_\theta \theta,$$

$$\frac{dr}{dt} = h_{32} q - c_r r - c_\psi \psi,$$

$$\frac{d\phi}{dt} = p - \tan\theta_0\, r, \quad \frac{d\theta}{dt} = q, \quad \frac{d\psi}{dt} = \sec\theta_0\, r, \tag{7.31}$$

where $h_{12} = [(C - E)/(AC - E^2)]h_{r_z}$, $h_{21} = (1/B)h_{r_z}$, $h_{23} = (1/B)h_{r_x}$, $h_{32} = [(A - E)/(AC - E^2)]h_{r_x}$.

The system is too involved to allow a general qualitative discussion. It requires numerical cases to be considered. To point out in a simple manner some aspects regarding the necessity of the various forms of artificial stabilizing and the effect of the gyroscopic terms, a few simplified cases are considered.

It is first assumed that the undisturbed angle of pitch is negligible. Such an assumption obviously applies to normal-attitude VTOL aircraft. The achievement of hovering (or vertical at quasi-zero speed) flight with $\theta_0 = 0$ is possible (i) by using tilting engines rotated in order that their thrust axis coinciding with the rotation axis of the engine rotors be vertical ($\sigma = 0$); (ii) by rotating these engines by an angle σ differing from zero and from $\pi/2$, combined with jet deflection aimed at obtaining a vertical thrust; (iii) or, finally, by fixed engines ($\sigma = \pi/2$) with jet deflection equal to $\pi/2$ (theoretical case). Three schemes will be considered here: scheme (a) $h_{r_x} = h_{r_z} = 0$, achievable by using an even number of engines, where the rotors of the corresponding ones on the opposite sides about the median plane of symmetry have the same rotation velocity and are opposite handed. This scheme is compatible with all three cases above, somehow constituting a standard case: the effect of the gyroscopic moments on the stability of the system can be made to stand out by comparison with it; scheme (b) $h_{r_x} = 0$, $h_{r_z} = J\Omega \neq 0$, corresponding to case (ii); and scheme (c) $h_{r_x} = J\Omega \neq 0$, $h_{r_z} = 0$, corresponding to the theoretical case (iii).

a. Standard scheme $h_{r_x} = h_{r_z} = 0$. System (7.31) is resolved—in the absence of the gyroscopic coupling terms—into three independent systems, corresponding to the three rotation components (about the body axes):

$$\frac{dp}{dt} = - c_p p - c_\phi \phi, \qquad \frac{d\phi}{dt} = p; \tag{7.32}$$

$$\frac{dq}{dt} = - c_q q - c_\theta \Delta\theta, \qquad \frac{d\Delta\theta}{dt} = q, \tag{7.33}$$

$$\frac{dr}{dt} = - c_r r - c_\psi \psi, \qquad \frac{d\psi}{dt} = r. \tag{7.34}$$

Since the systems are formally identical, only system (7.32), for example, will be considered, the conclusions being valid, *mutatis mutandis*, for the other two as well.

The roots of the characteristic equation being $-c_p/2 \pm [(c_p/2)^2 - c_\phi]^{1/2}$, for sufficiently efficient static stabilization also allowing acceleration of damping, it is required that $c_\phi > (c_p/2)^2$ (c_ϕ being expressed per second squared, sec^{-2}, and c_p per second, sec^{-1}). A damped oscillation is obtained in this case, characterized by a rate of damping equal to $-c_p/2$ and by a frequency $[c_\phi - (c_p/2)^2]^{1/2}$. If $c_\phi < (c_p/2)^2$, two convergent aperiodic modes are obtained, one of them having its rate of damping $-\{c_p/2 - [(c_p/2)^2 - c_\phi]^{1/2}\}$ lower than the oscillation in the preceding case.

It can be seen that to obtain uniform asymptotic stability of the zero solution of the three systems, all the controls represented by the equations (7.30) are necessary (and sufficient), all the static and damping coefficients being different from zero. As will be seen hereafter, this condition can be weakened—by renouncing part of the damping term—if the angular momentum of the spinning rotors or airscrews h_r is different from zero, wherefrom it will result that the gyroscopic moments have a damping effect.

b. Scheme $h_{r_x} = 0$, $h_{r_z} = J\Omega \neq 0$. As the coupling terms $h_{23}r$ and $h_{32}q$ of the rotations about the axes Oy and Oz vanish, and since the equations for dp/dt and $d\phi/dt$ (of (7.31)) do not contain terms in r and ψ, the equations representing the motion about axis Oz are separated. Two independent systems thus appear.

$$\frac{dr}{dt} = -c_r r - c_\psi \psi, \qquad \frac{d\psi}{dt} = r; \tag{7.35}$$

$$\frac{dp}{dt} = c_p p - h_{12} q - c_\phi \phi, \qquad \frac{dq}{dt} = h_{21} p - c_q q - c_\theta \Delta\theta,$$

$$\frac{d\phi}{dt} = p, \qquad \frac{d\Delta\theta}{dt} = q. \tag{7.36}$$

System (7.35) models the yawing motion, (7.36) the coupled rolling and pitching motions.

Regarding system (7.35), identical to (7.34), the conclusions concerning the ratio between the damping coefficient and the static stabilization coefficient expounded above are valid. The Hurwitz conditions for the characteristic equation

$$\lambda^4 + (c_p + c_q)\lambda^3 + (c_\phi + c_\theta + c_p c_q + h_{12}h_{21})\lambda^2 + (c_p c_\theta + c_q c_\phi)\lambda + c_\phi c_\theta = 0 \tag{7.37}$$

of system (7.36) are

$$c_p + c_q > 0, \tag{7.38a}$$

$$c_p c_\phi + c_q c_\theta + (c_p + c_q)(c_p c_q + h_{12} h_{21}) > 0, \tag{7.38b}$$

$$c_p c_q (c_\phi - c_\theta)^2 + (c_p + c_q)(c_p c_\theta + c_q c_\phi)(c_p c_q + h_{12} h_{21}) > 0, \tag{7.38c}$$

$$c_\phi c_\theta > 0. \tag{7.38d}$$

As can be seen, they are satisfied not only under all the controls (7.30a, b) with all the damping and static stabilization coefficients different from zero, but also if one of the damping coefficients (c_p or c_q) is zero. In the same conditions if h_{r_z} were equal to zero (resultant angular momentum of spinning rotors canceled) relation (7.38c) would turn into an equality. From condition (7.38d), simultaneous static stabilization about both axes is necessary.

From the foregoing, the gyroscopic moments represented in (7.36) by the terms $- h_{12}q$ and $h_{21}p$ have a damping effect, allowing us, at least theoretically, to renounce artificial damping in roll or in pitch. Obviously, they cannot have the effect of static stabilization.

It should be noted that the damping provided by the gyroscopic moments is not, practically, sufficient in evolutions requiring high accuracy, such as the ones along the vertical or when hovering, and it is therefore not recommended to forego simultaneous artificial damping in both roll and pitch; however, its effect will be stressed by the presence of the gyroscopic moments.

c. Scheme $h_{r_x} = J\Omega \neq 0$, $h_{r_z} = 0$. Analogous conclusions also stand out from the examination of this scheme, for which, however, the system corresponding to the motion about the longitudinal axis Ox is separated—as an independent system—from system (7.31). The effect of the vanishing of the artificial damping coefficient c_r on the pitch remains to be studied, the slow damping in yaw ($c_r = 0$ having as a main effect the slowing-up of this damping) involving fewer drawbacks regarding the safety and accuracy of flight than that of the other two components of rotation.

REFERENCES

1. B. Etkin, *Dynamics of Flight: Stability and Control.* Wiley, New York, 1959.
2. R. J. White, "Investigation of lateral dynamic stability in the XB-47 airplane," *J. Aeronaut. Sci.* 17(1960), 133–148.

3. W. J. Duncan, *The Principles of the Control and Stability of Aircraft.* Cambridge Univ. Press, London and New York, 1952.
4. R. Westerwick, "The roll coupling problem: a mathematical approach," *Aeronaut. Eng. Rev.* **16**(1957), 48–51.
5. A. W. Babister, *Aircraft Stability and Control.* Macmillan (Pergamon), New York, 1961.
6. H. C. Vetter, "Effect of turbojet engine on the dynamic stability of an aircraft," *J. Aerospace Sci.* **20**(1953), 797.

Chapter 8

STABILITY OF UNSTEADY MOTION: NONAUTONOMOUS SYSTEMS

8.1. INTRODUCTION

In Chapters 5, 6, and 7 the mathematical model of systems of autonomous ordinary differential equations, particularly linear equations with constant coefficients, were used on basis of the hypothesis of a strict constancy of the system parameters: strictly uniform, straight, and level régime of basic motion; constant mass, moments of inertia, and products of inertia about the body axes; invariable location of mass center; and so on. Rigorously speaking, such a hypothesis constitutes *eo ipso* an evident idealization, even in most cases of steady straight level flight. Indeed, to keep up such a steady motion in a resistant medium like the atmosphere, it is necessary to expend a certain amount of energy, which implies the variation of the mass, of the moment of inertia, and, generally, of the location of the center of gravity as well, all of which are undoubtedly essential parameters of the system. The model of autonomous systems differs still more from the actual situation in the cases, so frequent, of flight at variable speed or along paths differing from the level path.

Nevertheless, the model of autonomous systems, particularly that of linear systems with constant coefficients, is not to be set aside automatically as inadequate; we should only consider that its validity is not absolute or beyond discussion. We must know the bounds of this validity in order to be able to determine the restrictive conditions to be imposed on the system. It should be noted that, generally, in the literature, the linear model with constant coefficients continues to be used almost exclusively, its validity being accepted a priori or, sometimes, empirical criteria being resorted to. Thus, for example, the maximum relative variation (difference between the maximum and minimum values of the linear system coefficients versus their mean values) is compared with the order of magnitude of the error of determination (by experimental procedures and approximate computations).

The model of the linear system with constant coefficients is assumed to be acceptable if the maximum relative variation corresponding to actual flight conditions does not exceed the error involved in determining the value of the system coefficients. The values, often high, of the errors of determination, which reach, according to [1] (see page 482), up to 20%, prove rather comfortable in this respect.

The necessity is felt, however, to check the validity of the classical model in a more rigorous manner. (The errors of determination of the system parameters constitute for the system, viz., for the mathematical model, persistent disturbances. Hence, the model is valid (i) if the system is steady under persistent disturbances and (ii) if the respective errors do not exceed the admissible values of these disturbances.) This necessity becomes imperative in the case of nonconventional aircraft, in operating conditions differing considerably from the standard "flying qualities requirements" (in the case of missiles or spacecraft flying with high positive and negative accelerations or with rapid variations of altitude in the atmosphere, etc., as well as of military aircraft implying involved kinematics, etc.). In many situations, as will be apparent from the following two sections, a check of this kind, based on rigorous criteria, does not present noteworthy theoretical difficulties. Two classes of systems reducible to the linear model with constant coefficients will be considered, corresponding to a sufficiently small variation in the values of the linearized nonautonomous system coefficients and to a sufficiently slow variation of these values. In the case of the former class, the limit of validity will have to be sought in the form of an upper bound of the difference between the maximum and minimum values of the coefficients for the entire duration of the evolution, or of the ratio between this difference and a mean value of the coefficients considered. For the latter it is necessary to impose an admissible upper bound on the value of the time derivatives of these coefficients. For both cases the general theory includes theorems allowing us to establish limits that are rigorous in the framework of the adopted physical hypotheses. Within these limits the variation of the parameters may be neglected, an autonomous system being thus substituted as mathematical model for a system depending explicitly on t, and the nonlinear terms being eventually neglected. In other words, it is possible, from these theorems, to deduce validity criteria of the linear model with constant coefficients. These criteria generally represent sufficient conditions of validity and are thus liable to be continuously improved upon. The effort made to loosen the conditions is widely compensated by the possibility of applying the classical model, which is both well known and easy to handle. If, however, it is not possible to make the admissibility of the use of the autonomous system stand out by the existing processes, direct consideration of the respective case of the nonautonomous system becomes necessary.

The direct study of the stability of unsteady basic motions presents great difficulties in many respects. Since the mathematical models of these motions are equations depending explicitly on the independent variable (nonautonomous equations), that these equations do not generally integrate in a closed form is far from being the only difficulty. Another is connected with the universality of the procedures of approach. Unsteady motion does not mean a certain definite form of motion. On the contrary, the concept implies the absence of any individual characteristic feature of the motion and, as such, it is all-inclusive (as a sphere) and altogether poor in contents. Therefore, if steady motion, as a well-defined case, can be studied by generally applicable methods and processes, every unsteady motion requires specific methods and processes, an individual approach. Hence, a first problem in considering the stability of the unsteady basic motions consists in identifying those types of motion that are of the greatest practical interest, on the one hand by their frequency in current operation, and on the other by any discrepancy occuring in comparison with the cases considered in the classical theory.

Another difficulty when studying the stability of unsteady basic motion is linked with the individualized nature of the various cases of disturbed motion. In the case of a steady basic motion, generally valid stability criteria may be found; the various dynamic variables characterizing the disturbed motion have a qualitative behavior that is similar: all are stable, or all are unstable; all have, as a rule, the same number of similar modes; an asymptotically damped disturbed motion is concomitantly necessarily stable along the entire $t \geqslant 0$ half axis, and so on. In contrast, in the case of unsteady basic motions the conditions of stability found for a certain type of motion (or for a certain motion) are not generally valid for other motions or types of motion; there is no a priori certainty that the various variables of the disturbed motion system have similar qualitative behaviors; an acceptable asymptotic behavior (dying out of the variable for $t \to \infty$) does not exclude the possibility of large values of the induced deviations appearing at various stages of the disturbed motion, and so on.

Two important types of unsteady basic motions will be considered hereafter: variable speed flight (in Chapter 9) and evolutions in a medium of variable density (in Chapter 10).

8.2. CASES REDUCIBLE TO LINEAR SYSTEMS WITH CONSTANT COEFFICIENTS

8.2.1. Small Variation of the System Parameters

For small differences between the maximum and minimum values of the linearized system coefficients, neglecting their variation can be justified by means of the theory of stability by first approximation (see Section 3.4).

To this end the system of motion equations whose zero solution represent the basic motion (see Chapter 1, system (1.8)) will be written in the form

$$\frac{dx}{dt} = (A + B(t))x + N(t, x) \tag{8.1}$$

where the column vector x represents the deviations induced by a disturbance of the n dynamic variables, A is a constant $n \times n$ matrix, $B(t)$ a matrix function of t, bounded on the half axis $t \geqslant 0$ (let $||B(t)|| \leqslant \beta_0 = \text{const.}$), and $N(t, x)$ a column vector representing the nonlinear terms. Both $A + B$ and N depend on the system parameters. The system is defined in the domain $t \geqslant t_0$, $|x| < h$, h being a given constant, and it is assumed that in this domain the vector function $N(t, x)$ satisfies uniformly a Lipschitz condition with respect to t (the Lipschitz constant being independent of t). In these conditions, according to the theory of stability by the first approximation, if the zero solution of the linear system with constant coefficients

$$\frac{dx}{dt} = Ax$$

is uniformly asymptotically stable (in other words, if matrix A is a Hurwitz matrix), for sufficiently small values of β_0 and of Lipschitz's constant for function N, the basic motion $\tilde{x} = 0$, $\tilde{x}(t)$ being the solution of system (8.1), is likewise uniformly asymptotically stable. (See the theorem stated in Section 3.4.2.)

The upper bound β_0 of the variation of the coefficients of the linear system $B(t)$, or the error made by neglecting the addition $B(t)$, if β_0 is known, can easily be estimated by using, for example, one of the techniques described in Chapter 3. In the following two chapters β_0, and with its help the necessary lower bound of the spectrum of matrix A, defining dynamic stability margin in the sense shown, will be estimated for the cases of speed and flight altitude variation.

8.2.2. Slow Variation of the System Parameters: the Process of the "Freezing of Coefficients"

To any evolution with variable parameters there obviously corresponds a certain bounded region of the space of parameters. The maximum variation of the parameter values can therefore be estimated by means of the difference between the upper and lower bound of this region. If the estimate obtained is not sufficiently small in the sense of Section 8.2.1 (particularly if, for linear systems with coefficients depending on t, the estimate of β_0 is not sufficiently small), the use of a single linear system with constant coefficients (instead of a nonautonomous and nonlinear system) as a mathematical model is no longer justified. Nevertheless, in certain situations, the replacement of a nonautonomous and nonlinear system—considered as adequate in the framework of the physical hypotheses adopted—by a set of linear systems with constant coefficients (a procedure sometimes called the "freezing of coefficients") can be rigorously justified. This is the case, among others, for parameters with a slow variation (with $B(t)$ of (8.1), $||dB/dt||$ admits a sufficiently small upper bound). For example, in machines equipped with rocket engines, the mass, moments of inertia, and sometimes the center-of-gravity location undergo a noteworthy variation during the fulfillment of their mission. At the same time, however, the variation speed of these parameters may be rather small. The same may also be said, in certain cases, about the variation in density of the medium crossed by the machine, of the flying speed, and so on.

Thus, let the vectorial equation defined for $t \geqslant t_0$

$$\frac{dx}{dt} = X(t, x) \tag{8.2}$$

be the mathematical model of the (disturbed) motion corresponding to the physical hypotheses adopted. The autonomous systems

$$\frac{dy}{dt} = X(s, y) \tag{8.3}$$

corresponding to all the values of the parameter s satisfying the inequality $s \geqslant t_0$ are also considered.

Determine the bounds (or, at least, some bounds) within which the stability of the solutions $y = 0$ of all the (8.3) equations implies the stability of the trivial solution of equation (8.2). For this, corresponding theorems have been determined and estimating procedures suggested. (See, e.g., [2–5].)

Linear models will be used, with time-dependent coefficients of the disturbed motion corresponding to the unsteady basic motion, written in either the form

$$\frac{dx}{dt} = A(t)x \tag{8.4}$$

with the vector x and the matrix A as above, or the form

$$\frac{dx}{dt} = A(p(t))x \tag{8.5}$$

where p represents either the velocity V of the center of mass or the flying height h (or the air density ρ corresponding to the flight height), etc. The following sets of linear systems with constant coefficients, depending on the parameter s, $s \geqslant t_0$, can be substituted for these nonautonomous systems

$$\frac{dy}{dt} = A(s)y \tag{8.6}$$

or

$$\frac{dy}{dt} = A(p(s))y, \tag{8.7}$$

respectively, if the system parameters (hence matrix A or p in the case of the (8.5) systems) vary sufficiently slowly in the sense resulting, for instance, from the following lemma, due to Hale [4].

LEMMA. *Consider the nonautonomous system* (8.5) *and the set* (8.7) *of linear systems with constant coefficients, depending on parameter* s *through a given function* $p(s)$ *continuously differentiable for* $s \geqslant t_0$. *It is assumed that for set initial moment* $t_0 \geqslant 0$ *and function* $p(s)$, $s \geqslant t_0$, *there exist two functions* $\nu(p)$ *and* $K(p)$ *continuous and with continuous derivatives in a region containing the values of function* $p(s)$, *such that for any* s, $s \geqslant t_0$, *the solution* $y(t, p(s); t_0, y_0)$, $y_0 = y(t_0, p(s); t_0, y_0)$ *of system* (8.7) *satisfies the relation*

$$|y(t, p(s); t_0, y_0)| \leqslant K(p(s)) \exp[-\nu(p(s))(t-t_0)]|y_0|, \qquad t \geqslant t_0, \tag{8.8}$$

(i) *if* $\nu(p(s))$ *and* $K(p(s))$ *are constant,* $\nu > 0$, $K \geqslant 1$, *and there is a* T *for which*

$$\frac{1}{T}\int_t^{t+T} \left\| \frac{\partial A(p(s))}{\partial p} \right\| ds \leqslant \frac{\nu}{4(K-1)} - \mu \tag{8.9}$$

whatever $t \geqslant 0$ *is,* μ *being a positive quantity different from zero, then the zero solution of system* (8.5) *is asymptotically stable*; (ii) *if* $\nu(p(s))$ *and* $K(p(s))$ *are not constant but there is a* T *such that for any* $s \geqslant T$, $\nu(p(s)) \geqslant \mu_1 > 0$ *and*

$$\left\| \frac{\partial A(p(s))}{\partial p} \right\| \leqslant \frac{\nu^2(p(s)) - 2|\nu'(p(s))|\,|p'(s)| \ln K(p(s))}{4(K(p(s)) - 1)} - \mu_2$$

with μ_1 *and* μ_2 *positive constants differing from zero, then the solution of the nonautonomous system* (8.5) *is asymptotically stable.*

The estimation of the functions $\nu(s)$ and $K(s)$ is, in principle, possible (see, e.g., [5]) and it does not, generally, involve very important computing difficulties. The only valid objection is the one regarding the too stringent conditions that may result, particularly in the case of intricate systems (e.g., systems of an order higher than two). The objection does not, however, justify giving up the checking of the extent to which the substitution of a set of linear systems with constant coefficients for the nonautonomous system is valid, particularly if the flight behavior of an essentially new type of aircraft is studied.

The necessity, in principle, to check the admissibility of the substitution can be made evident by the following counterexample, borrowed from [6]. Let be matrix $A(t)$ of (8.4), of the form

$$A(t) = \begin{pmatrix} -1 + \frac{3}{2}\cos^2 t & 1 - \frac{3}{2}\cos t \sin t \\ -1 - \frac{3}{2}\cos t \sin t & -1 + \frac{3}{2}\sin^2 t \end{pmatrix}.$$

It can readily be seen that all the linear systems with constant coefficients (8.6) admit the same characteristic equation, independent of the parameter s:

$$\lambda^2 + \tfrac{1}{2}\lambda + \tfrac{1}{2} = 0,$$

having complex conjugate roots with a negative real part ($\operatorname{Re} \lambda = -\frac{1}{4}$). Hence, the zero solution of system (8.6), whatever the value of parameter s, is uniformly asymptotically stable. Concomitantly, the nonautonomous system (8.4) admits the unstable solution

$$x = \begin{pmatrix} -e^{t/2} \cos t \\ e^{t/2} \sin t \end{pmatrix}.$$

8.3. DIRECT INVESTIGATION OF UNSTEADY MOTION STABILITY

Often, a proper variation of the system parameters in the sense shown above cannot be pointed out, either because this variation is not actually *small* or *slow* in the sense required, or due to the imperfection of the investigation procedure (of the criterion used or of the estimations). The direct tackling of the problem of stability of the solution of nonautonomous systems then becomes necessary through specific procedures adequate to the particular form of the system considered. As already mentioned, it is not possible to obtain criteria that are generally valid, such as the algebraic criteria known in the case of autonomous systems, for systems depending explicitly on the independent variable. The criteria will be different for every type of dependency of the system on t, and will not, as a rule, be unique. Therefore, the aim is not to find any criteria whatever, but criteria that are as convenient as possible. (The Routh–Hurwitz criteria (see Chapter 5, Appendix I) derive from theorems providing necessary and sufficient conditions of stability. The theorems that furnish stability criteria for any nonautonomous systems whatever generally only indicate sufficient conditions of stability, wherefrom results, in principle, the possibility of improving the criteria.)

The method of the Liapunov function constitutes a classical source for such criteria, perhaps the only one of universal validity furnished so far by the general theory. But because of its very universal character, it does not immediately supply practical schemas, effective computing procedures for a given concrete case. For each case studied the establishing of as convenient as possible a Liapunov function becomes a matter of investigation. The choice of this function is often made considerably easier by a good knowledge of the physical phenomenon. Once found, Liapunov's function allows us not only to make stability stand out, but also to estimate the solution in a finite time interval. Thus, even in the case of instability, it can be used to obtain information about the behavior of the solution: character of the instability, potential efficiency of stabilization by controls, and so on.

After uniform asymptotic stability has been made evident, however, it is simpler and more accurate to obtain such information for the time interval in which the evolution considered takes place by the numerical integration of the equations of disturbed motion.

It is entirely clear that no asymptotic property of the solution of nonautonomous systems, and particularly its stability, can be made evident by numerical integration. However close the computed integral curve is to the exact one of the nonautonomous system considered, it only gives information for the time interval for which the computation has been made. Since the system is explicitly dependent on t, it is not generally possible to draw conclusions regarding the behavior of the integral curve in the following stages from its aspect for a finite time interval, and the less so when $t \to \infty$. For instance, an integral curve with a monotone decreasing aspect during a finite time interval $[t_0, T]$, but no matter how large otherwise, does not exclude, in principle, the possibility of divergence for $t > T$. Let, for example, $dx/dt = (-a + \varepsilon t)x$ be a scalar equation with a and ε positive constants. As its solution is $x = x(0) \exp[\int_0^t (-a + \varepsilon t)dt] = x(0) \exp[t(-a + (\varepsilon/2)t]$, for ε sufficiently small, the interval $[0, T]$ with $T = 2a/\varepsilon$, for which the integral curve has a decreasing aspect (for $x(0) > 0$), can be arbitrarily large. Nevertheless, any nontrivial solution diverges exponentially and hence the zero solution is strongly unstable.

REFERENCES

1. A. A. Lebedev and L. S. Chernobrovkin, *Flight Dynamics of Unmanned Flying Machines* (in Russian). Gostekhizdat-Oborongiz, Moscow, 1962.
2. L. Flatto and N. Levinson, "Periodic solutions of singularly perturbed differential equations," *J. Rat. Mech. Anal.* **4**(1955), 943–950.
3. J. K. Hale and A. P. Stokes, "Conditions for the stability of nonautonomous differential equations," *J. Math. Anal. Appl.* **3**(1961), 50–69.
4. J. K. Hale, "Functional-differential equations with parameters," *Contributions to Differential Equations.* (Wiley), **1**, 4, pp. 401–410.
5. Li Yueh-Sheng, "The boundedness, stability and error estimate for the solutions of nonlinear differential equations," *Chinese Math.* **3**(1963), 34–42.
6. L. Markus and H. Yamabe, "Global stability criteria for differential systems," *Osaka Math. J.* **12**(1960), 305–317.

Chapter 9

STABILITY OF UNSTEADY FLIGHT: VARYING UNDISTURBED VELOCITY

9.1. RAPID INCIDENCE ADJUSTMENT

9.1.1. Assumptions and Simplified Equations of Disturbed Motion

It is assumed that, by adequate actuation of the controls, variation of the undisturbed flight speed is assured according to a set program. Hence, if V represents the speed of the basic motion, $V(t)$ is a known time function.

If it is assumed that during the whole flight only these controls (called in Section 1.2 basic controls) act, the correction controls being identically zero, a scheme is obtained analogous to that of the locked controls used for steady-state motions.

The main parameters of the basic motion are functions of t assumed known as problem data or determined by means of the motion equations. In this manner, the basic motion is perfectly defined (and known). The following hypotheses are admitted for simplicity.

1. The mass, moments of inertia about the body axes, center-of-mass location, as well as air density, are constant.

2. Hypotheses 1, 3, and 4 of Section 2.2 are valid, hence the aircraft is considered to be a rigid symmetrical body and the disturbances are sufficiently small to allow linear approximation.

3. Hypothesis 5 of Section 2.2 becomes: Straight flight, level or sloping, is considered to be the basic motion, the value of the angles β, ϕ, and ψ corresponding to the basic motion being identically zero.

With these hypotheses the first-approximation system of the disturbed motion is separated into two groups, representing, respectively, the longitudinal and the lateral motion of the aircraft. Then, adopting a moving frame of reference fixed to the path of the center of mass (sometimes referred to as stability axes), we will describe the disturbed motion in the plane of symmetry—in a first approximation—by the linear system (2.15), whose coefficients appear as known functions of t.

Either the scheme corresponding to identically zero correction control or that of partially controlled motion will be utilized hereafter. In both the former and latter cases $\hat{M}_{c_\alpha} = \hat{M}_{c_q} = 0$ in the first-approximation system.

As in the case of uniform basic motion—and for the same reason, linked to the relatively high inertia of the aircraft—the induced deviations of the variables describing the motion of the center of mass develop slowly and their variation is slow in comparison with the induced deviations of the elements of motion about the center of mass. This allows either of the following two simplified schemes to be used.

(i) By extending for a first short stage of the disturbed motion the classic hypothesis $\Delta\hat{V} = 0$ to the case of unsteady basic motions (which amounts to the assumption that since the disturbed motion affects incidence, and to a certain extent the pitching angular velocity, it is practically exhausted within a time interval in which the induced deviation of flight speed remains negligible) system (2.15) is reduced to

$$\frac{d\Delta\alpha}{d\hat{t}} = -\left(\frac{C_{L_\alpha} + C_D}{2\mu}\hat{V} + \frac{1}{\hat{V}}\frac{d\hat{V}}{d\hat{t}}\right)\Delta\alpha + \Delta\hat{q},$$

$$\frac{d\Delta\hat{q}}{d\hat{t}} = \frac{1}{2i_B}\hat{V}^2\left[C_{m_\alpha} - C_{m_{\dot{\alpha}}}\left(\frac{C_{L_\alpha} + C_D}{2\mu}\hat{V} + \frac{1}{\hat{V}}\frac{d\hat{V}}{d\hat{t}}\right)\right]\Delta\alpha + \frac{1}{2i_B}\hat{V}^2(C_{m_q} + C_{m_{\dot{\alpha}}})\,\Delta\hat{q},$$

$$\frac{d\Delta\theta}{d\hat{t}} = \Delta\hat{q}. \tag{9.1}$$

This system was obtained by substituting $C_T\cos(\alpha - \varphi) - \mu\hat{g}\sin\gamma = \hat{V}^2C_D + \mu\, d\hat{V}/d\hat{t}$, according to the equation of forces along the direction of the tangent to the path written in nondimensional form. If the solution $\Delta\alpha = 0, \Delta\hat{q} = 0$, corresponding to the basic motion, of the first-approximation system formed by the first two equations is uniformly asymptotically stable, then, also assuming that the nonlinear parts depend only on t, $\Delta\alpha$, and $\Delta\hat{q}$ (but not on the other dynamic variables as well), the respective nonlinear system will also admit of a uniformly asymptotically stable trivial solution, according to the theorem of stability by the first approximation. The system will therefore be uniformly asymptotically stable with respect to the variables $\Delta\alpha$ and $\Delta\hat{q}$ and uniformly stable with respect to $\Delta\theta$ in accordance with the proposition relating to systems of the (3.26) type.

(ii) Considering the longitudinal motion as being partially controlled by constraining the variables $\Delta\hat{V}$ and $\Delta\theta$, we obtain the system formed of the first two equations of (9.1) as an auxiliary system. The uniform asymptotic stability of the solution $\Delta\alpha = 0$, $\Delta\hat{q} = 0$ will imply uniform stability with respect to $\Delta\alpha$ and $\Delta\hat{q}$, for the partially controlled system; if, in addition, the controls are sufficiently efficient to assure $\Delta\hat{V} \to 0$ and $\Delta\theta \to 0$ (practically, within a sufficiently short time), the uniform asymptotic stability of the trivial solution of the auxiliary system in $\Delta\alpha$ and $\Delta\hat{q}$ will entail the uniform asymptotic stability of the partially controlled basic motion (see Section 4.2).

Hence, for one scheme or the other, it is of interest to consider the system of the first two (9.1) equations, which will, for brevity, be transcribed in the form

$$\frac{d\Delta\alpha}{d\hat{t}} = -a\,\Delta\alpha + \Delta\hat{q}, \qquad \frac{d\Delta\hat{q}}{d\hat{t}} = -b\,\Delta\alpha - c\,\Delta\hat{q}, \tag{9.2}$$

where the time-dependent coefficients a, b, and c have the expressions

$$a = \frac{1}{2\mu} u(C_{L_\alpha} + C_D) + \frac{1}{u}\frac{du}{dt},$$

$$b = \frac{1}{2i_B} u^2 \left\{ -C_{m_\alpha} + C_{m_{\dot{\alpha}}} \left[\frac{1}{2\mu} u(C_{L_\alpha} + C_D) + \frac{1}{u}\frac{du}{dt} \right] \right\},$$

$$c = \frac{1}{2i_B} u^2 (-C_{m_q} - C_{m_{\dot{\alpha}}}).$$

Nondimensional flight speed is denoted by u: $u = \hat{V} = V/V_{\max}$ ($V_{\max}$ is the maximum flight speed of the aircraft), and t is the nondimensional time (ratio between time in seconds and unit of aerodynamic time: $2\bar{c}/V_{\max}$ with $\bar{c}$ in meters and $V_{\max}$ in meters per second).

9.1.2. Sufficient Conditions of Applicability for the Classical Model

The bounds of validity to replace system (9.2) by a linear system with constant coefficients will first be estimated in accordance with Section 8.2. To this end system (9.2), completed with the nonlinear terms, will be written in the form

$$\begin{aligned} \frac{dx}{dt} &= -a_1 x + y - \delta a(t) x + X(t, x, y), \\ \frac{dy}{dt} &= -b_1 x - c_1 y - \delta b(t) x - \delta c(t) y + Y(t, x, y), \end{aligned} \tag{9.3}$$

where $x = \Delta\alpha$, $y = \Delta\hat{q}$, $t = \hat{t}$, X and Y represent the respective nonlinear parts $(X(t, 0, 0) = 0, Y(t, 0, 0) = 0)$; a_1, b_1, and c_1 denote the respective values of the coefficients a, b, and c corresponding to the uniform basic motion at an undisturbed speed equal to the minimum speed actually reached; they are obtained by setting $u = u_1$ and $du/dt = 0$ in the expressions $a(t)$, $b(t)$, and $c(t)$ previously given. Obviously, $\delta a = a - a_1$, $\delta b = b - b_1$, $\delta c = c - c_1$. The minimum actual value of parameter u was chosen for u_1, as being the most unfavorable from the standpoint of dynamic stability margin of the linear system with constant coefficients

$$\frac{dx}{dt} = -a_1 x + y, \qquad \frac{dy}{dt} = -b_1 x - c_1 y. \tag{9.4}$$

As already mentioned, the value of the real part ν_1 of the roots

$$\lambda_{1,2}^{(1)} = \nu_1 \pm i\omega_1, \quad \nu_1 = -\frac{a_1 + c_1}{2},$$

$$\omega_1 = (b_1 + a_1 c_1 - \nu_1^2)^{1/2} = \left[b_1 - \frac{(a_1 - c_1)^2}{4}\right]^{1/2}$$

of the characteristic equation of the system may be considered as a measure of the stability margin.

For statically stable aircraft with a sufficiently high value of the derivative C_{m_α} it may be assumed that $b > (a - c)^2/4 > 0$ at any moment, inclusive for $\hat{V} = \hat{V}_1$. (This seems to be the case for all aircraft of current construction.) It is also assumed that for any t, $a > 0$, and hence also $a_1 > 0$. Function c and particularly its magnitude for V_1 $(c(\hat{V}_1) = c_1)$ are obviously positive. Then $\nu_1 < 0$ and hence the solution $x = y = 0$ of system (9.4) is uniformly asymptotically stable, which implies the uniform asymptotic stability of the trivial solution of system (9.3) for the sufficiently small additional terms $-\delta a.x$ and $-(\delta b.x + \delta c.y)$. Hereafter, these additional terms will be estimated in function of ν_1 (measuring the dynamic stability margin of the linear system with constant coefficients (9.4)) by the technique outlined in the preceding chapter, based on the method of the variation of parameters.

The general solution of system (9.4) may be set in the form

$$x = e^{\nu_1 t}(C^1 \cos \omega_1 t + C^2 \sin \omega_1 t),$$

$$y = e^{\nu_1 t}\{C^1[(\nu_1 + a_1) \cos \omega_1 t - \omega_1 \sin \omega_1 t] + C^2[(\nu_1 + a_1) \sin \omega_1 t + \omega_1 \cos \omega_1 t]\}$$

the integration constants C^1 and C^2 being determined from the initial conditions $x(0) = x_0$, $y(0) = y_0$

$$C^1 = x_0, \qquad C^2 = -\frac{\nu_1 + a_1}{\omega_1} x_0 + \frac{1}{\omega_1} y_0.$$

For system (9.3) the solution will be sought in the form

$$x = e^{\nu_1 t}(\eta^{(1)}(t) \cos \omega_1 t + \eta^{(2)}(t) \sin \omega_1 t),$$

$$y = e^{\nu_1 t}\{\eta^{(1)}(t) [(\nu_1 + a_1) \cos \omega_1 t - \omega_1 \sin \omega_1 t]$$
$$+ \eta^{(2)}(t) [(\nu_1 + a_1) \sin \omega_1 t + \omega_1 \cos \omega_1 t]\}.$$

It is assumed that the nonlinear parts satisfy the relations

$$|X(t, x, y)| \leqslant l_1(|x| + |y|), \qquad |Y(t, x, y)| \leqslant l_2(|x| + |y|),$$

l_1 and l_2 being constant. By using the procedure shown, we readily see that the following inequalities hold.

$$e^{-\nu_1 t}|x| \leqslant |C^1| + |C^2| + \frac{2}{\omega_1} \int_0^t \sigma(s) e^{-\nu_1 s}(|x| + |y|)\, ds,$$

$$e^{-\nu_1 t}|y| \leqslant \sqrt{b_1}(|C^1| + |C^2|) + \frac{2\sqrt{b_1}}{\omega_1} \int_0^t \sigma(s) e^{-\nu_1 s}(|x| + |y|)\, ds,$$

with $\sigma(t) = \max\{(l_1 \sqrt{b_1} + l_2 + \sqrt{b_1}|\delta a(t)| + |\delta b(t)|); \ (l_1 \sqrt{b_1} + l_2 + |\delta c(t)|)\}$, (by max is meant the highest of the two bracketed values) and hence

$$e^{-\nu_1 t}(|x| + |y|) \leqslant (1 + \sqrt{b_1})(|C^1| + |C^2|) + \frac{2(1 + \sqrt{b_1})}{\omega_1} \int_0^t \sigma(s) e^{-\nu_1 s}(|x| + |y|)\, ds; \tag{9.5}$$

or, by using Gronwall's inequality (see Section 3.6.3),

$$|x| + |y| \leqslant \exp(\nu_1 t)(1 + \sqrt{b_1})(|C^1| + |C^2|)$$
$$+ (|C^1| + |C^2|) \frac{2(1 + \sqrt{b_1})}{\omega_1} \int_0^t \sigma(s) \exp\left[\nu_1 t + \int_s^t \frac{2(1 + \sqrt{b_1})}{\omega_1} \sigma(\tau)\, d\tau\right] ds.$$

If function $\sigma(s)$ is known, the estimate of the additional terms represented by σ is obtained from the condition that the right-hand side of the last inequality be monotone decreasing along the entire semiaxis $t \geqslant 0$. A simple estimation is obtained if, in (9.5) the highest value of $\sigma(s)$ along the semiaxis

$s \geqslant 0$ is substituted for $\sigma(s)$. Let σ_m be this value. Then, on basis of Gronwall's lemma

$$e^{-\nu_1 t}(|x| + |y|) \leqslant (1 + \sqrt{b_1})(|C^1| + |C^2|)e^{\chi t}$$

is obtained, where

$$\chi = \frac{2(1 + \sqrt{b_1})\sigma_m}{\omega_1}$$

$$= \frac{2(1 + \sqrt{b_1})}{\omega_1}\{l_1 \sqrt{b_1} + l_2 + \operatorname*{lub}_{t \geqslant 0} \max[(\sqrt{b_1}|\delta a(t)| + |\delta b(t)|; (|\delta c(t)|)]\}$$

or, taking into account the dependence of the constants C^1 and C^2 on the initial conditions

$$|x| + |y| \leqslant \frac{1 + \sqrt{b_1}}{\omega_1}[(|\nu_1 + a_1| + \omega_1)|x_0| + |y_0|)] \exp[(\nu_1 + \chi)t].$$

The uniform asymptotic stability of the solution $x = y = 0$ of the system (9.3) obviously follows from

$$\chi < |\nu_1| = \frac{a_1 + c_1}{2}. \tag{9.6}$$

Let, for example, at any instant $\text{lub}_{t \geqslant 0} \max\{(\sqrt{b_1}|\delta a| + |\delta b|); (|\delta c|)\} = |\delta c|$. For simplicity, it is assumed that the nonlinear parts are negligible: $l_1 = l_2 \approx 0$ and that the rotary derivatives C_{m_q} and $C_{m_{\dot\alpha}}$ are independent of t. Then, from (9.6) the condition

$$\hat{V}_2^2 - \hat{V}_1^2 \leqslant \frac{i_B \omega_1}{(1 + \sqrt{b_1})(-C_{m_q} - C_{m_{\dot\alpha}})} \tag{9.7}$$

results, where $\hat{V}_2$ represents the maximum value actually reached by the undisturbed flight speed .($\hat{V}_1$ being the minimum actually reached).

Of $\text{lub}_{t \geqslant 0} \max\{(\sqrt{b_1}|\delta a| + |\delta b|); (|\delta c|)\} = \sqrt{b_1}|\delta a| + |\delta b|$, $\hat{V}$ will be considered, for illustration, of the form $\hat{V} = \hat{V}_{\min}(1 + \kappa e^{-t})$, $\kappa > 0$. Assuming, as above, $l_1 = l_2 \approx 0$ and, with the exception of $\hat{V}$ and $d\hat{V}/d\hat{t}$, the system parameters as constant, and admitting that the minimum speed actually reached $\hat{V}_1$ differs but little from $\hat{V}_{\min}$, it will be possible to estimate the variation of the parameters by means of the constant κ.

From $\hat{V} = \hat{V}_1(1 + \kappa e^{-t})$ there results $(1/\hat{V})\ (d\hat{V}/d\hat{t}) = -1 + \hat{V}_1/\hat{V}$ and hence, taking into account the expressions of the coefficients a and b (see (9.2)),

$$\delta a = a - a_1 = \kappa e^{-t}\left[\frac{C_{L_\alpha} + C_D}{2\mu} - \frac{1}{1 + \kappa e^{-t}}\right],$$

$$\delta b = b - b_1$$

$$= \kappa e^{-t}\,\frac{-\,C_{m_\alpha}\kappa e^{-t}/2 - C_{m_\alpha}\hat{V}_1 + C_{m_{\dot\alpha}}(C_{L_\alpha} + C_D)/4\mu + C_{m_{\dot\alpha}}/2(1 + \kappa e^{-t})}{i_B}.$$

In typical situations $\delta a(\hat{V})$ and $\delta b(\hat{V})$ are monotone increasing and positive functions; hence $\sqrt{b_1}|\delta a|_{\max} = (\kappa/2i_B)f_1$ and $|\delta b|_{\max} = (\kappa/2i_B)g_1$, with $f_1 = \hat{V}_1^2[(C_{L_\alpha} + C_D)/2\mu - 1][-C_{m_\alpha} + \hat{V}_1 C_{m_{\dot\alpha}}(C_{L_\alpha} + C_D)/2\mu]$ and $g_1 = -\hat{V}_1(\kappa + 2\hat{V}_1)\cdot C_{m_\alpha} + C_{m_{\dot\alpha}}[1 + (C_{L_\alpha} + C_D)]$. Now denoting $h_1 = [b_1 - (a_1 - c_1)^2/4]^{1/2}/(1 + \sqrt{b_1})$, condition (9.6) yields

$$\kappa \leqslant |\nu_1|\frac{i_B h_1}{f_1 + g_1}. \tag{9.8}$$

It should be noted that estimation (9.8) is only valid in cases where the restrictive conditions set above are fulfilled, among others if δa and δb are positive and increase with V.

When studying stability, inequality (9.6) supplies a sufficient condition, for example, in the form (9.7) or (9.8) permitting us to substitute system (9.4) with constant coefficients for system (9.3) with variable coefficients. The condition is given in the form of an upper bound imposed on the maximum variation of the parameters (characterized in the relations deduced above by the difference $\hat{V}_2^2 - \hat{V}_1^2$ or by the magnitude κ).

To check the procedure of the freezing of coefficients applied in the case of the "slow" variation of the system parameters, besides the linear system with variable coefficients (9.2), which will be written in the form

$$\frac{dx}{dt} = -\,a(u(t))x + y, \qquad \frac{dy}{dt} = -\,b(u(t))x - c(u(t))y, \tag{9.9}$$

the following set of linear systems with constant coefficients will also be considered.

$$\frac{d\xi}{dt} = -\,a(u(s))\xi + \eta, \qquad \frac{d\eta}{dt} = -\,b(u(s))\xi - c(u(s))\eta, \tag{9.10}$$

which correspond to all the fixed values of the parameter s on the semiaxis $s \geqslant 0$ or to all the corresponding (fixed) values of the function $u(s)$.† The nondimensional flight speed $\hat{V}$ was denoted by $u(s)$. It is assumed that, for any $s \geqslant 0$, the characteristic equation in λ

$$\begin{vmatrix} -a(u(s)) - \lambda & 1 \\ -b(u(s)) & -c(u(s)) - \lambda \end{vmatrix} = 0$$

has complex conjugate roots $\nu(u(s)) \pm i\omega(u(s))$, $i = \sqrt{-1}$, $\omega^2 = b - (a-c)^2/4 > 0$, $s \geqslant 0$, the real part $\nu = -(a+c)/2$ satisfying an inequality of the form

$$\nu(u(s)) \leqslant -\varepsilon_1 < 0 \tag{9.11}$$

for any positive s, ε_1 being constant. It is likewise assumed that for any s, $s \geqslant 0$, a, b, and c are positive.

The solution corresponding to the initial condition $\xi(0) = x_0$, $\eta(0) = y_0$ of a system whatever (9.10) may be written in the form

$$\xi = e^{\nu t}\left[x_0\left(\cos\omega t - \frac{\nu + a}{\omega}\sin\omega t\right) + y_0\frac{1}{\omega}\sin\omega t\right],$$

$$\eta = e^{\nu t}\left[-x_0\frac{b}{\omega}\sin\omega t + y_0\left(\cos\omega t + \frac{\nu + a}{\omega}\sin\omega t\right)\right].$$

Then, taking into account the obvious inequalities

$$\left|\cos\omega t \pm \frac{\nu + a}{\omega}\sin\omega t\right| \leqslant \sqrt{1 + \left(\frac{\nu + a}{\omega}\right)^2} = \frac{\sqrt{b}}{\omega}, \qquad \left|\frac{1}{\omega}\sin\omega t\right| \leqslant \frac{1}{\omega},$$

we can readily see that the following inequalities hold.

$$|\xi| \leqslant e^{\nu t}(|x_0| + |y_0|)\frac{1}{\omega}\max(\sqrt{b}, 1),$$

$$|\eta| \leqslant e^{\nu t}(|x_0| + |y_0|)\frac{1}{\omega}\max(b, \sqrt{b}),$$

† In contrast to the first-approximation system (9.4) representing the disturbed motion corresponding to a possible effective uniform basic motion, the linear systems with constant coefficients (9.10) obtained by fixing the coefficients of the nonautonomous linear system (9.9) constitute pure mathematical fictions in which, for instance, the parameters $\hat{V}$ and $d\hat{V}/dt$ (see (9.2)) are fixed (or "frozen") simultaneously at constant values different from zero, equal to various instantaneous values of these variables.

and further

$$|\xi| + |\eta| \leqslant k(u(s))e^{\nu(u(s))t}(|x_0| + |y_0|), \qquad s \geqslant 0$$

where $k = (1/\omega)[\max(\sqrt{b};\, 1) + \max(\sqrt{b};\, b)]$. Since ν satisfies relation (9.11), the solutions of all the systems with frozen coefficients (9.10) satisfy a relation of the form (8.8), and hence Hale's lemma, stated in Section 8.2.2, can be applied. For the case considered (8.9) becomes

$$\left|\frac{\partial a}{\partial u}\right| + \left|\frac{\partial b}{\partial u}\right| + \left|\frac{\partial c}{\partial u}\right| \leqslant \frac{\nu^2 + 2|\partial\nu/\partial u|\,|du/dt| \ln k(u(s))}{4[k(u(s)) - 1]}. \tag{9.12}$$

According to the lemma, inequality (9.12) must be satisfied, starting from a certain value T, for any fixed value of s, and hence of u.

Resuming the example above of decelerated flight at a flight speed varying according to the equation $u = u_1(1 + \kappa e^{-t})$ and using the shorthand notation $a = \alpha_1 u + (1/u)\, du/dt$, $b = \beta_1 u^2 - \beta_2 u^3 - \beta_3 u\, du/dt$, $c = \gamma_1 u^2$ (for the developed expressions of the positive constants α_1, β_1, β_2, β_3, and γ_1, see (9.2)), we have in (9.12): $|\partial a/\partial u| = |\alpha_1 - u_1/u^2|$, $|\partial b/\partial u| = |3\beta_2 u^2 - 2(\beta_1 + \beta_3)u + \beta_3 u_1|$, $|\partial c/\partial u| = 2\gamma_1 u$, $2|\partial\nu/\partial u| = |\partial(a + c)/\partial u| = |\alpha_1 - u_1/u^2 + 2\gamma_1 u|$, $|du/dt| = |u_1 - u| = u - u_1$; and for an accelerated flight with $u = u_0[1 + \kappa(1 - e^{-t})]$: $|\partial a/\partial u| = |\alpha_1 - (1 + \kappa)u_0/u^2|$, $|\partial b/\partial u| = |3\beta_2 u^2 - 2(\beta_1 + \beta_3)u + \beta_3 u_0(1 + \kappa)|$, $|\partial c/\partial u| = 2\gamma_1 u$, $2|\partial\nu/\partial u| = |\alpha_1 - (1 + \kappa)u_0/u^2 + 2\gamma_1 u|$, $|du/dt| = du/dt = (1 + \kappa)u_0 - u$.

9.1.3. Direct Approach Via the Liapunov Function Method

It will be shown that under common conditions, for statically stable aircraft, uniform asymptotic stability of motion about the lateral axis of the aircraft (rapid incidence adjustment) is also assured in the case of basic motion at variable flight speed, whatever the sense of this variation (sign of the acceleration).

In some earlier papers the problem was broached by using the method of the Liapunov function, some partial results being reached. Thus, in [1] it was shown that under certain hypotheses regarding the dependence of the coefficients of system (9.2) on undisturbed speed ($b'(u) \geqslant 0$), asymptotic stability is assured for accelerated basic motion. The same property is proved in [2] for decelerated flight if the function $u(t)$ satisfies the condition

$$\frac{d^2 \ln u}{dt^2} > -\frac{f}{u}, \tag{9.13}$$

f being a constant that is dependent on the system parameters. In both papers it was assumed that system (9.2) has bounded coefficients and that it preserves its structure; that is, the coefficients a, b, and c do not change their sign and do not vanish for $t \geqslant 0$, these conditions being, moreover, natural. However, the additional conditions $b'(u) \geqslant 0$ and (9.13) are restrictive.

Initially, some results of [3] will be shown. Namely, it will be proved, by choosing an adequate Liapunov function, that in the usual hypotheses, which are compatible with aircraft of current design, the unsteady basic motion represented by the zero solution of system (9.2) is uniformly asymptotically stable, whatever the variation sense of the flight speed. It is therefore proved that rapid incidence adjustment is assured whatever the basic motion of the aircraft in practice.

For this purpose, it will be assumed that the undisturbed flight speed $u(t)$ satisfies, along the entire semiaxis $t \geqslant 0$, the following conditions.

$$0 < u_1 \leqslant u(t) \leqslant u_2, \qquad u_1 = \text{const.}, u_2 = \text{const.} \tag{9.14}$$

$$\left|\frac{du}{dt}\right| < K, \qquad \left|\frac{d^2u}{dt^2}\right| < L, \tag{9.15a}$$

$$\lim_{t\to\infty}\left(\frac{du}{dt}\right) = 0. \tag{9.15b}$$

It is further assumed that the second derivative d^2u/dt^2 is uniformly continuous (or at least lower semicontinuous, uniformly along the semiaxis $t \geqslant 0$) wherefrom there results (see [4])

$$\lim_{t\to\infty}\left(\frac{d^2u}{dt^2}\right) = 0. \tag{9.16}$$

The inequalities (9.14) obviously amount to the existence of a maximum and a minimum speed, characteristic of any aircraft. Relations (9.15a) likewise express obvious physical realities, namely, that the values of the external forces along the tangent to the path and that of their variation in time are bounded. The conditions above are fulfilled, on the whole, if for example, the unsteady evolution, whatever its nature in the preceding stages, finally tends toward a steady-state motion.

The following hypotheses relating to the aerodynamic and design characteristics of the aircraft must be added to the ones regarding flight speed: the drag coefficient and the aerodynamic derivatives satisfy the relations

$$0 < C_D^{(1)} \leqslant C_D \leqslant C_D^{(2)}, \quad 0 < C_{L_\alpha}^{(1)} \leqslant C_{L_\alpha} \leqslant C_{L_\alpha}^{(2)}, \quad C_{m_\alpha}^{(1)} \leqslant C_{m_\alpha} \leqslant C_{m_\alpha}^{(2)} < 0,$$

$$C_{m_q}^{(1)} \leqslant C_{m_q} \leqslant C_{m_q}^{(2)} < 0, \quad C_{m_{\dot\alpha}}^{(1)} \leqslant C_{m_{\dot\alpha}} \leqslant C_{m_{\dot\alpha}}^{(2)} < 0, \tag{9.17}$$

and have time derivatives bounded in modulus, tending toward zero for $t \to \infty$ or, particularly, identically zero. In addition, the following inequalities hold.

$$C_{L_\alpha} + C_D > -\left(\frac{2\mu}{u^2}\right)\frac{du}{dt} + \varepsilon_1, \tag{9.18a}$$

$$-C_{m_\alpha} > -C_{m_{\dot\alpha}}\left[\frac{u(C_{L_\alpha} + C_D)}{2\mu} + \left(\frac{1}{u}\right)\frac{du}{dt}\right] + \varepsilon_2 \tag{9.18b}$$

where ε_1 and ε_2 are positive constants.

These relations are satisfied in current flight situations for all conventional types of aircraft that are statically stable for any usual régime. This statement will be illustrated hereafter by a typical numerical example.

The following conditions, imposed on the coefficients of system (9.2), result from the assumptions above.

I. For $t \geqslant 0$ the coefficients admit both lower and upper—negative—bounds:

$$0 < a_1 \leqslant a(t) \leqslant a_2,$$

$$0 < b_1 \leqslant b(t) \leqslant b_2,$$

$$0 < c_1 \leqslant c(t) \leqslant c_2.$$

II. The derivative $\dot b$ has zero as limit.

$$\lim_{t\to\infty} \dot b = 0.$$

Further, as consequences of relations (I) and (II) the following conditions result for the function $b(t)$.

III. For any positive quantity ε there is a function $T(\varepsilon)$ such that $t \geqslant T(\varepsilon)$ implies $|\dot b| < \varepsilon$.

IV. The derivative b is bounded in modulus, hence there is an $M > 0$ such that for an arbitrary positive t, $|\dot b| \leqslant M$.

The Liapunov function

$$\mathscr{V} = he^{-kt}(bx^2 + y^2) + x^2 + \left(\frac{1}{b}\right)y^2$$

will be used to prove the uniform asymptotic stability of the zero solution of system (9.2), which, for brevity, will be written in the form

$$\frac{dx}{dt} = -ax + y, \qquad \frac{dy}{dt} = -bx - cy. \tag{9.19}$$

It will be shown that there are an h and k, positive constants, for which this function satisfies the conditions required by Liapunov's theorem of uniform asymptotic stability. (See Section **3.4.3**, Theorem **2**.)

Indeed, in accordance with condition (I) referring to function $b(t)$, the relations $x^2 + (1/b_2)y^2 \leqslant he^{-kt}(bx^2 + y^2) + x^2 + (1/b)y^2 \leqslant (hb_2 + 1)x^2 + (h + (1/b_1))y^2$ stand, whatever the positive constants h and k are.

Further, h and k will be so determined that the derivative of function $\mathscr{V}$ with respect to t along the integral curves of system (9.19) will be negative definite; in other words, the following inequality should occur.

$$\left(\frac{d\mathscr{V}}{dt}\right)_{(9.19)} \equiv [-2a + he^{-kt}(\dot{b} - kb - 2ab)]\, x^2 + \left[-\frac{1}{b^2}(\dot{b} + 2bc) - he^{-kt}(2c + k)\right] y^2 \leqslant -\kappa_1 x^2 - \kappa_2 y^2, \tag{9.20}$$

κ_1 and κ_2 being positive constants.

Constant k is chosen equal to M/b_1. In accordance with relations I there immediately results

$$-2a + he^{-kt}(\dot{b} - kb - 2ab) \leqslant -2a_1 = -\kappa_1. \tag{9.21}$$

Let ε be a constant, $0 < \varepsilon < \min(2b_1^2 c_1/b_2;\, M)$ and let $T(\varepsilon)$ be such that $|\dot{b}| < \varepsilon$ for $t \geqslant T$. According to condition III, $T(\varepsilon)$ exists. Then, for $t \geqslant T$ we have

$$-\frac{1}{b^2}(\dot{b} + 2bc) - he^{-kt}(2c + k) \leqslant -\frac{2c_1}{b_2} + \frac{\varepsilon}{b_1^2} < 0, \tag{9.22}$$

the right-hand side being denoted by $-\kappa_2$. The constant h will be so determined that inequality (9.22) will also be verified for $0 \leqslant t < T$. Since, according to conditions I and IV, for $0 \leqslant t < T$

$$-\frac{1}{b^2}(\dot{b} + 2bc) - he^{-kt}(2c + k) \leqslant \frac{M}{b_1^2} - \frac{2c_1}{b_2} - he^{-kT}(2c_1 + k)$$

relation (9.22) is satisfied for $h \geqslant e^{kT}(M - \varepsilon)/[b_1^2(2c_1 + k)]$.

Then, based on relations (9.21) and (9.22), for

$$k = \frac{M}{b_1}, \qquad h = e^{kT}\frac{M - \varepsilon}{b_1^2(2c_1 + k)}$$

condition (9.20) is satisfied, κ_1 and κ_2 being $\kappa_1 = 2a_1$, $\kappa_2 = c_1/b_2 - \varepsilon/b_1^2$; hence the zero solution of system (9.19) is uniformly asymptotically stable. Since the hypotheses from (9.15) to (9.18) are generally valid in all cases of practical interest rapid incidence adjustment is therefore assured for variable-speed basic motions with practically the same likelihood as for the uniform basic motion.

Table 9.1

Gross weight during flight, considered constant	$W = 6900$ kg
Lifting surface area	$S = 21.7$ m^2
Reference chord length	$\bar{c} = 1.95$ m
Maximum flight speed	$V_{\max} = 296$ m/sec
Minimum flight speed	$V_{\min} = 78.75$ m/sec

The aircraft flies at a variable speed along a straight and level path at 2000 m altitude to which there corresponds, according to the data of standard atmosphere, an air density $\rho \approx 0.1027$ kg · sec^2/m^4. There result:

Relative density parameter	$\mu = 324$
Nondimensional moment of inertia in pitch (about axis Oy)	$i_B = 1186$
Unit of aerodynamic time related to the maximum speed	$[t] = 3.41 \times 10^3$ sec

Table 9.2

V (m/sec)	78.74 $= V_{\min}$	124	248	286 $= V_{\max}$
$u = V/V_{\max}$	0.275 $= u_1$	0.433	0.866	1.000 $= u_2$
C_D	0.100	0.027	0.020	0.073
C_{L_α}	4.56	4.56	5.12	4.30
C_{m_α}	−0.500	−0.488	−0.340	−0.746
C_{m_q}	−23.6	−23.6	−23.6	−23.6
$\max \dot{V}$ (9.18a)	$4.57g$	$11.5g$	$51g$	$57g$
$\max \dot{V}$ for $C_{m_{\dot{\alpha}}} = -11.8$ (9.18b)	$83g$	$127g$	$155g$	$484g$
$\max \dot{V}$ for $C_{m_{\dot{\alpha}}} = -4.3$ (9.18b)	$256g$	$396g$	$526g$	$1440g$

Finally, a typical numerical example will be given to illustrate the statements made with respect to the satisfaction of relations (9.18). An aircraft is considered, characterized by the design data in Table 9.1 [5, page 223]. Let it be assumed that the instantaneous aerodynamic characteristics in Table 9.2 correspond to the instantaneous speeds of 78.75 m/sec (minimum speed), 124 m/sec, 248 m/sec, and 286 m/sec (maximum speed). The longitudinal damping coefficient C_{m_q} was taken to be the same in the four instants of the flight, as was the value of the rotary derivative $C_{m_{\dot{\alpha}}}$, for which two alternatives have been considered. The last three rows of Table 9.2 represent the approximate maximum bound of the value of acceleration $\dot{V}$ expressed in multiples of g (acceleration of gravity), for which the (9.18) hypotheses remain valid. The results are conclusive. Inequality (9.18a) is satisfied in accelerated flight ($du/dt > 0$) whatever the value of acceleration, and for decelerated flight ($du/dt < 0$), if the absolute value of negative acceleration does not exceed 4.57 times the acceleration of gravity, for $V = V_{\min}$,† or up to 57 times the value of g, for $V = V_{\max}$. For inequality (9.18b) not to be satisfied, it is necessary, in accelerated flight, for the acceleration to exceed —in the alternative $C_{m_{\dot{\alpha}}} = -11.8$—from $83g$ (for $V = V_{\min}$) to $484g$ (for $V = V_{\max}$; see footnote), and if $C_{m_{\dot{\alpha}}} = 4.3$, from $256g$ (for $V = 0.433V_{\max}$) to 1440 times the acceleration of gravity (for $V = V_{\max}$; see footnote), these values being considerably higher than normal flight accelerations.

9.1.4. Comments

The hypotheses used above and particularly the relations (9.14) suggest some comments regarding the mathematical modeling of the variation of the basic motion parameters. To give consistency to the discussion, reference will be made to flight speed. The discussion will, however, be valid for any other parameter subjected—by the nature of the physical quantity it represents—to a similar constraint of bounds. Particularly, it will also concern the parameter ρ (air density at flight altitude). The effect of its variation on flight stability will be dealt with in Chapter 10.

The hypothesis expressed by the inequalities (9.14) reflects an actual situation and, as such, it must be taken into account when building the

† According to hypothesis (9.14), which, moreover, reflects the situation *de facto*, an actual acceleration at least equal to zero corresponds to the minimum flight speed and an actual acceleration at most equal to zero ($du/dt \leqslant 0$) corresponds to the maximum speed.

mathematical model of motion. In other words, the mathematical model must be consistent with the fact that aircraft speed can neither exceed a maximum value nor fall below a certain minimum. Maximum flight speed follows from design considerations (e.g., limited magnitude of thrust) or operational conditions (flight regulations or requirements), while the minimum is given by the stalling range of the lifting surface or that of the aerodynamic control surface. In the mathematical modeling of performance problems or even in estimating behavior in flight for a limited time interval, the fulfillment of condition (9.14) is assured without difficulty even by using some of the simplest mathematical models, such as uniform accelerated motion or uniform decelerated motion ($u = u_0 \pm jt$, u_0 and j being positive constants), the hyperbolic variation of speed ($u = u_0^2/(u_0 \pm j_0 t)$, u_0, j_0 being positive constants, for decelerated motion, eventually, $u = C/t$), the exponential variation ($u = u_0 e^{\pm t}$), and so on. Indeed, for certain bounded time intervals the variation of speed can be approximated by means of such functions. In those intervals, therefore, estimations of the solution of the system may be resorted to, with $u(t)$ having one of the foregoing forms, for example, by numerical integration and a posteriori ascertainment whether the duration required by the practical damping of the disturbance is or is not comprised in the interval in which the simplified model adopted for $u(t)$ is valid.

A difficulty occurs, however, because practical damping is required no matter what instant is considered as initial (no matter what the instant at which a disturbance appears); this makes it imperative to choose a function $u(t)$ compatible with the entire evolution of the basic motion, which obviously cannot be strictly accelerated or strictly decelerated (e.g., uniformly accelerated or decelerated, etc.) throughout. The property of stability, defined for arbitrarily large (unlimited) time intervals, is resorted to precisely in order to make evident an acceptable behavior, independent of the instant the disturbance appears. When operating with this mathematical notion, however, it is necessary for the mathematical model used to make sense for an unlimited time interval, too: the model should be compatible with the method.† From this standpoint

† Some examples of simple models of the variation of flight speed, acceptable from this standpoint and satisfying the conditions (9.14), are, for accelerated flight:

$$u = u_0[1 + \kappa(1 - e^{-t})] \quad \text{with} \quad u_0 \geqslant u_1, \quad \kappa \leqslant (u_2 - u_0)/u_0;$$

$$u = C_1/[1 + C_2/(t + C_3)] \quad \text{with} \quad C_1C_3/(C_2 + C_3) \geqslant u_1, \quad C_1 \leqslant u_2;$$

and for decelerated flight:

$$u = C(1 + \kappa e^{-t}) \quad \text{with} \quad C \geqslant u_1, \quad u_0 = C(1 + \kappa) \leqslant u_2;$$

$$u = C_1 + C_2/(t + C_3) \quad \text{with} \quad C_1 \geqslant u_1, \quad u_0 = C_1 + C_2/C_3 \leqslant u_2.$$

the models that allow or unlimited increase of the speed or the passing to negative speeds within a finite time interval (e.g., the model of uniform accelerated motion or that of uniform decelerated motion) are undoubtedly unacceptable. From the following there will result that a condition of the form $u(t) \geqslant 0$ is not sufficient either, even if the zero value appears as a limit for $t \to \infty$, and it is therefore not actually reached.

From the following it will stand out that if the norm of the matrix of system (9.19) is bounded, the inequality

$$u(t) \geqslant u_1 > 0, \qquad t \geqslant 0,$$

actually constitutes a necessary condition of stability.

The boundedness of the norm of the system matrix is a natural fidelity requirement of the model. For this reason the case of the vanishing of the function u for a finite t is excluded from the discussion hereafter since, in this case (see expression of coefficient a, system (9.2)), the coefficient a tends toward infinity for t finite.

Indeed, let $t = \tau$ be the instant where the function $u(t)$ vanishes. Then

$$\lim_{t\to\tau} a(t) = \lim_{t\to\tau} \frac{1}{u}\frac{du}{dt} = \lim_{t\to\tau} \frac{d}{dt} \log u$$

(by log is understood the natural logarithm) and since $u(\tau) = 0$, hence $\log u$ tends toward $-\infty$ in a finite time interval, the derivative $(d/dt) \log u$ cannot be bounded in this interval, whence $a \to \pm \infty$.

It will therefore be assumed that the function $u(t)$ satisfies the relations

$$0 < u(t) \leqslant u_2 < \infty, \tag{9.23}$$

$$\lim_{t\to\infty} u(t) = 0, \qquad \lim_{t\to\infty} \dot{u}(t) = 0, \tag{9.24}$$

and that the ratio $\dot{u}/u$ is bounded in modulus and admits a limit that is, obviously, negative, namely,

$$\left|\frac{\dot{u}}{u}\right| < R = \text{const.}, \qquad \lim_{t\to\infty} \frac{\dot{u}}{u} = -a_0 < 0. \tag{9.25}$$

Therefore, taking into account the expression of coefficients a, b, and c, we see that system (9.19) will tend toward

$$\frac{dx}{dt} = a_0 x + y, \qquad \frac{dy}{dt} = 0.$$

Since $a_0 > 0$, the characteristic equation of this system has a zero and a positive root.

Now system (9.19) is arranged in the form

$$\frac{dx}{dt} = a_0 x + y - (a + a_0)x, \qquad \frac{dy}{dt} = -bx - cy. \tag{9.26}$$

On basis of the assumptions (9.23) and (9.25) a, b, and c are bounded in modulus. If the respective upper bounds of the functions $|a(t) + a_0|$, $|b(t)|$, and $|c(t)|$ are sufficiently small, according to the theorem of instability by the first approximation (see Section 5.2) the zero solution of system (9.26) (hence of system (9.19)) will be unstable. Such a situation is possible based on relations (9.24) and (9.25), from which it results that for any $\varepsilon > 0$ there is $T(\varepsilon)$ such that $t \geqslant T$ implies $|b(t)| < \varepsilon$, $|c(t)| < \varepsilon$, $|a(t) + a_0| < \varepsilon$. Let ε be such that if the preceding inequalities hold, the instability of the trivial solution of system (9.26) results. Therefore, in order for this solution to be unstable it is sufficient for $t_0 \geqslant T(\varepsilon)$; in other words, for the disturbance to occur at an instant when the values of the functions $a + a_0$, b, and c, dependent on the undisturbed motion, are sufficiently small.

The aspect discussed regarding the modeling of the various physical quantities subjected to natural constraints will be illustrated by one more example. The system

$$\frac{dx}{dt} = y, \qquad \frac{dy}{dt} = -\beta_0 u^2(t)x - \gamma_0 u(t)y \tag{9.27}$$

is considered defined for $t \geqslant t_0 > 0$, where β_0 and γ_0 are positive constants and $u(t)$ is a monotone decreasing function in the definition range. It will be shown that the zero solution of this system is stable for

$$u(t) \geqslant u_1 > 0, \qquad t \geqslant t_0, \tag{9.28}$$

and uniformly stable for

$$0 < u_1 \leqslant u(t) \leqslant u_2 < \infty, \qquad t \geqslant t_0, \tag{9.29}$$

and may become unstable if $u(t)$ can, to any extent, draw closer to zero, namely, for

$$0 \leqslant u(t) < u_2 < \infty. \tag{9.30}$$

To prove the stability of the trivial solution in the case of the constraints (9.28) or (9.29), use will be made of the Liapunov function

$$\mathscr{V} = \beta_0 u^2 x^2 + y^2$$

positive definite for (9.28) since $\mathscr{V} \geqslant \beta_0 u_1^2 x^2 + y^2$, and satisfying, in addition, for (9.29) the inequality $\mathscr{V} \leqslant \beta_0 u_2^2 x^2 + y^2$. The derivative of this function with respect to t is, along the integral curves of system (9.27),

$$\left(\frac{d\mathscr{V}}{dt}\right)_{(9.27)} = 2u(\beta_0 \dot{u} x^2 - \gamma_0 y^2).$$

Function $u(t)$ being assumed monotone decreasing, $\dot{u} \leqslant 0$, the quadratic form on the right-hand side is negative semidefinite and hence the trivial solution of system (9.27) is stable for (9.28) and uniformly stable for (9.29) (see Section **3.4.3**, Theorem 1) whatever the positive constants β_0 and γ_0 are.

Now, let it be assumed that

$$\gamma_0 < 2\sqrt{\beta_0} \tag{9.31}$$

and let $u(t) = C/t$ with $C = (-\gamma_0 + 2\sqrt{5\beta_0 - \gamma_0^2})/4\beta_0 - \gamma_0^2$.

From relation (9.31), $C > 0$ and hence $u(t)$ satisfies (9.30) with $u_2 = C/t_0$, for $t \geqslant t_0 > 0$. It can readily be verified that for $t \geqslant 0$ the system

$$\frac{dx}{dt} = y, \qquad \frac{dy}{dt} = -\left(\frac{\beta_0 C^2}{t^2}\right)x - \left(\frac{\gamma_0 C}{t}\right)y,$$

obtained from (9.27), if we set $u = C/t$, admits the solution

$$\begin{aligned} x &= t^{\kappa} \cos \log t, \\ y &= t^{\kappa-1}(\cos \log t - \sin \log t) \end{aligned} \tag{9.32}$$

with $\kappa = \frac{1}{2}(1 - \gamma_0 C)$. Since on the strength of relation (9.31), we have $0 < \kappa < 1$,† the solution (9.32) is unstable with respect to the component x.‡ The appearance of the unstable solution is therefore linked to the nonfulfillment of the condition $u \geqslant u_1 > 0$. If $u(t)$ represents a physical mag-

† Indeed, $\gamma_0 C > 0$, hence $\kappa > 0$ and $\gamma_0(5\beta_0 - \gamma_0^2)^{1/2} = [\beta_0\gamma_0^2 - \gamma_0^2(\gamma_0^2 - 4\beta_0)]^{1/2} < [4\beta_0^4 - \gamma_0^2(\gamma_0^2 - 4\beta_0^2)]^{1/2} < 2\beta_0$ wherefrom $\gamma_0 C = [-\gamma_0^2 + 2\gamma_0(5\beta_0 - \gamma_0^2)^{1/2}]/(4\beta_0^2 - \gamma_0^2) < 1$ and hence $\kappa = (1 - \gamma_0 C)/2 < 1$.

‡ A system of the type (9.27), with $\beta_0 = 10/(3C)^2$, $\gamma_0 = 1/(3C)$ for $u = C/t$, hence $\kappa = 1/3$, was used in [6] as a counterexample to evidence the possible fallacy of the process of "freezing of coefficients." Indeed, the characteristic equations of the systems with constant coefficients, obtained from (9.27) by setting t at values higher or equal to $t_0 > 0$ ($s \geqslant t_0$), $\lambda^2 + (\gamma_0 C/s)\lambda + \beta_0 C/s^2 = 0$ are $\lambda_{1,2} = (C/2s)[-\gamma_0 \pm i(4\beta_0 - \gamma_0^2)^{1/2}]$, hence for $0 < \gamma_0 < 2\sqrt{\beta_0}$ complex with their real part negative. As shown, under the same conditions, the nonautonomous system (9.27) admits an unstable solution.

nitude, obviously admitting a lower bound differing from zero (e.g., undisturbed flight speed), the nonfulfillment of this condition amounts to the fact that the model used, in the given case $u = C/t$, does not reflect a property of the physical original essential for the problem of stability.

9.2. ESTIMATES OF VARYING SPEED EFFECT

9.2.1. Damping Velocity of Incidence Deviation

The damping velocity of the effect of a short-duration disturbance on the aircraft incidence will be estimated by applying to system (9.19) a technique outlined in Section 3.6. All the assumptions stated in the preceding section with regard to this system are assumed valid.

For a general guide the first step will be to use the Liapunov function $\mathscr{V} = he^{-kt}(bx^2 + y^2) + x^2 + (1/b)y^2$, introduced in Section 9.1. The information obtained in this way will then unable us to consider better-suited Liapunov functions for the various stages of the programmed motion, hence to obtain finer estimations. To estimate the damping velocity for finite time intervals, different Liapunov functions may be used, satisfying the condition required by the general theory only within the interval considered. This condition consists in the fact that the total time derivative of the Liapunov function $\mathscr{V}$ along the integral curves of the system be negative definite (for $\mathscr{V}$ positive definite), which ensures that, within the interval in which the condition is satisfied, the integral curve cuts the mobile ellipses $\mathscr{V}(t, x, y) = C$ in the plane (x, y) from the outside inward. In this manner the damping velocity of the variables x and y can be estimated by means of the variation in length of the semiaxes of this mobile ellipse.

As already seen, the function

$$\mathscr{V} = he^{-kt}(bx^2 + y^2) + x^2 + (1/b)y^2 \tag{9.33}$$

with h and k constants having the expressions given in the previous section, satisfies, for $t \geqslant 0$, the double inequality

$$x^2 + \frac{1}{b_2}y^2 \leqslant \mathscr{V} \leqslant (hb_2 + 1)x^2 + \left(h + \frac{1}{b_1}\right)y^2 \tag{9.34}$$

and its time derivative along the integral curves of (9.19), inequality (9.20), which is reproduced here for convenience.

$$\dot{\mathscr{V}} = \left(\frac{d\mathscr{V}}{dt}\right)_{(9.19)} \leqslant -\kappa_1 x^2 - \kappa_2 y^2$$

where $\kappa_1 = 2a_1$, $\kappa_2 = 2c_1/b_2 - \varepsilon/b_1^2$, with $0 < \varepsilon < \min[(2b_1c_1/b_2); M]$.

Further, let

$$\nu = \tfrac{1}{2} \min \left(\frac{\kappa_1}{hb_2 + 1} ; \frac{\kappa_2}{h + 1/b_1} \right). \tag{9.35}$$

Then, it can immediately be seen that the inequality

$$\dot{\mathscr{V}} \leqslant - 2\nu \mathscr{V}$$

occurs and hence, for $t_0 = 0$ and $\mathscr{V}(0, x(0), y(0)) = \mathscr{V}_0$

$$\mathscr{V} \leqslant \mathscr{V}_0 e^{-2\nu t}$$

or, further, taking (9.34) into account,

$$x^2 + \frac{1}{b_2} y^2 \leqslant \mathscr{V}_0 e^{-2\nu t}.$$

The figurative point of the integral curve in the plane of the dynamic variables x and y being, hence, constantly within the mobile ellipse

$$x^2 + \frac{1}{b_2} y^2 = \mathscr{V}_0 e^{-2\nu t},$$

the variable x will be dominated in modulus by the length of the corresponding semiaxis of this ellipse, that is, by the function $\mathscr{V}_0^{1/2} e^{-\nu t}$, and its damping velocity will be estimated by the quantity ν (9.35).

Let it be assumed that at the instant set as initial, $t_0 = 0$, there occurs a disturbance of the aircraft incidence

$$x(0) = x_0, \qquad y(0) = 0.$$

Then the induced deviation of the incidence for $t \geqslant 0$ will be estimated by means of relation

$$|x| \leqslant (hb_0 + 1)^{1/2} |x_0| e^{-\nu t} \tag{9.36}$$

where $b_0 = b(0)$.

It should be noted that this estimate, and particularly ν, depend on the choice of the positive constant ε, arbitrary in the limit of the interval $[0, \min((2b_1 c_1/b_2) ; M)]$. Hence ε will be chosen so that the right-hand side of inequality (9.36) results as small as possible. It seems a priori obvious that by dividing the total time interval considered into a number of subintervals, ε optimum will vary from one subinterval to the other. Thus, for the intervals $[t_i, t_{i+1}]$ with magnitudes of t_i and t_{i+1} below a certain limit, ε should be chosen such that $(hb_0 + 1)^{1/2}$, and therefore h should be as small as possible, while for the intervals where t_i and t_{i+1} exceed this limit, the preponderant factor becomes the exponential, ε having thus to be chosen such that ν results as large as possible. It can likewise be seen that too low a value of the constant ε implies both an excessively high value of h and a disadvantageous estimate of the damping velocity through a small value of κ_1 and therefore of ν. Values of ε close to $2b_1^2c_1/b_2$ are also to be avoided since they imply small values, close to zero, of κ_2, and therefore of ν.

The optimum value of the constant ε for each time interval considered is determined by effective computation. If, after this optimum value is reached, the products hb_1 and hb_2, hence implicitly also hb_0, are nevertheless large in comparison to 1, the estimation proves to be crude and the problem arises of finding, for this interval, a better-adapted Liapunov function.

For instance, an interval $[t_i, t_{i+1}]$ is considered, for which the basic motion is accelerated, u being such that in the whole interval

$$\dot{b} > m > 0.$$

As in the preceding cases, a, b, and c satisfy, in the interval considered, the relations

$$0 < a_1 \leqslant a \leqslant a_2,$$
$$0 < b_1 \leqslant b \leqslant b_2,$$
$$0 < c_1 \leqslant c \leqslant c_2. \tag{9.37}$$

As a Liapunov function, use is made of the quadratic form

$$\mathscr{V}^{(1)} = x^2 + \frac{1}{b} y^2$$

for which, obviously, occur the inequalities

$$x^2 + \frac{1}{b_2} y^2 \leqslant \mathscr{V}^{(1)} \leqslant x^2 + \frac{1}{b_1} y^2, \qquad t \geqslant 0. \tag{9.38}$$

The derivative of the function $\mathscr{V}^{(1)}$ along the integral curves of (9.19) being for $t_i \leqslant t \leqslant t_{i+1}$ dominated by a negative definite quadratic form with coefficients independent of t:

$$\left(\frac{d\mathscr{V}^{(1)}}{dt}\right)_{(9.19)} \equiv -2ax^2 - \left(\frac{\dot{b}}{b^2} + \frac{2c}{b}\right)y^2 \leqslant -2a_1x^2 - 2e_1y^2, \qquad t_i \leqslant t \leqslant t_{i+1}; \tag{9.39}$$

where $e_1 = m/2b_2^2 + c_1/b_2$, the integral curve will cut the mobile ellipse $\mathscr{V}^{(1)} = C$ in the plane of the variables x and y inward for $t \in [t_i, t_{i+1}]$.

From (9.38) and (9.39) there results, in the same manner as for the preceding case, $\mathscr{V}^{(1)} \leqslant \mathscr{V}_i^{(1)} \exp[-2\nu^{(1)}(t - t_i)]$, $t_i \leqslant t \leqslant t_{i+1}$, where $\nu^{(1)} = \min(a_1; b_1c_1/b_2 + mb_1/2b_2^2) \geqslant \min(a_1; b_1c_1/b_2)$ and hence the estimate of the induced deviation of incidence for the initial conditions $x(t_i) = x_0$, $y(t_i) = 0$

$$|x| \leqslant |x_0| \exp[-\nu^{(1)}(t - t_i)], \qquad t_i \leqslant t \leqslant t_{i+1}. \tag{9.40}$$

Now, let $[t_j, t_{j+1}]$ be an interval in which the basic motion is decelerated and u varies so that

$$\dot{b}(t) < -m < 0, \qquad t_j \leqslant t \leqslant t_{j+1}$$

and a, b, and c satisfy, in this interval, the inequalities (9.37). Use will be made of a Liapunov function of the form [2]

$$\mathscr{V}^{(2)} = bx^2 + y^2.$$

By repeating the reasoning above, we get $\mathscr{V}^{(2)} \leqslant \mathscr{V}_j^{(2)} \exp[-2\nu^{(2)}(t - t_j)]$, $t_j \leqslant t \leqslant t_{j+1}$, with $\nu^{(2)} = \min[(a_1b_1/b_2 + m/2b_2); c_1] \geqslant \min(a_1b_1/b_2; c_1)$, and for $x(t_j) = x_0$, $y(t_j) = 0$, the following estimate of the variable x is obtained.

$$|x| \leqslant |x_0| \left(\frac{b_0}{b_1}\right)^{1/2} \exp[-\nu^{(2)}(t - t_j)], \qquad t_j \leqslant t \leqslant t_{j+1}, \tag{9.41}$$

with $b_0 = b(t_j)$.

A certain indication of the fineness of the estimate above can be obtained by comparison with the case of the uniform basic motion. Hence, let $a = a_1 = a_2 = a_0$, $b = b_1 = b_2 = b_0$, $c = c_1 = c_2 = c_0$, $\dot{b} \equiv 0$. The constants M, m, k, and ε, and hence also h, will likewise be zero. Then, $\nu = \nu^{(1)} = \nu^{(2)} = \min(a_0, c_0) = \nu^{(0)}$ and the estimates (9.36), (9.40), and (9.41) come down to only one:

$$|x| \leqslant |x_0| \exp[-\nu^{(0)}(t - t_0)], \qquad t_0 \leqslant t \leqslant t_f;$$

$[t_0, t_f]$ being the interval in which the basic motion is uniform.

For a uniform basic motion the system describing the rapid longitudinal disturbed motion will be

$$\frac{dx}{dt} = -a_0 x + y, \qquad \frac{dy}{dt} = -b_0 x - c_0 y.$$

The amplitude of the component x of the solution of this system, corresponding to the initial condition $x(t_0) = x_0$, $y(t_0) = 0$, is in the interval $t_0 \leqslant t \leqslant t_f$, $|x_0| \exp[-\nu_0(t - t_0)]$, and hence

$$|x| \leqslant |x_0| \exp[-\nu_0(t - t_0)]$$

where the rate of damping is $\nu_0 = (a_0 + c_0)/2$. (As already shown in the previous chapter, the characteristic equation of the system has complex roots with negative real part: $\lambda_{1,2} = -\nu_0 \pm i\omega_0$, $\omega_0 = [4b_0 - (a_0 - c_0)^2]^{1/2}/2$.)

Generally the coefficients a and c are of a relatively close order of magnitude. In the numerical example of Section 9.1, by considering $u = 1$, $\dot{u} = 0$, $C_{m_{\dot{\alpha}}} = -4.3$, in the four cases $\frac{1}{2}(a_0 + c_0)$ varies approximately between 0.0135 and 0.0140, and $\min(a_0, c_0)$ between 0.007 and 0.008; hence the ratio between the exact and the estimated rate of damping is smaller than 2 (1.7–1.9).

9.2.2. Influence of Flight Speed Variation on Frequency of Phugoid Oscillations

System (9.19) was deduced as an auxiliary system for the partially controlled longitudinal motion, corresponding to the constraint of the variables $\Delta\hat{V}$ and $\Delta\theta$. Hence, implicitly, the controls achieving this constraint were assumed to be sufficiently efficient, in the sense pointed out in Section 5.7.

The assumption implies, as was shown, the existence of intervals in which the constrained variables maintain their signs or are monotone time functions, sufficiently long in comparison with the response time of the respective controls. In the case of linear systems with constant coefficients the interval in which the variables keep their sign or vary monotonously obviously coincides with the semiperiod of oscillation, the aspect of the solution being either monotone or periodic. In the case of systems with variable coefficients the behavior of the solution depends on the manner in which these coefficients vary in time. Therefore, to determine an efficiency criterion of the controls, a lower bound will be sought for the entire duration of the evolution of the intervals during which the variable considered maintains its sign, that is, of the interval between two successive zeros of the respective component of the solution of the system of the free disturbed motion equations. The criterion is then obtained by comparing this lower bound with the time of response to the actuation of controls.

To study the aspect shown, we will consider the system of equations of the pure phugoid motion, corresponding to the assumption that after the practical consumption of the rapid mode of disturbed motion (i.e., after rapid incidence adjustment), the deviation of the incidence remains negligible for a certain time interval including the duration of effecting the control and the aircraft response. In other words, it is admitted that in the interval of free flight subsequent to the consumption of the rapid mode, the deviation of incidence and that of pitching velocity may be neglected and hence system (2.15) is reduced to

$$\frac{d\varDelta\hat{V}}{d\hat{t}} = a_{11}(\hat{t})\,\varDelta\hat{V} + a_{13}(\hat{t})\,\varDelta\gamma, \qquad \frac{d\varDelta\theta}{d\hat{t}} = a_{31}(\hat{t})\,\varDelta\hat{V} + a_{33}(\hat{t})\,\varDelta\gamma \tag{9.42}$$

where the a_{jk} coefficients have the expressions

$$a_{11} = \frac{1}{\mu}(-uC_D - \tfrac{1}{2}u\mathbf{M}C_{D_\mathbf{M}} + C_{T_V}\cos(\alpha-\varphi)),$$

$$a_{13} = -\hat{g}\cos\gamma,$$

$$a_{31} = \frac{1}{\mu}\left(C_L + \tfrac{1}{2}\mathbf{M}C_{L_\mathbf{M}} + \frac{1}{u}C_{T_V}\sin(\alpha-\varphi) - \frac{\mu}{u}\frac{d\gamma}{dt}\right),$$

$$a_{33} = \frac{\hat{g}\sin\gamma}{u},$$

with $u = \hat{V} = V/V_{\max} \geqslant u_1 > 0$. Let the basic motion be horizontal; hence $\gamma = d\gamma/dt = 0$ and $a_{33} = 0$. Generally the derivative C_{T_V} is negligible.
With these simplifying assumptions the coefficients become

$$a_{11} \approx -\frac{u}{\mu}(C_D + \tfrac{1}{2}\mathbf{M}C_{D_\mathbf{M}}), \qquad a_{13} = -\hat{g} < 0,$$

$$a_{31} \approx \frac{1}{\mu}(C_L + \tfrac{1}{2}\mathbf{M}C_{L_\mathbf{M}}), \qquad a_{33} = 0.$$

By substituting

$$\varDelta\hat{V} = \xi\cos\eta, \qquad \varDelta\theta = \xi\sin\eta \tag{9.43}$$

we get, from system (9.42),

$$\begin{aligned} \dot{\xi}\cos\eta - \dot{\eta}\xi\sin\eta &= a_{11}\xi\cos\eta + a_{13}\xi\sin\eta, \\ \dot{\xi}\sin\eta + \dot{\eta}\xi\cos\eta &= a_{31}\xi\cos\eta + a_{33}\xi\sin\eta. \end{aligned} \tag{9.44}$$

If $\xi = 0$, the variables $\Delta\hat{V}$ and $\Delta\gamma$ with their time derivatives vanish simultaneously. This situation is to be excluded as improbable till the complete damping of the disturbance, hence assuming that ξ does not vanish on the half axis $\hat{t} \geqslant 0$. It is likewise assumed that function $\eta(t)$ is monotone on the half axis and let, for definiteness, $\dot{\eta} \geqslant 0$. Then the interval between two successive zeros of any of the variables $\Delta\hat{V}$ and $\Delta\gamma$ coincides with the interval between two successive vanishings of the function $\cos\eta$. Hence, if t_i is an instant in which $\cos\eta = 0$, the next instant $\hat{t}_i + T$ in which $\cos\eta$ vanishes is obtained from the equation in T

$$\eta(\hat{t}_i + T) - \eta(\hat{t}_i) = \pi. \tag{9.45}$$

Let $T_i = T(\hat{t}_i)$ be the lowest value of the unknown T satisfying this equation and let, as above, $[\hat{t}_0, \hat{t}_f]$ be the time interval within which the evolution considered takes place. In particular, $\hat{t}_0$ may be zero, and $\hat{t}_f$ may tend toward infinity. Then the minimum duration T between two successive vanishings of the variables $\Delta\hat{V}$ and $\Delta\theta$—components of any solution whatever of system (9.42)—will be

$$\tilde{T} \geqslant \operatorname*{glb}_{\hat{t}_0 \leqslant \hat{t} \leqslant \hat{t}_f} T(\hat{t})$$

(glb denotes the greatest lower bound).

The reason for the transformation (9.43) consists in the fact that the quantity $\tilde{T}$ can be estimated, in the hypothesis of the monotone variation of function $\eta(t)$, with no necessity for a numerical integration of the systems (9.42) or (9.44). Indeed, by multiplying the (9.44) equations by $-\sin\eta$ and $\cos\eta$, respectively, and by summation, we get

$$\dot{\eta} = a_{31}\cos^2\eta - (a_{11} - a_{33})\cos\eta\sin\eta - a_{13}\sin^2\eta. \tag{9.46}$$

In accordance with the assumption that $\dot{\eta} \geqslant 0$, the eigenvalues $\Lambda_1(\hat{t})$ and $\Lambda_2(\hat{t})$ of the matrix

$$\begin{pmatrix} a_{31} & -\dfrac{a_{11} - a_{33}}{2} \\ -\dfrac{a_{11} - a_{33}}{2} & -a_{13} \end{pmatrix}$$

of the quadratic form in $\cos\eta$ and $\sin\eta$ in the right-hand side of equation (9.46) are, for $t \geqslant 0$, positive or zero. Let $\Lambda_2(\hat{t}) \geqslant \Lambda_1(\hat{t}) \geqslant 0$. Then, the obvious relation

$$\dot{\eta} = a_{31}\cos^2\eta - (a_{11} - a_{33})\cos\eta\sin\eta - a_{13}\sin^2\eta \leqslant \Lambda_2(\hat{t}), \qquad \hat{t} \geqslant 0$$

yields

$$\int_{\hat{t}}^{\hat{t}+T} \Lambda_2(s)\,ds \geqslant \eta(\hat{t}+T) - \eta(\hat{t}) = \pi;$$

$\Lambda_2(\hat{t})$ being a known function, the estimate of $\tilde{T} = \text{glb}_{\hat{t}_0 \leqslant \hat{t} \leqslant \hat{t}_f} T(\hat{t})$ is obtained in the form

$$\tilde{T} \geqslant \operatorname*{glb}_{\hat{t}_0 \leqslant \hat{t} \leqslant \hat{t}_f} T_2(\hat{t}) \tag{9.47}$$

where $T_2(\hat{t})$ is the solution of the equation in T_2

$$\int_{\hat{t}}^{\hat{t}+T_2} \Lambda_2(s)\,ds = \pi.$$

This equation can be solved graphically. The graphic process is suitable not only for its simplicity or because an accurate estimation is not required, but particularly since the variation in time of the system parameters (functions $a_{ik}(\hat{t})$, and hence $\Lambda_2(\hat{t})$), are usually also furnished graphically. It should be noted that if $a_{33} = 0$ and $a_{31}(\hat{t})$ does not alter its sign in the domain considered, relation (9.47) also estimates the minimum interval of monotone variation of the deviation in the inclination of the flight path, since $\dot{\gamma} = a_{31}\xi\cos\eta$.

It remains to be seen in what conditions the hypothesis $\dot{\eta} \geqslant 0$ holds. Since

$$\dot{\eta} = a_{31}\cos^2\eta - (a_{11} - a_{33})\cos\eta\sin\eta - a_{13}\sin^2\eta \geqslant \Lambda_1(\hat{t}), \qquad \hat{t} \geqslant 0$$

the condition sought is obtained from the inequality

$$\Lambda_1(\hat{t}) = \tfrac{1}{2}\{a_{31} - a_{13} - [(a_{31} - a_{13})^2 + 4a_{13}a_{31} + (a_{11} - a_{33})^2]^{1/2}\} > 0$$

in the form of the set of inequalities

$$a_{31} - a_{13} \geqslant 0, \qquad 4a_{13}a_{31} \leqslant -(a_{11} - a_{33})^2. \tag{9.48}$$

For straight level basic motion, taking into account that $\hat{g} \approx C_L/2\mu$ (see Section 2.3), we see that these relations change into

$$\mathbf{M}C_{L_\mathbf{M}} \geqslant -3C_L, \qquad \mathbf{M}C_{L_\mathbf{M}} \geqslant -2C_L + u^2\frac{(C_D + \frac{1}{2}\mathbf{M}C_{D_\mathbf{M}})^2}{C_L}.$$

It can be seen that such inequalities might not hold in the range of the supercritical Mach numbers ($C_{L_M} < 0$). In this zone, however, for many aircraft, because of the very sudden and apparently disorderly variations of the aerodynamic coefficients, the validity of some basic hypotheses regarding the system parameters and the use of a linear first-approximation system is questionable. Therefore, any procedure based on such hypotheses might, within the transonic range, be fallacious.

REFERENCES

1. T. Hacker, "On the stability of an aeroplane in unsteady, straight level flight," *Rev. Méc. Appl. Acad. R. P. R.* **1**(1956), 23–36.
2. T. Hacker, "Quelques problèmes non classiques de la stabilité du vol," *Rev. Roum. Sci. Techn.-Méc. Appl.* **11**(1966), 363–381.
3. T. Hacker, "Le rétablissement rapide de l'incidence lors des vols à vitesse et altitude variables," *Rev. Roum. Sci. Techn.-Méc. Appl.* **12**(1967), 779–796.
4. I. Barbălat, "Systèmes d'équations différentielles d'oscillations non linéaires," *Rev. Math. Pures et Appl.* **4**(1959), 267–270.
5. I. V. Ostoslavskiy and G. S. Kalatchev, *Longitudinal Stability and Controlability of Aircraft* (in Russian). Oborongiz, Moscow, 1951.
6. A. R. Collar, "On the stability of accelerated motion: some thoughts on linear differential equations with variable coefficients," *Aeronaut. Quart.* **8**(1957), 309–330.

Chapter 10

STABILITY OF UNSTEADY FLIGHT: VARYING UNDISTURBED HEIGHT

10.1. INTRODUCTION: UNDISTURBED HEIGHT AS A TIME-DEPENDENT PARAMETER OF THE SYSTEM

The theory expounded in the preceding chapter is generally applicable in the case of nonstationarity of any kind; nevertheless, most of the conclusions were formulated considering the flying speed as the basic parameter characterizing the undisturbed régime. Assuming the flight program as defined in the form of the law of the variation of speed, it was admitted that the other parameters (e.g., undisturbed incidence) vary as a function of the speed variation, or (as regards, e.g., local air density and mass) are constant. In this chapter undisturbed flight height will be considered as a basic parameter. The importance of this parameter is obvious. To this end it is sufficient to recall that the aerodynamic forces and moments for set values of the kinematic parameters (speed, incidence, angular velocity, etc.) are proportional to air density at the flight altitude,† or that the Mach number for a given flight speed is inversely proportional to sound velocity at that altitude. The thrust developed by the engines, the efficiency of the airscrew, and so on are also dependent, to a noticeable extent, on altitude.

In contrast with the case of level flight considered in Section 6.4, for inclined flight at a steep slope (with a considerable vertical component of speed) or vertical flight—the only cases that warrant discussion in this chapter—generally, stability in relation to flight height is not significant; therefore it is not useful to include the deviation of the latter as a dynamic variable in the equations of disturbed motion. To make evident the effect of height variation, it is sufficient to consider it as a variable parameter of

† Thus, according to the data of standard atmosphere, atmospheric density, and with it the value of the aerodynamic forces and moments, drops to approximately one third at the altitude of 10 km and drops approximately 68 times at the altitude of 30 km, in comparison with its value on the ground.

the system of equations. In other words, it is adequate to study stability with respect to the conventional dynamic variables (flight speed, incidence, flight path angle, etc.) in the conditions where flight height varies. The corresponding model will therefore be a nonautonomous system of equations, with respect to the usual variables (the same as the ones considered so far), also (or only) dependent on the independent variable through the intermediary of the function $h(t)$, h being the flight height (e.g., $\dot{x} = X(t, h(t), x)$ or the linearized model $\dot{x} = A(t, h(t))x)$; as before, the column vector x represents the dynamic variables of disturbed motion, X a nonlinear vector function, and A a matrix dependent on the independent variable.

The adoption of such a model involves the assumption that the deviations in value of function $h(t)$ due to disturbance do not induce noticeable deviations of the external forces. In other words, if $\mathcal{F}$ represents the resultant of the external forces and $\mathcal{M}$ that of the external moments acting on the aircraft, the terms $(\partial\mathcal{F}/\partial h)\Delta h$ and $(\partial\mathcal{M}/\partial h)\Delta h$ of the variational system are negligible in comparison with $(\partial\mathcal{F}/\partial V)\Delta V$, $(\partial\mathcal{F}/\partial\alpha)\Delta\alpha, \ldots, (\partial\mathcal{M}/\partial V)\Delta V$, and so on. Such an assumption, debatable in the case of small or slow variations in height of the basic motion and certainly inadequate in the case of level flight, seems perfectly justified in the case of evolutions with important vertical components (steep dive or climb, etc.), on basis of the following simple heuristic reasoning: for evolutions (basic motions) such as steep dive or steep climb, the variation in flight height induced by a disturbance is small in comparison with the one corresponding to the basic motion itself, and consequently the effect of the former may be considered negligible in comparison with that of the latter. Such a neglect ceases to be justified when the variations become comparable in magnitude (in the case of small or low variation of undisturbed flight height) and it loses its sense for constant undisturbed altitude (for level flight). (In the latter case, however, the variation of the external forces induced by the disturbance of flight height must still be compared with the variation of the same forces caused by the deviations of the other régime parameters (V, α, etc.) in relation to their undisturbed values. As shown in Section 6.4, the relative magnitude of the variation induced by height disturbance is sufficient to justify the investigation of this effect by introducing an additional dynamic variable.)

10.2. EQUATIONS TO BE CONSIDERED

Only the longitudinal motion of the aircraft will be dealt with. In the hypothesis of a zero undisturbed angle of bank ($\phi_0 = 0$), the lateral asymmetrical motion is independent, in linear approximation, of the induced deviations of vertical displacements (of Δh, $\Delta \dot{h}$, etc.). Indeed, these deviations only have, in the hypotheses mentioned, components differing from zero along the Ox and Oz axes, but not along Oy, of the reference frame fixed in the aircraft (body axes) since the direction cosines, referred to the body axes, of a vertical vector (viz., $\sin \theta$, $-\sin \phi \cos \theta$, $\cos \phi \cos \theta$; see Appendix to Chapter 2, Table A.1) for $\theta = \theta_0 + \Delta\theta$ and $\phi = \Delta\phi$ become $\sin \theta_0 + O(\Delta\theta)$, $0 + O(\Delta\phi) + O(\Delta\theta^2)$, $\cos \theta_0 + O(\Delta\theta^2, \Delta\phi^2)$, respectively. The influence of height variation on the characteristics of the asymmetrical disturbed motion will only result in the change in value of some parameters of the system depending on ρ and $\mathbf{M}$.

The considering of ρ and a as time-dependent parameters makes imperative a change of the nondimensional system. It will thus be necessary to renounce relation $\hat{\rho}\hat{S} = 1$ as well as the use of some quantities dependent on air density or sound velocity as dimensional units, excepting the dimensional units for the aerodynamic forces and moments ($\rho V^2 S/2$ and $\rho V^2 S l/2$, respectively), which are deeply rooted in practice. In general, the use of variable dimensional units should be avoided since, on the one hand, it necessitates making additional assumptions of boundedness, noncancellation, and so on, and on the other hand, it often makes it inconvenient to interpret the results at the instant when the original dimensional physical quantities are resumed. For this reason, in the preceding chapter the quantity $\bar{c}/2V_0$, $V_0 = \text{const.}$, was used as unit of aerodynamic time instead of $\bar{c}/2V$, $V = V(t)$.

To bring system (2.3) with $M_c \equiv 0$ (since free flight is considered) to a suitable nondimensional form, in system (2.13)—also valid in this case—instead of the additional relations $\hat{\rho}\hat{S} = 1$ and $\hat{l} = 1$, we will choose

$$\hat{S} = 1, \qquad \hat{V}_0 = 1, \qquad \hat{\rho}_0 = 1, \tag{10.1}$$

which is tantamount to adopting the dimensional units for area, speed, and density as a system of basic units, instead of length, time, and mass.

From the first relation of (10.1) there results $[S] = S$ and $[l] = S^{1/2}$; from the second, $[V] = V_0$, $[l]/[t] = V_0$; therefore the unit of aerodynamic time will be $[t] = S^{1/2}/V_0$; and, finally, from the third relation $[\rho] = \rho_0$, $[\rho] = [m]/S^{3/2} = m/\mu_0 S^{3/2} = \rho_0$ and hence the relative density parameter will be $\mu_0 = m/\rho_0 S^{3/2}$, the nondimensional moment of inertia about the lateral axis $i_{B_0} = \mu_0 \hat{k}_y^2 = \mu_0 k_y^2/S$ and $\hat{g}_0 = g\sqrt{S}/V_0^2$.

With these transformations system (2.13) becomes

$$\frac{d\Delta\hat{V}}{d\hat{t}} = a_{11}\Delta\hat{V} + a_{12}\Delta\alpha + a_{13}\Delta\theta,$$

$$\frac{d\Delta\alpha}{d\hat{t}} = a_{21}\Delta\hat{V} + a_{22}\Delta\alpha + a_{23}\Delta\theta + \Delta\hat{q},$$

$$\frac{d\Delta\theta}{d\hat{t}} = \Delta\hat{q},$$

$$\frac{d\Delta\hat{q}}{d\hat{t}} = a_{41}\Delta\hat{V} + a_{42}\Delta\alpha + a_{43}\Delta\theta + a_{44}\Delta\hat{q}, \tag{10.2}$$

where

$$a_{11} = -\frac{1}{\mu_0}\left[\sigma u(C_D + \tfrac{1}{2}\mathbf{M}C_{D_\mathbf{M}}) - C_{T_V}\cos(\alpha - \varphi)\right]$$

$$a_{12} = -\frac{1}{\mu_0}(\tfrac{1}{2}\sigma u^2 C_{D_\alpha} + C_T\sin(\alpha - \varphi)) + \hat{g}_0\cos\gamma$$

$$a_{13} = -\hat{g}_0\cos\gamma$$

$$a_{21} = -\frac{1}{\mu_0}\left[\sigma(C_L + \tfrac{1}{2}\mathbf{M}C_{L_\mathbf{M}}) + \frac{\sin(\alpha - \varphi)}{u}C_{T_V}\right] + \frac{1}{u}\frac{d\gamma}{d\hat{t}}$$

$$a_{22} = -\frac{1}{\mu_0}(\tfrac{1}{2}\sigma u C_{L_\alpha} + C_T\cos(\alpha - \varphi)) + \frac{\hat{g}_0\sin\gamma}{u}$$

$$a_{23} = -\frac{\hat{g}_0\sin\gamma}{u}$$

$$a_{41} = \frac{1}{i_{B_0}}\left[\sigma u(C_m + \tfrac{1}{2}\mathbf{M}C_{m_\mathbf{M}}) - C_{T_V}\hat{z}_T + \sigma u^2 C_{m_{\dot{\alpha}}}a_{21}\right]$$

$$a_{42} = \frac{\sigma u^2}{i_{B_0}}(\tfrac{1}{2}C_{m_\alpha} + C_{m_{\dot{\alpha}}}a_{22})$$

$$a_{43} = \frac{\sigma u^2}{i_{B_0}}C_{m_{\dot{\alpha}}}a_{23}$$

$$a_{44} = \frac{\sigma u^2}{i_{B_0}}(C_{m_q} + C_{m_{\dot{\alpha}}}).$$

According to the equations of basic motion along the tangent and the normal to the flight path, $-\sigma u^2 C_L - 2\mu_0\hat{g}_0\cos\gamma + 2\mu_0 u\, d\gamma/d\hat{t}$ and $\sigma u^2 C_D + 2\mu_0\hat{g}_0\sin\gamma + 2\mu_0\, du/d\hat{t}$ can be substituted for $C_T\sin(\alpha - \varphi)$ and $C_T\cos(\alpha - \varphi)$, respectively.

10.3. RAPID INCIDENCE ADJUSTMENT

It can be admitted, in this case also, that the induced deviation of the variables describing the motion of the center of mass have a slow evolution allowing us to consider the two simplified schemes previously used. That is, (i) it can be assumed, for a first stage of the disturbed motion, that $\Delta\hat{V} = 0$ (conventional scheme) and the last three equations of (10.2) can be considered, with $\Delta\hat{V} \equiv 0$, or (ii) the model of partially controlled motions under flight speed and angular pitch constraint can be resorted to, in which case we take into consideration the second and fourth equations of (10.2), with $\Delta\hat{V} \equiv 0$ and $\Delta\theta \equiv 0$, viz., the system

$$\frac{d\Delta\alpha}{d\hat{t}} = a_{22}\Delta\alpha + \Delta\hat{q}, \quad \frac{d\Delta\hat{q}}{d\hat{t}} = a_{42}\Delta\alpha + a_{44}\Delta\hat{q}, \tag{10.3}$$

a_{jk} being as in the preceding section. As for the cases previously treated (e.g., in Chapter 9), investigation of the stability of system (10.3) is essential for both schemes. That is, it is required for the trivial solution of this system to be uniformly asymptotically stable. It can readily be shown, by extending the results in Section 9.1, that this requirement is fulfilled for all current cases.

Three additional time-dependent parameters (i.e., σ, a,† and γ) appear in (10.3) in comparison with the case treated in Section 9.1. Among them only the parameter $\sigma(t)$ requires additional restrictive hypotheses of the type used in the preceding chapter for $u(t)$. That is, the assumptions (9.14) and (9.15) are completed by the following, numbered (10.4), (10.5), and (10.6):

$$0 < \sigma_1 \leqslant \sigma(t) \leqslant \sigma_2 < \infty, \qquad t \geqslant 0, \tag{10.4}$$

with σ_1 and σ_2 constant. If ρ_0 denotes the air density at the altitude zero, $\sigma_2 = 1$. The constant σ_1 corresponds to the theoretical or practical ceiling altitude of the aircraft, being obviously different from zero;

$$\left|\frac{d\sigma}{d\hat{t}}\right| < N = \text{const.}, \qquad \left|\frac{d^2\sigma}{d\hat{t}^2}\right| < P = \text{const.}, \tag{10.5}$$

$$\lim_{\hat{t}\to\infty} \frac{d\sigma}{d\hat{t}} = 0, \qquad \lim_{\hat{t}\to\infty} \frac{d^2\sigma}{d\hat{t}^2} = 0. \tag{10.6}$$

† Sound velocity does not appear explicitly in (10.3) but only through the intermediary of the Mach number, which ceases to be proportional with the flight speed: $\mathbf{M} = V(t)/a(t)$.

The last relation implies uniform continuity along the entire semiaxis $\hat{t} \geqslant 0$ of $d^2\sigma/d\hat{t}^2$. The first and second derivatives of function σ with respect to $\hat{t}$ depend essentially on the vertical component of the flight speed (i.e., on the sinking or climbing velocity), and therefore in a way, relations (10.5) are implied by (9.15) and (9.16). Conditions (10.6) are fulfilled if the aircraft flight path finally tends toward level flight. Therefore, as for those relating to parameter u, the conditions (10.4), (10.5), and (10.6) are obviously real.

Further, relations (9.18) are replaced by

$$\begin{aligned} C_{L_\alpha} + C_D &> -\left(\frac{2\mu_0}{\sigma u^2}\right)\left(\frac{du}{d\hat{t}} + \hat{g}_0 \sin\gamma\right) + \varepsilon_1, \\ -C_{m_\alpha} &> -C_{m_{\dot{\alpha}}}\left[\frac{\sigma u(C_{L_\alpha} + C_D)}{2\mu_0} - \frac{\hat{g}_0 \sin\gamma}{u} - \frac{1}{u}\frac{du}{d\hat{t}}\right] + \varepsilon_2, \end{aligned} \tag{10.7}$$

ε_1 and ε_2 being positive constants, or by the following conditions, providing a safety margin.

$$\begin{aligned} C_{L_\alpha} + C_D &> \frac{2\mu_0(\hat{g}_0 + K)}{\sigma_1 u_1^2} + \varepsilon_1, \\ -C_{m_\alpha} &> -C_{m_{\dot{\alpha}}}\left[\frac{\sigma_2 u_2(C_{L_\alpha} + C_D)}{2\mu_0} + \frac{\hat{g}_0 + K}{u_1}\right] + \varepsilon_2, \end{aligned} \tag{10.8}$$

while (9.17) is accepted *tale quale.*†

On basis of these hypotheses, for the coefficients of system (10.3) the conditions I–IV of Section 9.1 are satisfied, where $a = -a_{22}$, $b = -a_{42}$, $c = -a_{44}$, $a_1 = |a_{22}^{(1)}|$, $a_2 = |a_{22}^{(2)}|$, $b_1 = |a_{42}^{(1)}|$, and so on, and therefore the zero solution of this system is uniformly asymptotically stable.

It can easily be verified that asymptotic stability with respect to motion about the center of mass also occurs in the case of aircraft whose mass variation is not negligible, function $m(t)$ or $\mu(t)$ satisfying conditions similar to $u(t)$ (conditions (9.15)–(9.17)) or $\sigma(t)$ (see (10.4)–(10.6)).

† These additional assumptions are also fulfilled in all current cases by statically stable aircraft. Again taking the numerical example in Section 9.1 with $\sigma_1 = 0.3$ (corresponding to tropo-pause altitude), $\sigma_2 = 1$, $u_1 = 0.275$, $u_2 = 1$, $\gamma = \pi/2$ (vertical climb), $K = 3g_0$ (maximum acceleration three times larger than gravitational acceleration), $C_{m_{\dot{\alpha}}} = -11.8$—hence, by using entirely unfavorable data and resorting to relations (10.8), which provide a safety margin, there results approximately 0.67 (the minimum effective value being 4.37) as the acceptable lower bound for $C_{L_\alpha} + C_D$ and for $-C_{m_\alpha}$ with $C_{L_\alpha} + C_D = 5.14$, the remainder as above, 0.118 (the minimum effective value being 0.34).

Therefore, the rapid restoration of the undisturbed value of incidence is not proper only in the case of the steady basic motion, but actually occurs whatever the basic motion of the conventional aircraft.

Finally, we make a remark similar to that made for the case of basic motion at variable speed, regarding the compatibility of the model of time variation of density, with the method used. Here, too, the method used requires a faithful modeling of the basic motion along the entire interval (i.e., a model describing the basic motion satisfactorily within any time interval taken into consideration). This observation concerns not so much the choice of functions $\sigma(h)$ and $a(h)$, which can be the ones corresponding to standard atmosphere, as of the variation of the kinematic variables in time. Indeed, adopting for σ the model of dependency on h, defined for $h \geqslant 0$ according to standard atmosphere, (or a more accurate law based on local measurements, etc.), that is (h is in meters)

$$\sigma = \begin{cases} \left(\dfrac{1-\alpha_T h}{T_0}\right)^{-1+1/\alpha_T R} \approx (1-0.00002255)^{5.26} \\ \quad \text{for} \quad 0 \leqslant h \leqslant h_{\text{tr}} \approx 11{,}000\,\text{m}, \\ (gRT_{\text{str}})^{-1} \exp\left(\dfrac{h-h_{\text{tr}}}{-RT_{\text{str}}}\right) \approx 0.0001605 \exp[-0.001575(h-h_{\text{tr}})] \\ \quad \text{for} \quad h_{\text{tr}} \leqslant h \leqslant h_2 \approx 25{,}000\,\text{m}, \end{cases}$$

the constraints of bounds (10.4) imposed on function $\sigma(t)$ are fulfilled if the range of altitude variation is correspondingly narrowed. This constraint of function $h(t)$ is, in turn, achieved by conditions imposed to the kinematic parameters V and γ. Thus, for example, the basic motion may contain a straight vertical dive or climb at constant speed, but it obviously cannot be limited to this evolution.

Accordingly, the mathematical model should reflect the fact that the kinematic parameters γ and V cannot be constant all the time. For example, when studying the stability of the evolution containing diving, the pull-out from the dive should necessarily be included.

However, after the stability of the entire evolution has been made to stand out, the quantitative estimation of the various characteristics of the disturbed motion may be made piecewise. This is even advisable in order to obtain greater accuracy. Thus, in the case of the example mentioned, the disturbed motion will be estimated separately for the portion of straight dive, the pull-out, and so on.

10.4. INFLUENCE OF UNDISTURBED FLIGHT HEIGHT VARIATION ON PHUGOID MOTION

The above conclusion regarding the rapid restoration of undisturbed incidence—no matter what the basic motion—justifies the adoption of the conventional assumption (also used in the case of flight speed variation, Section 9.2), namely, that after the rapid phase of the longitudinal disturbed motion is over, the undisturbed motion is not completely restored with regard to speed and flight path angle, but the induced incidence deviation remains negligible until the transitory process is damped. Therefore, by setting $\Delta\alpha = d\Delta\alpha/d\hat{t} = 0$, the following simplified linear system of the phugoid motion is separated from (10.2).

$$\frac{d\Delta\hat{V}}{d\hat{t}} = a_{11}\,\Delta\hat{V} + a_{13}\,\Delta\gamma, \qquad \frac{d\Delta\gamma}{d\hat{t}} = -\,a_{21}\,\Delta\hat{V} - a_{23}\,\Delta\gamma. \tag{10.9}$$

As in the case of level flight at variable speed considered in the preceding chapter, the first point of interest—and for the same reasons—is the frequency with which the phugoid variables (i.e., the induced deviations of speed and flight path angle) change signs. Before passing on to the estimation of this quantitative characteristic of disturbed motion, a remark will be made regarding phugoid stability in vertical flight. For vertical flight, system (10.9) reduces to

$$\frac{d\Delta\hat{V}}{d\hat{t}} = a_{11}\,\Delta\hat{V}, \qquad \frac{d\Delta\gamma}{d\hat{t}} = -\,a_{21}\,\Delta\hat{V} - a_{23}\,\Delta\gamma. \tag{10.10}$$

Since a_{11} is practically always negative, it is the sign of coefficient a_{23}, hence that of the function $\sin\gamma$, that is decisive as regards stability. Indeed, there results from the first equation, with the variable separated, that whatever the sense of the vertical motion (climb or dive), it is stable in the phugoid phase with respect to flight speed, since

$$\Delta\hat{V} = \Delta\hat{V}(\hat{t}_0)\exp\left(\int_{\hat{t}_0}^{\hat{t}} a_{11}(s)\,ds\right) \tag{10.11}$$

and if $\lim_{\hat{t}\to\infty}\int_{\hat{t}_0}^{\hat{t}} a_{11}(s)\,ds = -\,\infty$, stability is asymptotic.

Let $f(\hat{t}) = -\,a_{21}(\hat{t})\,\Delta\hat{V}(\hat{t})$. On the basis of stability with respect to $\Delta\hat{V}$, and because function $a_{21}(\hat{t})$ is bounded, for a sufficiently small initial value $\Delta\hat{V}(\hat{t}_0)$, $f(\hat{t})$ will be arbitrarily small. From the second equation of system (10.10),

$$\frac{d\Delta\gamma}{d\hat{t}} = - a_{23}\Delta\gamma + f(\hat{t}),$$

there results that the basic motion is stable or unstable with respect to $\Delta\gamma$ according to the sign of the coefficient $- a_{23} = \hat{g}_0 \sin\gamma/u$. Since $\hat{g}_0/u > 0$, the flight direction will be stable in the case of a vertical dive ($\gamma = - \pi/2$) and is unstable in the case of vertical climb ($\gamma = \pi/2$), no matter how flight speed, local air density, and sound velocity vary in time. For stability in a dive to be asymptotic it is necessary that $f(t) \to 0$ as $t \to \infty$, hence that stability with respect to flight speed also be asymptotic and hence, finally, that $\lim_{\hat{t}\to\infty} \int_{\hat{t}_0}^{\hat{t}} a_{11}(s)\, ds = - \infty$.

It can be seen that within the limits of validity of the assumptions that have led to system (10.9), the induced deviation of flight speed mainly varies monotonously (i.e., the variation in time of its dominant mode is monotone) and for sufficiently small $f(\hat{t})$ the conclusion also extends to the variation of the angular deviation of the flight path from the undisturbed vertical path. On the basis of the continuity of the solution with respect to the system parameters, the significant mode of the phugoid motion will also be non-oscillatory for evolutions sufficiently close to the vertical (with $|\pi/2 - \gamma|$ sufficiently small). If, in this case, the motion is divergent with respect to γ, the only requirement for good efficiency of the stabilizing controls in the sense shown in Section 5.7 is that the divergence not be too severe (i.e., that the rate of increase of deviation $\Delta\gamma$ not be too high). Divergence is estimated by means of the coefficient $- a_{23} = \hat{g}_0 \sin\gamma/u > 0$, and it will be the slower (and hence control efficiency the greater) the smaller the climbing path inclination and the larger the flight speed. It should be noted that below a certain value of the undisturbed flight path angle γ a change in the entire picture of the phugoid motion may (but need not necessarily) occur, rendering it oscillatory.

Consequently, for steep undisturbed flight paths the parameter σ does not (qualitatively) influence the aspect of the phugoid motion, while it only acts quantitatively in determining the rate of decay of deviation $\Delta\hat{V}$, function a_{11} under the integral sign in formula (10.11) being dependent on σ. As results from the expression of coefficient a_{11} (see (10.2)), high values of function $\sigma(t)$ as well as of undisturbed speed $u(t)$ favor the damping of induced deviation if the quantity $C_D + \mathbf{M}C_{D_\mathbf{M}}/2$ is positive. Usually this is indeed the case.

Consider now the oscillatory phugoid motion corresponding to values of the angle γ to which the foregoing considerations do not apply, but for which the (9.48) conditions written for system (10.9) are satisfied for suitable aerodynamic and régime characteristics. These conditions will be

$$a_{13} + a_{21} \leqslant 0, \tag{a}$$

$$4a_{13}a_{21} \geqslant (a_{11} + a_{23})^2. \tag{b}$$

That is, γ will be assumed sufficiently small and the aerodynamic and régime characteristics such that (i) $a_{21} < 0$, this amounting to $\sigma(C_L + \mathbf{M}C_{L_\mathbf{M}}/2) - (\mu_0/u)d\gamma/d\hat{t} + C_{T_V}\sin(\alpha - \varphi)/u > 0$ or, neglecting C_{T_V} for simplicity and considering the basic flight as rectilinear, $C_L + \mathbf{M}C_{L_\mathbf{M}}/2 > 0$; (ii) $-4a_{13}a_{21} < -(a_{11} + a_{23})^2$ (i.e., $4\mu_0\hat{g}_0\cos\gamma\,[\sigma(C_L + \mathbf{M}C_{L_\mathbf{M}}/2) - (\mu_0/u)\,d\gamma/d\hat{t} + C_{T_V}\sin(\alpha - \varphi)/u] > [\sigma u(C_D + \mathbf{M}C_{D_\mathbf{M}}/2) - C_{T_V}\cos(\alpha - \varphi) + \mu_0\hat{g}_0\sin\gamma/u]^2$) or, with $\gamma = \text{const.}$ and $C_{T_V} = 0$, $4\mu_0\hat{g}_0\cos\gamma(C_L + \mathbf{M}C_{L_\mathbf{M}}/2)\sigma > [\sigma u(C_D + \mathbf{M}C_{D_\mathbf{M}}/2) + \mu_0\hat{g}_0\sin\gamma/u]^2$.

In the particular case of inclined rectilinear flight at constant speed, for $\gamma = \text{const.}$, $u = 1$, $\mu_0\hat{g}_0\cos\gamma = \sigma C_L/2 + C_T\sin(\alpha - \varphi)$, and $\mu_0\hat{g}_0\sin\gamma = -\sigma C_D/2 + C_T\cos(\alpha - \varphi)$.

Assuming the thrust to be zero (e.g., rectilinear descent in gliding flight at constant speed), the last inequality becomes $2C_L(C_L + \mathbf{M}C_{L_\mathbf{M}}/2) \geqslant (C_D + \mathbf{M}C_{D_\mathbf{M}})^2/4$.

There results that, in the above assumption, the conditions (9.48) being satisfied, the results found in Section 9.2. for system (9.42) may be applied to system (10.9) corresponding to the case where σ and a are variable. Therefore, the minimum interval T' in which a component of any solution whatever of system (10.9) keeps its sign will be estimated by means of the relation

$$T' \geqslant \operatorname*{glb}_{\hat{t}_0 \leqslant \hat{t} \leqslant \hat{t}_f} T_2'(\hat{t})$$

where $\hat{t}_f - \hat{t}_0$ is the duration of the evolution considered, and $T_2'(\hat{t})$ is the solution of the equation in T_2'

$$\int_{\hat{t}}^{\hat{t}+T_2'} \Lambda_2'(s)\,ds = \pi, \tag{10.12}$$

$\Lambda_2'(t)$ being at every instant the largest eigenvalue of the matrix

$$\begin{pmatrix} -a_{21} & \dfrac{-(a_{11}-a_{23})}{2} \\ \dfrac{-(a_{11}-a_{23})}{2} & -a_{13} \end{pmatrix}$$

and being positive in sign. Function $\Lambda_2'(s)$ and, through it, the estimation T' sought will obviously be dependent on functions $\sigma(s)$ and $a(s)$.

The expression of Λ_2' as a function of the coefficients of system (10.9) is

$$\Lambda_2' = \frac{-a_{13}-a_{21}+[(a_{13}-a_{31})^2+(a_{11}+a_{23})^2]^{1/2}}{2}, \tag{10.13}$$

which, for $\sigma = \text{const.}$, $C_T = C_{T_V} = 0$, $u = 1$, becomes

$$\Lambda_2' = \sigma \frac{3C_L + \mathbf{M}C_{L_\mathbf{M}} + [(C_L + \mathbf{M}C_{L_\mathbf{M}})^2 + (C_D + \mathbf{M}C_{D_\mathbf{M}})^2]^{1/2}}{4\mu_0}.$$

The analysis of the influence of parameter σ on the basis of the latter relation is made more difficult by the fact that generally all the magnitudes of this expression except μ_0 depend on h, and therefore vary together with σ. Thus, in the hypotheses stated, the products σC_L and σC_D appear as constant, $\mathbf{M}$ in inverse proportion to sound velocity, and so on. To obtain qualitative information notwithstanding this difficulty, a range of flight régimes will be considered in which the products $\mathbf{M}C_{L_\mathbf{M}}$ and $\mathbf{M}C_{D_\mathbf{M}}$ may be assumed constant since they are less affected by the variation in flight height than σ. It is then seen that Λ_2' increases and hence, according to equation (10.12), T_2' decreases with the increase of air density. Hence, for the minimum interval T' in which the phugoid variables keep their sign there results a probable increase with the mean altitude of the evolution.

Quantitative—and more certain—results are obtained by graphically solving the equation (10.12) in T_2', where Λ_2' is given by the formula (10.13). (Usually, the quantities on which the coefficients a_{jk} (C_L, C_D, $C_{L_\mathbf{M}}$, $C_{D_\mathbf{M}}$) depend as functions of h and, through the relation $h = Vt \sin\gamma$, as functions of $\hat{t}$, are also obtained in the form of graphs.)

Chapter 11

EFFECT OF TIME LAG ON THE STABILITY OF CONTROLLED FLIGHT

11.1. THE MATHEMATICAL MODEL

11.1.1. Closed-Loop Systems with Time Lag

As was recollected in Chapter 1, the controlled flight can be modeled by means of certain closed-loop (or feedback) systems of the form (1.1) in which, from a mathematical point of view, the state variables and control variables play the same part of system variables. The variables $u(t)$ represent correction controls released by the appearance of a disturbance. The trivial solution of system (1.1) represents the basic motion, containing as a component the program of basic controls (represented by $u \equiv 0$).

Generally, the dependence of the control variables u on the state variables is supposed to be known in the form of an integral and differential, not necessarily linear, relation. In what follows, simplified cases will be considered in which the control variables will be approximated through known finite linear functions of the state variables. Thus, it will be possible to model the controlled motion by means of certain systems of the type (1.2):

$$\frac{dx}{dt} = X(t; x, u(x)), \tag{11.1}$$

or

$$\frac{dx}{dt} = A(t)x + B(t)u(x) \tag{11.2}$$

with X and A having the same meanings as in Chapter 1 and B another time-dependent $n \times n$ matrix.

Such schemes were used in Section 7.3 to study the stability of the hovering flight of certain VTOL aircraft with ideal rotation dampers and ideal static stabilizers. Unlike our earlier approach, we will take into account here that the time of response of the control is not zero, and we will be interested in the effect of this time lag on the efficiency of stabilization. Hence, as will further be specified, we will use systems of the form

$$\frac{dx}{dt} = X(t, x(t), u(x(t - \tau))) \tag{11.3}$$

or

$$\frac{dx}{dt} = A(t)x(t) + B(t - \tau)u(x(t - \tau)) \tag{11.4}$$

instead of the systems (11.1) or (11.2), respectively, the vector τ representing the delay of response for the various components u_i of the vector u.

11.1.2. Ideal Controls and Controls with Time Lag

In Section 7.3 it was admitted that the connection between the appearance of a disturbance and the counteraction of the controls is instantaneous. The control variables were each supposed to be proportional to one of the six state variables considered, hence connected through simple relationships of the form $u_i = - c_i x_i$, $c_i = \text{const.}$, whereby the time of response was considered negligible.

In what follows, the control models in which the time of response is taken as equal to zero will be named *ideal controls*. This chapter will consider the effect of the delay of response on the stability of controlled flight. There will implicitly result certain limits of applicability of the theory constructed with the aid of the models of ideal controls. To this end, we will use, beside the *ideal control* model, another model that should only add to the former another attribute of the actual control process, namely, that of its duration. Hence, as in the case of the ideal control, leaving the physical nature of this process aside, it will be assumed that the output is connected with the input through a known functional relationship, but the variation with time of the input is followed by that of the output lagging by an interval of time equal to the duration of the process of control.

Some of the qualitative aspects of the effect of the response delay on the stabilizing efficiency of the controls were considered in Section 5.7, but in a somewhat roundabout manner, by comparing the time lag value with certain quantitative characteristics of the disturbed motion in free flight (the period of oscillation, time required to halve or double the magnitude of a disturbance). In this chapter the question will be approached more directly by considering the effect of time lag under the specific conditions of the controlled flight. Differential equations with time lag will be used as the mathematical model.

II.1.3. Equation System with Time Lag

Ordinary differential equations with time lag or retarded argument are those ordinary differential equations in which: (i) the unknown functions and their time derivative appear with different values of the argument, and (ii) the highest-order derivative of this function occurs for but one value of the argument, which is not less than all arguments of the unknown functions and their derivatives occurring in the equations [1]. (Such equations were considered as early as 1728 by Johann Bernoulli, and are also met as a sporadic concern in Euler's work. They became the object of systematic concern only in about the last two decades, as a result of the requirements of certain branches of technical science, primarily control theory. Of recent years, certain synthetical expositions have also been published (see, e.g., [1–6]).)

In what follows, we will make use of systems of equations with retarded argument of the form

$$\frac{dx}{dt} = X(t, x(t), x(t - \tau)) \tag{11.5}$$

where the vector τ ($\tau \geqslant 0$) represents the duration of the various control loops of the system $\tau_1, \tau_2, \ldots$. The time lag τ will be assumed constant.

Let $\varphi_0(t)$ be a function given on a set E_{t_0} consisting of the values $t - \tau_j$, $t \geqslant t_0$, smaller than or equal to t_0; E_{t_0} will be named initial set and $\varphi_0(t)$ the initial function of the system (11.5). The function $\varphi_0(t)$ is assumed continuous over the interval $[t_0 - \tau, t_0]$. A solution of system (11.5) is defined for an initial function, just as a solution of a system of conventional differential equations is defined for an initial condition expressed by the value of the unknown function at the initial moment. The existence and unity of the solution of systems of the form (11.5) has been proved for the initial function having the stated property if the function $X(t, x, y)$ is continuous in all arguments and verifies a Lipschitz condition with respect to x for any y. In other words, under these conditions the system (11.5) admits of one and only one solution that coincides with the initial function $\varphi_0(t)$ over the interval $[t_0 - \tau, t_0]$.

II.I.4. Stability of Solution

The definitions in Chapters 3 and 4 are adapted to the case of systems with time lag, keeping in mind that the initial conditions are not given for a certain initial moment t_0, but for a certain interval $[t_0 - \tau, t_0]$ by means of the initial function $\varphi_0(t)$ with which the solution must coincide in this interval.

We denote by $x(t; t_0, \varphi_0)$ the solution of the system (11.5) defined for $t \geqslant t_0 - \tau$, which for $t_0 - \tau \leqslant t \leqslant t_0$ coincides with the initial function $\varphi_0(t)$. It is assumed that $X(t, 0, 0) \equiv 0$, hence system (11.5) admits the trivial solution.

DEFINITION 1. The trivial solution $x(t; t_0, 0) \equiv 0$ of system (11.5) is stable whenever for any $\varepsilon > 0$ there exists an $\eta(t_0, \varepsilon) > 0$ for which $|x(t; t_0, \varphi_0)| < \varepsilon$ for $t \geqslant t_0$ if $|\varphi_0(t)| < \eta$ for $t_0 - \tau \leqslant t \leqslant t_0$.

The zero solution is uniformly stable if η is independent of t_0.

DEFINITION 2. The solution $x = 0$ of system (11.5) is uniformly asymptotically stable if there exists a number δ and two functions $\eta(\varepsilon)$ and $T(\varepsilon)$ such that $|\varphi_0(t)| < \eta$ for $t_0 - \tau \leqslant t \leqslant t_0$ implies $|x(t; t_0, \varphi_0)| < \varepsilon$ for $t \geqslant t_0$, while the inequalities $|\varphi_0(t)| < \delta$ and $t \geqslant t_0 + T$ imply $|x(t; t_0, \varphi_0)| < \varepsilon$.

We will further consider the partially controlled system with time lag

$$\frac{dy}{dt} = Y(t, y(t), y(t - \tau), z), \quad \frac{dz}{dt} = Z(t, y(t), y(t - \tau), z) \tag{11.6}$$

where the k vector y represents the uncontrolled or free variables, and the m vector z the constrained ones (in the sense pointed out in Section 4.2) and with

$$Z(t, y(t), y(t - \tau), 0) \equiv 0.$$

Let φ_1 and φ_2 be two components, k- and m-dimensional, respectively, of the initial function φ_0 which is a $(k + m)$ vector. Hence the solution of the system (11.6) corresponding to the initial function $\varphi_0(t)$ satisfies, identically over the interval $[t_0 - \tau, t_0]$, the relations $y(t; t_0, \varphi_0) = \varphi_1(t)$, $z(t; t_0, \varphi_0) = \varphi_2(t)$.

DEFINITION 3. The trivial solution ($y = 0$, $z = 0$) of the system (11.6) is uniformly stable with respect to the variables y if for any $\varepsilon > 0$ there exist two functions $\eta(\varepsilon)$ and $\delta(\varepsilon)$ such that $|\varphi_0(t)| < \eta$ for $t_0 - \tau \leqslant t \leqslant t_0$ and $|z(t)| < \delta$ for $t \geqslant t_0$ imply $|y(t; t_0, \varphi_0)| < \varepsilon$ for $t \geqslant t_0$ and z_0 arbitrary within the indicated limit for $t \geqslant t_0$.

Definition 4. The zero solution of system (11.6) is uniformly asymptotically stable with respect to the free variables y if it is uniformly stable with respect to these variables in agreement with the previous definition and, in addition, $\lim_{t\to\infty} y(t; t_0, \varphi_0) = 0$.

Now, in addition to system (11.6) we will consider the corresponding auxiliary system

$$\frac{dy}{dt} = Y(t, y(t), y(t-\tau), 0). \tag{11.7}$$

The propositions stated in Section 4.2 in the case of partially controlled systems with time lag become:

1. If the trivial solution of the auxiliary system is uniformly asymptotically stable according to Definition 2, then the solution $y = 0$, $z = 0$ of system (11.6) is uniformly stable with respect to the free variables y (according to Definition 3).

2. If the trivial solution of the auxiliary system (11.7) is uniformly asymptotically stable according to Definition 2, and in addition $\lim_{t\to\infty} z(t; t_0, \varphi_0) = 0$, then the zero solution of system (11.6) is uniformly asymptotically stable with respect to the variables y (in the sense of Definition 4).

11.1.5. A Relation between the Stability of an Ideally Controlled Motion and That of a Motion Controlled with Time Lag

Time lags in control systems being generally small in comparison with the main quantitative characteristics of the disturbed motion (considered in Section 5.7), the stability of controlled flight was investigated for the most part under the assumption of their being disregarded. Therefore the question of substantiating this neglect and of establishing its validity limit appears to be natural. Thus, we must establish whether the conclusions arrived at in considering the solution of the system

$$\frac{dx}{dt} = \tilde{X}(t, x) \tag{11.8}$$

where $\tilde{X}(t, x) = X(t, x, x)$, particularly those concerning the stability, can be extended to the system (11.5) and up to what values of the time lag τ this can be done. The fact that the uniform asymptotic stability of the trivial solution of system (11.8) implies for small values of the time lag the uniform asymptotic stability of the zero solution of system (11.5) appears intuitively obvious for reasons of continuity. Moreover, it follows immediately by particularizing a theorem concerning the conservation of the stability of systems with time lag for small variations of the time lag [7, Section 3], or it can be inferred directly by the method of Liapunov's function [1, Chapter III]. Nothing remains but to establish by estimates the values of the time lags τ up to which the model of the ideal control, represented by the system (11.8), is still conclusive. In the next section certain estimates will be given for the case of the longitudinally controlled motion of a conventional machine, and in Section 11.4 we will give estimates for hovering flight of VTOL aircrafts.

11.2. ARTIFICIAL STABILIZATION OF LONGITUDINAL MOTION ALLOWING FOR SMALL TIME-LAGS; VALIDITY LIMITS OF THE IDEAL CONTROL MODEL; ESTIMATES

11.2.1. Mathematical Aids: Lemmata Concerning Differential Inequalities with Time Lag

In the estimates to be presented in this section for establishing the validity limit of neglecting the time lag, we will have recourse to the following lemma, the proof of which we owe to Halanay [6, Section 4.5].

LEMMA. *Let $f(t)$ be a positive function and α, β, γ, δ positive constants. If the inequality*

$$\frac{df}{dt} \leqslant -\alpha f(t) + \beta \operatorname*{lub}_{t-\tau \leqslant s \leqslant t} f(s) + \gamma \exp[-\delta(t - t_0)]$$

holds for $t \geqslant t_0$ and $\alpha > \beta$, then there exist $\mu > 0$ and $f_0 > 0$ such that

$$f(t) \leqslant f_0 \exp[-\mu(t - t_0)]$$

for $t \geqslant t_0$, whatsoever γ and δ might be.

Similarly, it can be shown that if

$$f(t) \leqslant \beta \int_{t_0}^{t} \exp[-\alpha(t - \sigma)] \operatorname*{lub}_{\sigma-\tau \leqslant s \leqslant \sigma} f(s)\, d\sigma + \gamma \exp[-\alpha(t - t_0)]$$

and $\alpha > \beta$ there exist $\mu > 0$ and $f_0 > 0$ for which

$$f(t) \leqslant f_0 \exp[-\mu(t - t_0)]$$

for $t \geqslant t_0$ and γ arbitrary.

11.2.2. Removal of Phugoid Divergence Through Ideal Controls in the Case of a Steady Basic Motion

It is assumed that in the case of locked controls the straight level uniform flight of the considered aircraft is unstable in the phugoid phase. Considering system (6.1) with $a_{23} = a_{43} = 0$ to be the mathematical model of the disturbed motion, we will assume, therefore, that the solution corresponding to an arbitrary initial disturbance contains a divergent aperiodic mode, in other words, that the free term a_4 of the characteristic equation (6.2) is slightly negative;

$$a_4 = -a_{13}(a_{22}a_{41} - a_{21}a_{42}) < 0.$$

We seek to remove this inherent instability by using an automatic control reacting to a deviation of the pitch angle. We assume, namely, that this control induces a longitudinal correction control moment proportional to the magnitude of the angular attitude deviation, admitting for the present that the response is instantaneous ($\tau = 0$). Then, denoting the nondimensional time by t, the linear system of disturbed motion becomes

$$\frac{d\Delta\hat{V}}{dt} = a_{11}\,\Delta\hat{V} + a_{12}\,\Delta\alpha + a_{13}\,\Delta\theta,$$

$$\frac{d\Delta\alpha}{dt} = a_{21}\,\Delta\hat{V} + a_{22}\,\Delta\alpha + \Delta\hat{q},$$

$$\frac{d\Delta\theta}{dt} = \Delta\hat{q},$$

$$\frac{d\Delta\hat{q}}{dt} = a_{41}\,\Delta\hat{V} + a_{42}\,\Delta\alpha - c_\theta\,\Delta\theta + a_{44}\,\Delta\hat{q}, \tag{11.9}$$

where t denotes nondimensional time, a_{jk} having the same expressions as in system (6.1) with $\gamma = 0$, while c_θ is a positive constant characterizing the control. The system admits the following characteristic equation.

$$\lambda^4 + a_1\lambda^3 + (a_2 + c_\theta)\lambda^2 + [a_3 - c_\theta(a_{11} + a_{22})]\lambda + a_4 + c_\theta(a_{11}a_{22} - a_{12}a_{21}) = 0 \tag{11.10}$$

where a_i, $i = 1, 2, 3, 4$, represent the coefficients of the characteristic equation (6.2) corresponding to the fixed-control condition. As the coefficients a_2 and a_3 are, in typical cases, positive and the sum $a_{11} + a_{22}$ negative, it can be seen that using the correction term with $c_\theta > 0$ does not alter the sign of the coefficients of λ^2 and λ.

For the free term with $a_4 < 0$ to become positive, a condition called for by the Routh–Hurwitz criterion, it is requisite that

$$c_\theta > -\frac{a_4}{a_{11}a_{22} - a_{12}a_{21}}. \tag{11.11}$$

The order of magnitude of the coefficient c_θ of the correction term is that of the group of small coefficients of equation (6.2) and, at any rate, it is much lower than that of the coefficient a_2. Consequently, as follows from equation (11.10), a value of the coefficient c_θ large enough to ensure a positive free term, hence negative real roots, will not essentially affect the large roots of the equation. In what follows we will admit that both equation (6.2) and (11.10) have a pair of complex roots with negative real parts of relatively large value and two real roots of contrary signs in the case of equation (6.2), both negative for (11.10).

Inequality (11.11) is a sufficient condition of uniform asymptotic stability for the case of an ideal control and, consequently (in accordance with what has been shown in Section 11.1), for the case of a small time of response, also. Further, we will estimate the magnitude of the response time for which condition (11.11) remains valid.

11.2.3. Estimation of the Validity Limit of Condition (11.11)†

If the assumption is maintained that the longitudinal correction control moment is proportional to the deviation of angular pitch, but account is taken of the time of response, the additional term representing the effect of the control in the system of equations will correspond, not to the instant t for which the related variables and their derivatives specifying the motion of the aircraft are written, but to a previous instant $t - \tau$, τ representing the time lag of the control. For simplicity, the time lag τ is considered constant. Then the system of equations with time lag of the disturbed longitudinal motion with automatic attitude control will be written as follows.

$$\frac{d\Delta\hat{V}}{dt}(t) = a_{11}\Delta\hat{V}(t) + a_{12}\Delta\alpha(t) + a_{13}\Delta\theta(t),$$

$$\frac{d\Delta\alpha}{dt}(t) = a_{21}\Delta\hat{V}(t) + a_{22}\Delta\alpha(t) + \Delta\hat{q}(t),$$

† See [8, Sections 6–8].

$$\frac{d\Delta\theta}{dt}(t) = \Delta\hat{q}(t),$$

$$\frac{d\Delta\hat{q}}{dt}(t) = a_{41}\Delta\hat{V}(t) + a_{42}\Delta\alpha(t) - c_\theta\Delta\theta(t-\tau) + a_{44}\Delta\hat{q}(t). \qquad (11.12)$$

To make system (11.9), corresponding to the ideal control, stand out in this system, we note

$$Q = c_\theta(\Delta\theta(t) - \Delta\theta(t-\tau))$$

and (10.2.4) becomes

$$\frac{d\Delta\hat{V}}{dt} = a_{11}\Delta\hat{V} + a_{12}\Delta\alpha + a_{13}\Delta\theta,$$

$$\frac{d\Delta\alpha}{dt} = a_{21}\Delta\hat{V} + a_{22}\Delta\alpha + \Delta\hat{q},$$

$$\frac{d\Delta\theta}{dt} = \Delta\hat{q},$$

$$\frac{d\Delta\hat{q}}{dt} = a_{41}\Delta\hat{V} + a_{42}\Delta\alpha - c_\theta\Delta\theta + a_{44}\Delta\hat{q} + Q. \qquad (11.13)$$

The last system differs from (11.9) only by the term Q, added to the right-hand side of the last equation. The admissible limit of the time lag value will be estimated through the agency of function Q.

Two estimation techniques will be used, as in Section 3.6, one based on Lagrange's method of the variation of parameters and the other on the method of the Liapunov function. Estimations will be obtained depending on system parameters.

11.2.4. Estimation Based on the Method of the Variation of Parameters

For the first estimation technique we will write the solution of system (11.9) (it will be recalled that equation (11.10) was assumed to have one pair of complex-conjugate roots and two real negative roots, the orders of magnitude of whose values are lower than the modulus of the complex roots)

$$\Delta\hat{V} = C_1\gamma_1(\lambda_1)e^{\lambda_1 t} + C_2\gamma_1(\lambda_2)e^{\lambda_2 t} + e^{\nu t}[C_3(\gamma_1'(\nu,\omega)\cos\omega t$$
$$- \gamma_1''(\nu,\omega)\sin\omega t) + C_4(\gamma_1'(\nu,\omega)\sin\omega t + \gamma_1''(\nu,\omega)\cos\omega t)],$$

$$\begin{aligned}
\Delta\alpha &= C_1\gamma_2(\lambda_1)e^{\lambda_1 t} + C_2\gamma_2(\lambda_2)e^{\lambda_2 t} + e^{\nu t}[C_3(\gamma_2'(\nu, \omega)\cos\omega t \\
&\quad - \gamma_2''(\nu, \omega)\sin\omega t) + C_4(\gamma_2'(\nu, \omega)\sin\omega t + \gamma_2''(\nu, \omega)\cos\omega t)], \\
\Delta\theta &= C_1\gamma_3(\lambda_1)e^{\lambda_1 t} + C_2\gamma_3(\lambda_2)e^{\lambda_2 t} + e^{\nu t}[C_3(\gamma_3'(\nu, \omega)\cos\omega t \\
&\quad - \gamma_3''(\nu, \omega)\sin\omega t) + C_4(\gamma_3'(\nu, \omega)\sin\omega t + \gamma_3''(\nu, \omega)\cos\omega t)], \\
\Delta\hat{q} &= C_1\gamma_4(\lambda_1)e^{\lambda_1 t} + C_2\gamma_4(\lambda_2)e^{\lambda_2 t} + e^{\nu t}[C_3(\gamma_4'(\nu, \omega)\cos\omega t \\
&\quad - \gamma_4''(\nu, \omega)\sin\omega t) + C_4(\gamma_4'(\nu, \omega)\sin\omega t + \gamma_4''(\nu, \omega)\cos t)]. \qquad (11.14)
\end{aligned}$$

Integration constants C_1, C_2, C_3, and C_4 are determined from the initial conditions, and $\gamma_i(\lambda_j)$ from the system

$$\begin{aligned}
(a_{11} - \lambda_j)\gamma_1 + a_{12}\gamma_2 + a_{13}\gamma_3 &= 0, \\
a_{21}\gamma_1 + (a_{22} - \lambda_j)\gamma_2 + \gamma_4 &= 0, \\
-\lambda_j\gamma_3 + \gamma_4 &= 0, \\
a_{41}\gamma_1 + a_{42}\gamma_2 - c_\theta\gamma_3 + (a_{44} - \lambda_j)\gamma_4 &= 0. \qquad (11.15)
\end{aligned}$$

λ_j being the roots of the characteristic equation of system (11.9) cancels the determinant of system (11.15), hence three out of four unknown quantities γ_i are determined from three equations of this system, for instance, from the first three, depending on the fourth unknown quantity taken arbitrarily. For instance, let $\gamma_3 = 1$. Then

$$\gamma_1(\lambda_j) = \frac{(a_{12} + a_{13})\lambda_j - a_{12}a_{22}}{\lambda_j^2 - (a_{11} + a_{22})\lambda_j + a_{11}a_{22} - a_{12}a_{21}},$$

$$\gamma_2(\lambda_j) = \frac{\lambda_j^2 - a_{11}\lambda_j + a_{13}a_{21}}{\lambda_j^2 - (a_{11} + a_{22})\lambda_j + a_{11}a_{22} - a_{12}a_{21}},$$

$$\gamma_3(\lambda_j) = 1, \qquad \gamma_4(\lambda_j) = \lambda_j.$$

The roots of equation (11.10) being complex conjugate, $\lambda_{3,4} = \nu \pm i\omega$, the corresponding coefficients γ_i will generally be complex conjugate, also, for example $\gamma_i(\lambda_{3,4}) = \gamma_i'(\nu, \omega) \pm i\gamma_i''(\nu, \omega)$. Obviously $\gamma_3' = 1$, $\gamma_3'' = 0$.

As in Section 3.6, to obtain the estimates we seek, (11.12) is converted into a system of integral equations

$$\Delta\hat{V} = g_1(t) + \int_0^t k_1(t, s)Q(s)\,ds,$$

$$\Delta\alpha = g_2(t) + \int_0^t k_2(t, s)Q(s)\, ds,$$

$$\Delta\theta = g_3(t) + \int_0^t k_3(t, s)Q(s)\, ds,$$

$$\Delta\hat{q} = g_4(t) + \int_0^t k_4(t, s)Q(s)\, ds, \qquad (11.16)$$

with $g_i(t)$, $k_i(t, s)$ known functions obtained by simple transformations.

The function $Q = c_\theta(\Delta\theta(t) - \Delta\theta(t - \tau))$ is small as τ is small. This dependence will be made conspicuous by writing

$$Q = \tau c_\theta \frac{d\Delta\theta}{dt}(t - \tau_m) = \tau c_\theta \Delta\hat{q}(t - \tau_m)$$

with $0 \leqslant \tau_m \leqslant \tau$.

Now the last equation in (11.13) will be transcribed in the following form.

$$\frac{d\Delta\hat{q}}{dt} = a_{44}\Delta\hat{q}(t) + \tau c_\theta \Delta\hat{q}(t - \tau_m) + a_{41}\Delta\hat{V} + a_{42}\Delta\alpha - c_\theta\Delta\theta.$$

Multiplying both sides with $\Delta\hat{q}(t)$ and taking account of (11.16) from this equation we obtain

$$\Delta\hat{q}(t)\frac{d\Delta\hat{q}}{dt}(t) = a_{44}\Delta\hat{q}^2(t) + \tau c_\theta \Delta\hat{q}(t - \tau_m)$$
$$+ \Delta\hat{q}(t)g(t) + \Delta\hat{q}(t)\int_0^t k(t, s)Q(s)\, ds,$$

where

$$g(t) = a_{41}g_1(t) + a_{42}g_2(t) - c_\theta g_3(t),$$
$$k(t, s) = a_{41}k_1(t, s) + a_{42}k_2(t, s) - c_\theta k_3(t, s).$$

Let $\delta = \min(|\lambda_1|, |\lambda_2|, |\nu|)$. Equation (11.10) has been assumed Hurwitzian, hence $\lambda_1 < 0$, $\lambda_2 < 0$, $\nu < 0$. There exists then $g_0 > 0$ such that $|g(t)| \leqslant g_0 e^{-\delta t}$ and K such that $\int_0^t |k(t, s)|\, ds < K$. Denoting $f = |\Delta\hat{q}|$, from the last equation we obtain

$$\Delta\hat{q}\frac{d\Delta\hat{q}}{dt} = \frac{1}{2}\frac{d\Delta\hat{q}^2}{dt} = \frac{1}{2}\frac{df^2}{dt}$$
$$\leqslant a_{44}f^2 + \tau c_\theta \operatorname*{lub}_{t-\tau\leqslant s\leqslant t} f(s) + f|g(t)|$$
$$+ f\int_0^t |k(t,s)|\tau c_\theta \operatorname*{lub}_{t-\tau\leqslant\sigma\leqslant t} f(\sigma)\,ds$$
$$\leqslant f[a_{44}f + \tau c_\theta \operatorname*{lub}_{t-\tau\leqslant s\leqslant t} f(s)(1+K) + g_0e^{-\lambda t}]$$

whence

$$\frac{df}{dt} \leqslant a_{44}f + \tau c_\theta(1+K) \operatorname*{lub}_{t-\tau\leqslant s\leqslant t} f(s) + g_0e^{-\delta t}.$$

From this inequality there follows, according to the lemma stated at the beginning of this section, that there exists $\mu > 0$; namely, $\mu = \delta$, and $f_0 > 0$ such that

$$f(t) = |\Delta\hat{q}(t)| \leqslant f_0e^{-\mu t} = f_0e^{-\delta t} \tag{11.17}$$

for $t \geqslant 0$, therefore, system (11.13) is exponentially stable with respect to $\Delta\hat{q}$ if

$$-a_{44} > \tau c_\theta(1+K)$$

or

$$\tau < \frac{-a_{44}}{c_\theta(1+K)}\,. \tag{11.18}$$

It can readily be shown that in this case the system is exponentially stable with respect to the other variables as well. Indeed, taking into account (11.17), from (11.16) we have

$$|\Delta\hat{V}| \leqslant A_{11}e^{\lambda_1 t} + A_{12}e^{\lambda_2 t} + A_{13}e^{\nu t} + A_{14}e^{-\delta t},$$
$$|\Delta\alpha| \leqslant A_{21}e^{\lambda_1 t} + A_{22}e^{\lambda_2 t} + A_{23}e^{\nu t} + A_{24}e^{-\delta t},$$
$$|\Delta\theta| \leqslant A_{31}e^{\lambda_1 t} + A_{32}e^{\lambda_2 t} + A_{33}e^{\nu t} + A_{34}e^{-\delta t}, \tag{11.19}$$

with A_{ik} known functions of the system parameters and of the time lag τ.

Consequently, if the magnitude of the time lag τ satisfies the inequality (11.18), the trivial solution of system (11.13) is uniformly asymptotically stable. Thus (11.18) is the estimate looked for. The numerator of the second term represents a measure of the aerodynamic damping: $a_{44} = (C_{m_q} + C_{m_{\dot{\alpha}}})/2i_B$. The admissible time lag will be so much the greater as this damping, and therefore the incidence adjustment, is more rapid. The aerodynamic damping constitutes the main component of the dynamic stability margin. The presence in the denominator of the gearing ratio is also natural: the stronger the requisite artificial stabilization (greater c_θ), the smaller is the admissible time lag for realizing this stabilization (for $c_\theta = 0$, hence for an aircraft having an inherent asymptotic stability, the admissible τ is infinite).

The constant K can readily be estimated depending on the parameters of the system. Nevertheless, its dependence on these functions is too complicated to permit a discussion for the general case. It appears obviously necessary to use certain numerical computations for each case investigated.

Using the same technique, another estimation can be obtained by considering the last equation of system (11.16). As above, let $\delta = \min(|\lambda_1|, |\lambda_2|, |\nu|)$. Since $\lambda_1 < 0$, $\lambda_2 < 0$, $\nu < 0$, there exists $\gamma_0 > 0$ such that

$$|g_4(t)| \leqslant \gamma_0 e^{-\delta t}$$

and β_0 such that

$$|k_4(t, s)| \leqslant \beta_0 \exp[-\delta(t-s)];$$

β_0 is, for instance, equal to $|\lambda_1|\Gamma_1 + |\lambda_2|\Gamma_2 + 2(\nu^2 + \omega^2)^{1/2}\Gamma_3$, with Γ_1, Γ_2, and Γ_3 known positive constants, depending on the system parameters. The function $Q(s)$ can be estimated as follows.

$$|Q(s)| = c_\theta \tau f(s - \tau_m) \leqslant c_\theta \tau \operatorname*{lub}_{s-\tau \leqslant \sigma \leqslant s} f(\sigma)$$

with $f(t) = |\Delta \hat{q}(t)|$ and $0 \leqslant \tau_m < \tau$, as above. Then from (11.16) we obtain

$$f(t) \leqslant \beta \int_0^t \exp[-\delta(t-s)] \operatorname*{lub}_{s-\tau \leqslant \sigma \leqslant s} f(\sigma)\, ds + \gamma_0 e^{-\delta t}$$

with $\beta = \beta_0 c_\theta \tau$, and therefore according to the lemma stated in this section there exists $f_0 > 0$ and $\mu > 0$ such that the inequality (11.17) holds for $t \geqslant 0$, if $\beta < \gamma$, or

$$\tau < \frac{\delta}{c_\theta \beta_0} = \frac{\min(|\lambda_1|, |\lambda_2|, |\nu|)}{c_\theta[|\lambda_1|\Gamma_1 + |\lambda_2|\Gamma_2 + 2(\nu^2+\omega^2)^{1/2}\Gamma_3]}. \tag{11.20}$$

As already stated, the exponential damping of component $\Delta \hat{q}$ of the solution of system (11.13), according to inequality (11.17), implies the exponential damping, according to relations (11.19), of the other components, namely, of $\Delta\hat{V}$, $\Delta\alpha$, and $\Delta\theta$, and therefore (11.20) represents a sufficient condition of uniform asymptotic stability.

The numerator specifies, as in relation (11.18), the dynamic stability margin of the ideal control system, but in a less advantageous form, δ representing the value of the minimum rate of damping. As was to be expected, the coefficient c_θ measuring the control gearing appears in the denominator here, too: the weaker the control requisite for stabilization, the greater the admissible time lag. Instead of the constant K, β_0 appears. In the last instance a comparison of the two estimates is possible by computations only, retaining obviously that which gives a greater admissible τ.

II.2.5. An Estimate Based on the Method of the Liapunov Function

We denote $\Delta\hat{V} = x_1$, $\Delta\alpha = x_2$, $\Delta\theta = x_3$, $\Delta\hat{q} = x_4$. It is attempted to determine a quadratic form

$$\mathscr{V} = \sum_{j,k=1}^{4} \alpha_{jk} x_j x_k$$

with $\alpha_{jk} = \alpha_{kj}$, whose time derivative along the integral curves of system (11.12) be identically equal to a given negative definite quadratic form $-\mathscr{U}$. Let, for instance,

$$-\mathscr{U} = -2(x_1^2 + x_2^2 + x_3^2 + x_4^2).$$

The characteristic equation (11.10) being assumed Hurwitzian, the function $\mathscr{V}$, uniquely determined, will be positive definite, according to Liapunov's well-known theorem (see [9, Section 20]).

For a sufficiently small addition Q, the derivative of function $\mathscr{V}$ thus determined along the integral curves of system (11.13) will be negative definite. The admissible value of the time lag τ will result from an estimate of the addition Q for which $(d\mathscr{V}/dt)_{(11.13)} < 0$.

Let $x_i(t)$, $i = 1, 2, 3, 4$, be an arbitrary solution of system (11.13) and denote $\mathscr{V}(x_1(t), x_2(t), x_3(t), x_4(t)) = f^2(t)$. Then

$$\frac{1}{2}\left(\frac{d\mathscr{V}}{dt}\right)_{(11.13)} = f\frac{df}{dt} = -\sum_{1}^{4} x_i^2 + Q\sum_{i} \alpha_{i4} x_i. \qquad (11.21)$$

Let further $\alpha_4 = \max_i |\alpha_{i4}|$ and Λ_M and Λ_m, respectively, be the greatest and least eigenvalues of the matrix of quadratic form $\mathscr{V}(x_1, x_2, x_3, x_4)$. Then, considering the obvious relations $|x_4| \leqslant \sum_i |x_i| \leqslant 2(\sum_i x_i^2)^{1/2}$ and $f^2(t)/\Lambda_M \leqslant$

$\sum_i x_i^2 \leqslant f^2(t)/\Lambda_m$ and the expression of Q given above, it will readily be seen that on the basis of relation (11.21) the following inequalities will hold successively.

$$f\frac{df}{dt} \leqslant -\sum_i x_i^2 + \tau\alpha_4 c_\theta \sum_i |x_i| \operatorname*{lub}_{t-\tau\leqslant\sigma\leqslant t} |x_4(\sigma)|$$

$$\leqslant -\sum_i x_i^2 + 4\tau\alpha_4 c_\theta (\sum_i x_i^2)^{1/2} \operatorname*{lub}_{t-\tau\leqslant\sigma\leqslant t} (\sum_i x_i^2(\sigma))^{1/2}$$

$$\leqslant -\frac{f^2(t)}{\Lambda_M} - 4\tau\alpha_4 c_\theta \frac{f(t)}{\Lambda_m} \operatorname*{lub}_{t-\tau\leqslant\sigma\leqslant t} f(\sigma)$$

and hence

$$\frac{df}{dt} \leqslant -\frac{1}{\Lambda_M} f(t) - \frac{4\alpha_4 c_\theta}{\Lambda_m} \tau \operatorname*{lub}_{t-\tau\leqslant\sigma\leqslant t} f(\sigma).$$

On the basis of the lemma stated at the beginning of this section,

$$f(t) \leqslant f_0 e^{-\mu t}$$

with f_0 and μ positive constants, hence the trivial solution of system (11.13) is uniformly asymptotically stable if the time lag τ satisfies the inequality

$$\tau < \frac{\Lambda_m}{\Lambda_M} \frac{1}{4\alpha_4 c_\theta}.$$

This estimating technique also applies, obviously, to nonautonomous systems with time lag, if we succeed in finding a convenient Liapunov function for the ideal system ($\tau = 0$). The procedure will be described in outline in what follows.

11.2.6. The Case of Unsteady Basic Motion

Let (11.9) with a_{jk} known functions of t and $c_\theta =$ const., be the linear system of the longitudinal disturbed motion with ideal control that detects the deviation of longitudinal attitude, corresponding to a given unsteady basic motion, and (11.12) or (11.13) with $a_{jk} = a_{jk}(t)$, and $Q = c_\theta(\Delta\theta(t) - \Delta\theta(t-\tau))$ be the related linear system with time lag. (Henceforth by (11.9), (11.12), and (11.13) we will denote only the systems corresponding to an unsteady basic motion, hence with a_{jk} known functions of t.)

It is assumed that the trivial solution of the system corresponding to the ideal control is uniformly asymptotically stable, and that a convenient Liapunov function $\mathscr{V}(t; x_1, x_2, x_3, x_4)$ has been found that proves this property.

In other words, there exist Λ_m and Λ_M, positive constants such that

$$\Lambda_m \sum_i x_i^2 \leqslant \mathscr{V}(t, x_1, x_2, x_3, x_4) \leqslant \Lambda_M \sum_i x_i^2 \tag{11.22†}$$

and its derivative with respect to t along the integral curves of system (11.9) is negative definite, hence there exists a positive constant M such that

$$\left(\frac{d\mathscr{V}}{dt}\right)_{(11.9)} \leqslant -M \sum_i x_i^2. \tag{11.23}$$

It is assumed that the partial derivative of function $\mathscr{V}$ with respect to x_4 satisfies the inequality

$$\left|\frac{\partial \mathscr{V}}{\partial x_4}\right| \leqslant N \sum_i |x_i|. \tag{11.24}$$

The addition Q and, implicitly, the magnitude of the time lag τ, will be estimated so that the derivative of function $\mathscr{V}$ along the integral curves of system (11.13) should remain negative definite.

We have

$$\left(\frac{d\mathscr{V}}{dt}\right)_{(11.13)} = \left(\frac{d\mathscr{V}}{dt}\right)_{(11.9)} + \frac{\partial \mathscr{V}}{\partial x_4} Q = \left(\frac{d\mathscr{V}}{dt}\right)_{(11.9)} + \frac{\partial \mathscr{V}}{\partial x_4} c_\theta \tau x_4(t - \tau_m)$$

with $0 \leqslant \tau_m \leqslant \tau$ and hence

$$\begin{aligned}\left(\frac{d\mathscr{V}}{dt}\right)_{(11.13)} &\leqslant \left(\frac{d\mathscr{V}}{dt}\right)_{(11.9)} + \left|\frac{\partial \mathscr{V}}{\partial x_4}\right| c_\theta \tau \operatorname*{lub}_{t-\tau\leqslant\sigma\leqslant t} |x_4(\sigma)| \\ &\leqslant \left(\frac{d\mathscr{V}}{dt}\right)_{(11.9)} + \left|\frac{\partial \mathscr{V}}{\partial x_4}\right| c_\theta \tau \operatorname*{lub}_{t-\tau\leqslant\sigma\leqslant t} \sum_i |x_i(\sigma)|\end{aligned}$$

or, taking account of (11.23) and (11.24),

$$\begin{aligned}\left(\frac{d\mathscr{V}}{dt}\right)_{(11.13)} &\leqslant -M \sum_i x_i^2 + \tau c_\theta N \sum_i |x_i| \operatorname*{lub}_{t-\tau\leqslant\sigma\leqslant t} \sum_i |x_i(\sigma)| \\ &\leqslant -M \sum_i x_i^2 + \tau 4 c_\theta N \left(\sum_i x_i^2\right)^{1/2} \operatorname*{lub}_{t-\tau\leqslant\sigma\leqslant t} \left(\sum_i x_i^2(\sigma)\right)^{1/2}.\end{aligned}$$

Denoting $\mathscr{V}(t; x_1(t), x_2(t), x_3(t), x_4(t)) = f^2(t)$, where $x_i(t)$, $i = 1, 2, 3, 4$, represents a solution of system (11.13) and taking account of (11.22), we obtain, finally

† If, for instance, $\mathscr{V}$ is a quadratic form, $\mathscr{V} = \sum_{j,k} \alpha_{jk}(t) x_j x_k$ with $\alpha_{jk} = \alpha_{kj}$, Λ_m and Λ_M represent the lower bound of the least, and the upper bound of the greatest, eigenvalues for $t \geqslant 0$, of the matrix of the quadratic form $\mathscr{V}$, respectively.

$$\frac{df}{dt} \leqslant -\frac{M}{2\Lambda_M} + \frac{2Nc_\theta}{\Lambda_m}\tau \underset{t-\tau\leqslant\sigma\leqslant t}{\text{lub}} f(\sigma).$$

From this inequality it follows, according to the known lemma, that the trivial solution of the nonautonomous system (11.13) is uniformly asymptotically stable if system (11.9) corresponding to the ideal control is uniformly asymptotically stable and the magnitude of time lag τ does not exceed the value provided by the relation

$$\tau < \left(\frac{\Lambda_m}{\Lambda_M}\right)\left(\frac{M}{4Nc_\theta}\right). \tag{11.25}$$

Here, a measure of the stability margin, that depends on the method (on the function $\mathscr{V}$) is provided by the factor M, the number $M/2$ being an estimate of the minimum damping velocity of the modes of the solution of the system without time lag.

11.3. ARTIFICIAL STABILIZATION OF LONGITUDINAL MOTION IN THE CASE OF ARBITRARY TIME LAG

11.3.1. Mathematical Aids

For the mathematical modeling of the disturbed motion with time lag in the case of steady and quasi-steady basic motions, we will have recourse to homogeneous linear systems with constant coefficients of the form

$$\frac{dx}{dt} = Ax + Bx(t-\tau) \tag{11.26}$$

with A and B constant matrices. Seeking the solution in a manner similar to that of the case $\tau = 0$, in the form $x = \alpha e^{\lambda t}$ with $\alpha = (\alpha_1, \alpha_2, \ldots, \alpha_n)'$ (the prime denotes transpose), we obtain a system of linear and homogeneous algebraic equations in α_i. This system will admit roots different from zero if λ satisfies the transcendental equation

$$\det(A + e^{-\tau\lambda}B - \lambda I) = 0, \tag{11.27}$$

which is referred to as characteristic by analogy. Here I denotes the identity matrix. Equation (11.27) admits an infinity of roots, each having a corresponding solution of system (11.26). It was but natural to seek an extension of the analogy with linear systems having constant coefficients without time lag also as regards the necessary and sufficient condition for asymptotic stability expressed by means of the real parts of the roots of the characteristic equation. Indeed, although the result was not obvious, it was proved, the same as for the case of zero time lag, that fulfillment of the inequalities

$$\operatorname{Re} \lambda_i < 0, \tag{11.28}$$

this time in an infinite number, λ_i representing all roots of the characteristic equation (11.27), is the necessary and sufficient condition of asymptotic stability for the system with time lag (11.26) (see [1, Chapter 3, Section 2]). The respective theorems of stability by first approximation, warranting the use of the linear model, were also proved. (It appears that the first results belong to Bellman [10] and were subsequently extended by other authors, such as Wright [11] and Krasovskiy [4].)

To investigate the fulfillment of conditions (11.28), criteria analogous to those of the Routh–Hurwitz type were sought that would permit us to draw certain conclusions valid for the whole infinite set of roots λ_i by means of a finite number of relations (see [12]). Likewise, criteria derived from the theorem of the variation of the argument as, for instance, the Nyquist criterion (giving additional indications, too; e.g., the number of roots with real positive parts, in case they exist) apply the same as in the case of $\tau = 0$. Since only the region of stability in the space of parameters has to be determined, a simpler procedure will be used in what follows.

Considering the quantities λ_i as functions of the system parameters, inequalities (11.28) define the region of uniform asymptotic stability in the hyperspace of these parameters. (Unlike the case $\tau = 0$, where the condition $\operatorname{Re} \lambda_i$ represents a finite number of inequalities, (11.28) refers to the whole infinity of roots of the transcendental equation (11.27).) The boundary of the region will obviously be formed by the totality of the points of the hyperspace in which at least one characteristic root λ_i vanishes. Denoting by $F(\lambda; p_1, p_2, \ldots, p_m)$ the characteristic quasi-polynomial depending on the m parameters p_j of the system, all the points of the boundary will satisfy therefore the equation†

$$F(i\omega; p_1, p_2, \ldots, p_m) = 0$$

† It can be said, by using a geometrical language, that the boundary of the uniform asymptotic stability region in the hyperspace of the parameters is a hypersurface mapping in this space the imaginary axis $\nu = 0$ in the complex plane, obtained through the transformation defined by the equation $F(\lambda; p_1, p_2, \ldots, p_m) = 0$.

where $i\omega = \text{Im}\,\lambda$, $(\lambda = \nu \pm i\omega)$. Thus the boundary is described by the parametric equations

$$\begin{aligned} \text{Re}\,F(i\omega; p_1, p_2, \ldots, p_m) &= 0, \\ \text{Im}\,F(i\omega; p_1, p_2, \ldots, p_m) &= 0 \end{aligned} \tag{11.29}$$

(having ω as a parameter).

Actually, the boundary of the asymptotic stability region is constructed on the basis of equations (11.29). To determine which portion of the space of the parameters delimited by this boundary represents the region of stability, we can proceed by giving an increment to one of the coordinates of one point of the boundary (or of several points if only one point is not conclusive) and by finding the sign of the corresponding increment of ν (whose value in the point on the boundary is zero). For instance, let the time lag τ be the chosen coordinate. Obviously, to an increment $\delta\tau$ there corresponds an increment of ν equal to $\delta\nu = -\,\text{Re}\,\{(\partial F/\partial\tau)/(\partial F/\partial\lambda)\}_{\nu=0}\,\delta\tau + O(\delta\tau^2)$. The increment $\delta\tau$ will be taken sufficiently small for the sign of $\delta\nu$ to be given by the term of the first degree in $\delta\tau$.

To establish if a certain region of the hyperspace of the parameters delimited by a portion of the boundary belongs or not to the region of asymptotic stability, we can proceed in a simpler manner, by verifying the inequality (11.28) for an arbitrary point of the respective region. If among the parameters there is one single time lag τ, a point in the hyperplane $\tau = 0$ can often be chosen to advantage, because in this way the characteristic equation degenerates into an algebraic equation to which the usual criteria for the sign of the real parts of the roots can be applied (Hurwitz, etc.).

It is to be noted that in the usual practical cases an asymptotically stable solution with a sufficient margin for preventing the generally destabilizing effect of the time lag corresponds to the ideal control ($\tau = 0$). The verification mentioned above is therefore tantamount to the investigation of stability of the respective ideal control system.

If the existence of the stability of the solution in this ideal case is proved in another manner, all the verifications exposed are superfluous, the region of asymptotic stability in the space of the parameters being, a priori, that which contains (entirely) the hyperplane $\tau = 0$. (Existence of stability for $\tau = 0$, at least theoretically, is not necessary in order to have stability for certain values different from zero of the time lag.)

11.3.2. The Case of Stationary and Quasi-Stationary Basic Motions (Autonomous Systems)

What was said above will be applied to system (11.12) (see [8, Section 9]). The characteristic equation of this system can be put in the form

$$\lambda^4 + a_1\lambda^3 + (a_2 + c_\theta e^{-\tau\lambda})\lambda^2 + [a_3 - (a_{11} + a_{22})c_\theta e^{-\tau\lambda}]\lambda$$
$$+ a_4 + (a_{11}a_{22} - a_{12}a_{21})c_\theta e^{-\tau\lambda} = 0 \tag{11.30}$$

where a_1, a_2, a_3, and a_4 are the coefficients of the characteristic equation (e.g., (6.2)) of the system corresponding to the free motion (i.e., with $c_\theta = 0$). Taking into account that the magnitudes a_i depend on the system parameters, the parametric equations (11.29) that define the stability boundary in the space of these parameters will be

$$\omega^4 - a_2\omega^2 + a_4 - c_\theta[(a_{11} + a_{22})\omega \sin \tau\omega - (a_{11}a_{22} - a_{12}a_{21} - 1) \cos \tau\omega] = 0,$$
$$- a_1\omega^3 + a_3\omega - c_\theta[(a_{11} + a_{22})\omega \cos \tau\omega + (a_{11}a_{22} - a_{12}a_{21} - 1) \sin \tau\omega] = 0. \tag{11.31}$$

With the help of these equations by removing the parameter ω, it is possible to determine—for instance, graphically—the stability region in the space of the system parameters on which the quantities a_i, a_{jk}, c_θ, and τ are dependent. To facilitate the analysis, it is fitting to determine the stability regions in plane sections of the hyperspace of the parameters by choosing one pair of parameters at a time, the rest being considered constant, if such an assumption is compatible with the variation of the selected pair. In an attempt to bring out clearly in the least roundabout way the influence of time lag, one of the variable parameters considered in what follows will be the time lag τ itself. The proportional attitude control factor c_θ will be taken as the second variable parameter. The boundary thus defined will indicate the admissible values, from the viewpoint of flight stability, of the time lag depending on the values of the correction factor c_θ, specifying the control.

The set of parametric equations of the asymptotic stability boundary in the plane (c_θ, τ) is then easily obtained from (11.31) in the following form.

$$c_\theta = \frac{\Omega_1(1 + \Omega^2)^{1/2}}{(\Omega\Omega_2 - k)},$$
$$\tau = \frac{1}{\omega} \tan^{-1} \Omega(\omega), \tag{11.32}$$

where $\Omega_1, \Omega_2, \Omega$, are functions of the parameter ω; that is, $\Omega_1 = \omega^4 - a_2\omega^2 + a_4$, $\Omega_2 = (a_{11} + a_{22})\omega$, $\Omega = (\Omega_2\Omega_3 - k\Omega_1)/(\Omega_1\Omega_2 + k\Omega_3)$ with $\Omega_3 = a_1\omega^3 + a_3\omega$, and constant k, depending on the coefficients of system (11.12) is $k = a_{11}a_{22} - a_{12}a_{21} - 1$.

In the case of the ideal control the system is assumed to be uniformly asymptotically stable for $c_\theta \geqslant c_0 > 0$. Then the semistraight line $\tau = 0$ in the plane (c_θ, τ) located to the right of point c_0 will be contained in the region of asymptotic stability. It belongs to the boundary since, for $\tau < 0$ (system with advanced argument), the trivial solution of a system of the type considered is always unstable. The other branch of the boundary supplying useful information in the given case (the curve described by equations (11.32) has, obviously, an infinity of branches) can be constructed graphically from the parametric equations (11.32). The part of the plane included between this branch and the axis of abscissas beginning from the abscissa $c_\theta = c_0$ is the zone of the stability region of interest. The boundary curve of this zone indicates the variation of admissible time lag as a function of the quantity c_θ specifying the intensity of the control. As was to be expected, it can readily be seen that for a very powerful control, that is, for a very great ("infinite") control moment required, the admissible time lag tends toward zero.

In the next section the foregoing procedure will be applied for the investigation of the admissible time lag of the controls depending on the characteristics of the control outfit of the machine in the case of hovering flight of vertical takeoff and landing aircraft.

11.3.3. Note Concerning Nonautonomous Systems

In Section 11.2 we considered the case of small time lags via Liapunov's direct method, giving in the form of inequality (11.25) an estimate of the admissible time lag. As specified, however, this estimate depends on the method (selection of the function $\mathscr{V}$), and on many occasions may be pessimistic.

To investigate the stability of unsteady control motions with arbitrary time lag, recourse could be had to the existing mathematical theory based on the use of Liapunov's functionals (see [4]). But, although this theory is elaborated under all its aspects of principle, there are few examples of its application up to the present and consequently we cannot speak of any significant experience in this field. Its application to the problem of stability of controlled flight is still outstanding and should be undertaken as development of the attempts to apply direct methods in the investigation of the stability of unsteady motions of the uncontrolled aircraft, such as were presented in Chapters 8–10.

11.4. HOVERING FLIGHT OF VTOL AIRCRAFT

We will consider the case of the simultaneous use of a roll damper and the stabilizer that induces a restoring moment to counter angular deviations in roll (referred to as a lateral static stabilizer). In Section 7.3 this case was studied by assuming the neglect of the time lag. In what follows precisely the effect of this time lag will be investigated by maintaining the assumptions corresponding to the scheme previously referred to as standard. Namely, assumption will be made that the terms of gyroscopic gearing are vanishing ($h_{r_x} = h_{r_z} = 0$). In this case the disturbed motion about the center of mass is separated into three independent rotations about the three body axes. For exemplification, here too, the motion about the longitudinal axis (rolling motion) will be considered [13].

For $\tau = 0$ the most efficient stabilization was shown to be obtained if the static stabilization is sufficient for ensuring an oscillating variation of the induced deviations ($c_\phi > (c_p/2)^2$) maintaining nevertheless a high enough value of the coefficient c_p to ensure a sufficiently heavy damping. In this section the admissible time lag will be determined as a function of the ratio $\kappa = c_\phi/c_p$, chosen as the second parameter.

System (7.32) in the case of nonzero response time, assumed for simplicity to be the same for both the roll damper and the lateral static stabilizer (the largest of the actual times of response will be taken for τ), is modified as follows.

$$\frac{dp}{dt} = -c_p p(t-\tau) - c_\phi \phi(t-\tau), \quad \frac{d\phi}{dt} = p(t). \tag{11.33}$$

To facilitate the analysis, we will consider parameter c_ϕ set. The characteristic equation of system (11.33) being

$$\lambda^2 + c_p e^{-\tau\lambda}\lambda + c_\phi e^{-\tau\lambda} = 0,$$

the stability boundary in the plane (κ, τ) will be described by the parametric equations

$$-\omega^2 + c_\phi(\kappa\omega \sin \tau\omega + \cos \tau\omega) = 0,$$

$$c_\phi(\kappa\omega \cos \tau\omega - \sin \tau\omega) = 0,$$

or, making κ and τ explicitly stand out,

$$\kappa = \frac{(\omega^4 - c_\phi^2)^{1/2}}{|c_\phi \omega|}, \qquad \tau = \frac{1}{\omega} \tan^{-1} \kappa\omega. \tag{11.34}$$

The parametric equations (11.34) define an infinite set of curves issued from the origin in the plane (κ, τ). The system with ideal controls (7.32) is asymptotically stable in theory for any pair of values c_ϕ, c_p (both of them being strictly positive); hence c_ϕ being set, for any c_p or κ. For the reason shown in the previous section the semiaxis ($\kappa > 0, \tau = 0$) will belong to the boundary of asymptotic stability and the admissible time lag will be given by the branch nearest to the abscissa axis from the curves defined by (11.34). Figure 11.1 illustrates a numerical example computed for the dimensionless c_ϕ equal to

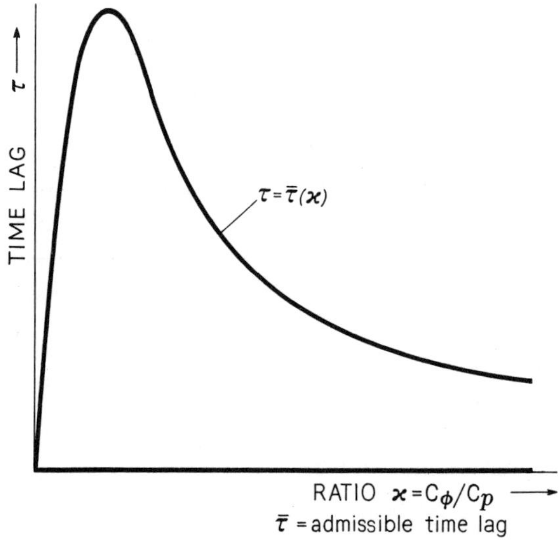

Figure 11.1.

1. There results a rapid increase of the admissible value of time lag with the increase of ratio κ from zero to some optimal value (for which the admissible time lag is maximum). There follows a decrease of the admissible τ that is steep at first, then gradually smoother, tending toward zero for $\kappa \to \infty$. In the numerical case computed, the optimum nondimensional ratio $\hat{\kappa}$ has resulted approximately of 1.25, for c_ϕ and c_p dimensionless, equal to $(k_x g)^{-1} c_\phi$ and $(k_x g)^{-1/2} c_p$, respectively, hence $\hat{\kappa} = (k_x g)^{1/2} \kappa$, k_x being the radius of gyration of the aircraft about the longitudinal axis (Ox) and g the acceleration

due to gravity. (To bring the equations describing the disturbed motion of hovering flight to a nondimensional form, the Glauert transformation, based on physical considerations in regard to aerodynamic forces, is obviously inadequate. It is also meaningless to express the dimensional unit of time, of the angular velocity disturbance, and so on by means of a velocity of the center of mass, e.g., of the maximum horizontal velocity, because it has no significance for the problem studied. For constructing the dimensional units of forces, moments, time, velocities, angular velocities, in addition to the quantities selected as fundamental units, i.e., the mass and a specific length, we can use the gravitational constant g. Then, if m is the mass of the machine, l a specific length (e.g., the semispan, semichord, or radius of gyration about one of the body axes), the dimensional units could be $(l/g)^{1/2}$ for time, $(g/l)^{1/2}$ for angular velocity disturbances, $(gl)^{1/2}$ for the induced deviations of the velocity of the center of mass, mg for forces, mgl for moments, etc.)

From (11.34) thus the admissible time lag is obtained as a function of the ratio κ, which in turn depends on the parameter c_p. To obtain a more conclusive estimate of the effect of time lag, the operation can be repeated for a set of fixed values of the parameter c_ϕ, covering the range of usual values. Nevertheless, the conclusion drawn for an arbitrary c_ϕ appears to be sufficient to provide a qualitative image of the phenomenon. Anyhow, for computations required in designing it is desirable to have recourse to more complex models than that represented by system (11.33) and to use a greater number of parameters of the system. In this case however, it becomes difficult if not impossible to draw general conclusions, and the only possibility that remains is to use a computer of acceptable performances.

REFERENCES

1. L. E. El'sgol'ts, *Introduction to the Theory of Differential Equations with Deviating Arguments.* Holden-Day, San Francisco, 1966.
2. A. D. Myshkis, *Linear Differential Equations with Retarded Arguments.* Gordon and Breach, New York, 1966.
3. E. Pinney, *Ordinary Difference-Differential Equations.* Univ. of California Press, Berkeley, 1959.
4. N. N. Krasovskiy, *Some Problems in the Theory of Stability of Motion.* Stanford Univ. Press, Stanford, California, 1963.
5. R. Bellman and K. L. Cooke, *Differential-Difference Equations.* Academic Press, New York, 1963.
6. A. Halanay, *Differential Equations: Stability, Oscillations, Time Lags.* Academic Press, New York, 1966.

7. Yu. M. Repin, "On the stability of solutions of equations with retarded argument" (in Russian), *Prikl. Mat. Mekh.* **21**(1957), 253–261.
8. T. Hacker, "On the longitudinal stability of an aircraft with automatic pilot," *Rev. Méc. Appl. Acad. R. P. R.* **7**(1962), 673–692.
9. A. M. Liapunov, *Stability of Motion.* Academic Press, New York, 1966.
10. R. Bellman, "On the existence and boundedness of solutions of nonlinear differential-difference equations," *Ann. of Math.* **50**(1949), 347–355.
11. E. M. Wright, "The stability of solutions of non-linear difference-differential equations," *Proc. Roy. Soc.* (*Edinburgh*) **A63**(1950).
12. N. N. Meyman and N. G. Chebotarev, "The Routh–Hurwitz problem for polynomials and entire functions" (in Russian), *Steklov Mathematics Inst. Works* **26**(1949).
13. T. Hacker, "Problems of stability of VTOL aircrafts in hovering flight," *Rev. Méc. Appl. Acad. R. P. R.* **7**(1962), 225–239.

Chapter 12

OPTIMIZATION PROBLEMS

12.1. INTRODUCTION

12.1.1. Object

The present chapter is connected with the previous ones in two ways. Sections from 12.4 to 12.7 will seek the law of optimal control for the passage from one flight régime to another that is not in its vicinity. (The problem was stated in Section 3.3.) For definiteness, the problem of optimal climb and optimal acceleration will be treated, hence the optimization of the transition from a régime marked by a given height and speed to another defined by a higher flight altitude and velocity. Further, the question of optimal stabilization will be considered in Sections 12.8–12.11. This problem is directly connected with several topics treated in the previous chapters. In particular, it develops the case of the transition from a given flight régime to another close one outlined in Section 3.3; this chapter can, in a way, be considered as a development of Section 3.3.

What was said above implies that no attempt will be made to present here the numerous results (see, e.g., [1–4]) obtained so far in the theory of optimization of terrestrial flight. Furthermore, such an attempt could not fit into the framework of this book. The object is a supplementation I have considered necessary of the matter exposed, in a direction in which particular efforts are being made at present.

Finally, it is to be noted that a comparative study of the efficiency of the various mathematical methods in existence will not be attempted. The use of one or another of these methods is not an implicit recommendation of it as most adequate for the type of problem to which it is applied.

However, although we will treat individual technical problems in this chapter, by using them as patterns, we will be able to approach other optimization and optimal stabilization problems without having to supplement the matters of principle.

12.1.2. Problem Formulation; the Physical Model

The notion of optimal programming occurs almost spontaneously. A flight mission is generally given by fixing only certain requirements, so that it can be achieved by different means, namely, by various sequences of controls or evolutions. In other words, when a flight mission is given, the aggregate movement is not generally uniquely determined; there is a certain freedom left in its choice, which is used for achieving the aim set forth in the flight mission in the most advantageous way relative to a certain additional requirement. Therefore, a certain performance index expressing a criterion of optimality is to be set in addition to the aim prescribed by the flight mission. Formally, both the objective sought and the quality of performance could be included in formulating the flight mission. There is, however, an essential difference between the two categories of problem data, which lies in their degree of definiteness: the objective is defined with accuracy, as a rule quantitatively; the quality of the achievement described by the performance index may be expressed only in vague terms, being subordinated to the necessity of achieving the aim exactly. Thus, it is stated: the aircraft *must* reach the height h_1, so that—by using the total fuel consumption as a performance index—a *minimal* amount of fuel should be consumed. Or: the flying machine (for instance a missile) provided with a *set* fuel and oxidizer capacity should climb so that, if the height of powered flight path is chosen as the performance index, this height h_p would be *maximal*. In both cases it is a question of climbing flight, and as main parameters for both, quantities of the same physical nature are used: the altitude and the fuel capacity. In the first case, however, an altitude to be reached is set with certainty for which a minimum but not yet determined fuel consumption is required; whereas in the second case the amount of fuel (together with the amount of oxidizer if a missile is involved) is given from the outset and a flight program (e.g., prescribing the trajectory, law of fuel consumption per unit time, etc.) is required such that the height reached in powered flight be maximal (although not yet determined, resulting at the most as a consequence of the solution of the problem). Summing up, therefore, the flight program is determined on the basis of the following data.

1. Purpose of the flight mission: in what follows, it will be given in the form of the initial and final states of motion, the mission consisting in a transfer from the former to the latter.

2. Performance index expressing an optimality criterion: this is given in Sections 12.4–12.7, in the form of either the total duration of the evolution or the total fuel consumption, both having to be minimized.

3. System of constraints imposed upon the movement of the machine and the controls: indeed, a number of limitations to be imposed upon the system appear by the physical nature itself of the aircraft's movement and of the operation of the controls, as well as by the necessity of taking into consideration other facts that are external with respect to the process under study, such as the structural strength of the machine, physiological endurance of the crew, and so on. Hence, to study the means of achieving the mission's aim by minimizing the performance index, it is compulsory to take into account certain conditions of constraint on the machine's movement and the operation of the controls.

Since the flight program is secured by an appropriate selection of the control policy, the essential problem to be treated will concern the determination of this policy so as to satisfy the requirements implied by points 1, 2 and 3.

12.2. THE MATHEMATICAL MODEL

12.2.1. State Variables and Control Variables; Assumption of Piecewise Continuity of the Control Variables

As has been mentioned before, three types of variables are involved in the mathematical model of the motion of a flying machine considered as a controlled process, namely, an *independent variable*, which is the time measured from a moment t_0 considered initial; the variables describing the state of motion (in short, the *state variables*) (denoted by the n vector x); and the variables that impose the development of the process, termed *control variables* (denoted by the r vector u). Therefore, given the equation of motion set up by taking into account the modeling assumptions, and the initial conditions (or another equivalent system of conditions), the solution is uniquely determined if the control variables are given as time functions over the time interval considered. The mathematical model, therefore, expresses the following situation: in order to determine the development of the controlled process over a given time interval, it is sufficient to set the succession of commands to be transmitted to the system.

Thus, in contrast with the state variables, the control variables are characterized by the fact that their magnitude may be imposed arbitrarily at any instant. The possibility of arbitrarily choosing at any instant the magnitude of the control variable is assured by the hypothesis concerning the lack of inertia of the controls, which in the language of the mathematical model used is expressed by the fact that the admissible control variables will be piecewise continuous functions of the independent variable. By the piecewise continuous time function is designated a continuous function at any point t with the possible exception of a finite number of points at which it may have discontinuities of the first kind (finite discontinuities, i.e., points in which the function has finite lateral limits. Thus, if $u(t)$ is the control function and t_k a point of discontinuity, we will have $u(t_k - 0) = \lim_{t \to t_k, t < t_k} u(t) < \infty$, $u(t_k + 0) = \lim_{t \to t_k, t > t_k} u(t) < \infty$.) Such functions are mathematical models describing physical controls that can pass from one value to some other value within a negligible time interval.

Let us assume, for exemplification, that the only control variable considered is the elevator setting. Then the vector function $u(t)$ reduces to the scalar function $\delta_e(t)$ $(r = 1)$. The function $\delta_e(t)$ varies in the range $-\delta_e^{(1)} \leqslant \delta_e(t) \leqslant \delta_e^{(2)}$, where $\delta_e^{(1)}$ is the maximum value of negative elevator angles, and $\delta_e^{(2)}$ of the positive ones. The assumption concerning the piecewise continuity of the function $\delta_e(t)$ expresses the fact that by a sufficiently sudden actuation of the stick the time required for changing the elevator setting from the value $-\delta_e^{(1)}$ to $\delta_e^{(2)}$ (or the other way round) is negligible in comparison with the process considered.

12.2.2. Control Variables

A first step in making the mathematical model of a complex process as a rule consists in selecting the variables considered to be essential for the studied aspect. In the given case of modeling the controlled flight, as could be seen from foregoing, some of the chosen variables, namely, those of state, will characterize the mechanical motion of the machine, while the others the control policy. As in the rest of the book, the models in this chapter will also be built by leaving the processes, whatever their nature (mechanical, electrical, psychophysiological, and others), that take place along the chain of the control system aside and replacing them with assumptions securing the functional relation between the input and output of this system. In this way, the state and control variables will generally be quantities of the same kind (geometrical, kinematic, or dynamic). In agreement with what was said earlier, however, unlike the state variables, those of control will not appear in equations as dynamic variables but as external constraints, that is, as time functions imposed on the system (forcing functions).

Therefore, we have to deal with control variables that can be imposed arbitrarily on the one hand, and with state variables that result uniquely determined if the equations (corresponding to the assumptions), a system of initial conditions (defined by the problem under investigation), and the law of variation of the control variables are given, on the other hand. However, since both classes of variables represent quantities of the same physical nature, there will always be a certain degree of arbitrariness about the selection of the control variables from among all the essential variables of the problem. This selection will be prescribed, among other things, also by formal considerations in connection with the task of facilitating the actual solution of the problem proposed. However, before attempting to examine this aspect of particular significance for the object of the present chapter, in what follows a classification of the control variables will be proposed.

As already stated, in the mathematical models used in this book the processes taking place within the control system will be disregarded. These processes connect the control movement carried out by the pilot or the servomotor with the effect of this movement on the value of the "control variable," which is a magnitude of the same kind as certain state variables. If, by replacing the control motion proper with a "control variable" in the sense shown, we cause the disregarded process to be independent of the remaining (state and control) variables of the system, the related control variable will be termed a *direct* one. The direct control variables include, among others, the elevator angle, throttle setting, or the thrust itself, if it is assumed to be independent of the speed and of the altitude.

However, certain kinematic parameters, whose value is determined by the control movement through an alteration of the whole dynamical picture of the controlled object, may also be used as control variables. In other words, the chain that in this case connects the control movement performed by the pilot or servomotor with the magnitude represented by the control variables contains elements depending on at least one of the state or control variables of the equations of motion. In such cases we will say that we have to deal with *indirect* control variables. Obviously, imposing the value of the indirect control variables arbitrarily is not generally consistent with the complete set of equations constructed on the basis of the physical assumptions admitted. The necessity arises to adopt additional assumptions permitting us to remove the incompatibility of imposing certain arbitrary values on the related control variable by restraining the generality of the problem.

For instance, in order to use the incidence or pitch angle as a control variable it must be assumed that (i) the normal and the tangential external forces acting on the machine are independent of the related angular velocities ($\dot{\alpha}$ and q, respectively); and (ii) the variation region of all the variables (α, $\mathbf{M}$, q, $\dot{\alpha}$, etc.) on which the longitudinal control moment depends enables it to take a sufficiently large value at any instant.

12.2.3. Constraints; Admissible Controls

In the previous section the compulsory imposition of certain constraints upon the state and control variables was included among the data that determine flight programming. As to the state variables, in what follows we will assume that the conditions implied by the physical nature of the magnitudes on which the motion depends (natural constraints) do not require a separate mathematical modeling. Such are the constraints expressed by: the existence of a maximum velocity; the nonexistence of negative velocities; the boundedness of external forces (hence of accelerations); and other factors that are generally implied by the equations of motion. The solution of the mathematical problem is appreciably complicated by the imposition of bounds on the state variables (see [5]). This problem will be resumed in Section 12.7 by means of a simple system exemplified. Obviously, if in formulating the mathematical problem the additional conditions mentioned are abandoned, it will be mandatory to check if they are satisfied by the actual solution resulting from the computation.

With control variables, the imposition of constraints, whether natural or called for by collateral reasons, is most often compulsory;† since they are imposed arbitrarily on the system no constraints whatever can be implied in the equations. The constraints concerning the control variables will be given in the form of a certain range of their values.

Let r be the number of control variables chosen. Consider the r-dimensional Euclidean space, and let U be a closed region of this space. By an admissible control is meant any piecewise continuous vector function $u(t)$ defined for

† In discussing optimal stabilization in Sections 12.8–12.11 we will renounce the imposition of certain constraints on the control variables. The reasons permitting this renunciation will be sketched in the contents of the respective section; they will include the intuitive ground that slight control movements are sufficient for correcting small disturbances.

the interval $[t^0, t^1]$ over which the process is considered and which takes for this interval only values belonging to the region U. (In this chapter, Arabic numeral subscripts will be reserved for the vector components: $(x_1, x_2, \ldots, x_n)$, $(u_1, u_2, \ldots, u_r)$, etc. The magnitude corresponding to the initial instant t^0 will be designated by the superscript 0 (e.g., $x_j^0 = x_j(t^0)$), those corresponding to the final instant t^1 by the superscript 1 (e.g., $x_j^1 = x_j(t^1)$). Superscripts in parentheses will generally be used to denote the lower and upper bounds (superscripts (1) and (2), respectively), except in Appendix I, where they will indicate the order of iteration.)

In what follows the region U will be given in the form of a set of inequalities imposed on the components of the vectorial function $u(t)$:

$$u_i^{(1)} \leqslant u_i(t) \leqslant u_i^{(2)}, \qquad i = 1, 2, \ldots, r$$

where u_i are the magnitudes measuring the controls used.

In the example mentioned, in which the only control variable is the elevator angle δ_e, the region U reduces to the interval $-\delta_e^{(1)} \leqslant \delta_e(t) \leqslant \delta_e^{(2)}$, with the remark that $\delta_e^{(1)}$ and $\delta_e^{(2)}$ need not necessarily be the constructively admitted bounds of the elevator angle, as they can—at option—be lower.

12.2.4. Proposed Objective (Purpose of the Flight Mission)

In formulating the flight mission, it is required that the machine be transferred from an initial state of motion S^0, which is not necessarily completely defined by a full set of initial conditions, to a final state of motion S^1, also defined only to the extent of the requirements considered. For generality, they will be defined parametrically.

Let p be a point of the open set P in an s-space. Then S^0 is defined as the set of points in the n-space with the coordinates $\omega_1^0(p), \omega_2^0(p), \ldots, \omega_n^0(p)$ and S^1 is defined as the set of points in the same space with the coordinates $\omega_1^1(p), \omega_2^1(p), \ldots, \omega_n^1(p)$.

NOTE. In what follows, S^0 and S^1 will be given in the form of a certain number m^0 ($m^0 \leqslant n$) of initial conditions, and m^1 ($m^1 \leqslant n$) of terminal conditions, respectively. Therefore S^0 and S^1, defining the purpose of the mission, will appear as $(n - m^0)$-dimensional and $(n - m^1)$-dimensional manifolds, respectively, in the n-space of the state variables. If, for example, only the altitude h^1 and the velocity V^1 are specified as final conditions, then m^1 will be equal to 2. The values of the other state variables at the final instant t^1 will be determined by actually solving the optimum problem considered.

12.2.5. Performance Index

The performance considered will be described by an integral functional of the form

$$J = \int_{t^0}^{t^1} f_0(x(t), u(t))\, dt \tag{12.1}$$

the minimization of which is sought. In other words, it is required to determine the function $u(t)$ so that the corresponding solution of the system should minimize the functional J.

Obviously, the problem can also be defined for performance indices K the magnitude of which it is required to maximize (K, e.g., can measure the range or endurance, the ceiling height of an aircraft, or the altitude of the powered flight path of a missile, etc.). If this be the case, in order to leave the definition unchanged, we will take $J = -K$.

In Sections 12.4–12.7 the performances measured by the functional J will be the flying time or the aggregate fuel consumption. In the case of the minimum time problem, the function f_0 will obviously be identically equal to unity, and $J = t^1 - t^0$. For the minimum consumption problem, f_0 represents a magnitude proportional to the fuel consumption per unit of time. It is assumed that the thrust depends only on the fuel supply. Between these two magnitudes there is a biunique correspondence. Through the rate of fuel supply the throttle lever setting (which is the "natural" control variable) may then be replaced by the thrust, which if assumed to be independent of the flight velocity appears as a direct control variable. (If the dependence of the thrust upon the flight velocity could not be neglected, the magnitude of the thrust would appear as an indirect control variable.) Consequently, f_0 will depend only on one control variable and will be independent of the state variables, namely, $f_0 = f_0(\tau)$. It is to be noted that f_0 being independent of the state variables makes it appreciably easier to obtain qualitative conclusions without actual integration.

12.2.6. Mathematical Formulation of the Optimum Problem

Henceforth we will assume the meaning of the *control system* to be the following set of equations (only autonomous systems will be considered)

$$\frac{dx}{dt} = f(x, u) \tag{12.2}$$

together with the boundary conditions

$$x(t^0) \in S^0, \qquad x(t^1) \in S^1. \tag{12.3}$$

Generally, in the set of admissible controls $u(t) \in U$ there can exist several that transfer the system from a state $x \in S^0$ to another state $x \in S^1$. The question is to select one of them that achieves this purpose, and at the same time realizes the best performance from a given point of view. This where the optimality criterion expressed by means of the functional (12.1) comes in. The problem will thus consist in determining the admissible control $u(t)$ that transfers the system from a state $x \in S^0$ to another state $x \in S^1$ so that the functional J is minimal. The function $u(t)$ having these properties will be referred to as optimal control and will be denoted by $\tilde{u}(t)$, and the corresponding solution of the control system (12.2), (12.3) as the optimal solution will be denoted by $\tilde{x}(t)$.

Simultaneously with finding the optimal solution, the initial and final conditions are completed, and the corresponding points in the manifolds S^0 and S^1 determined. The respective point of the point set P will be denoted by $\tilde{p}$ (e.g., $\tilde{x}(t^0) = \omega^0(\tilde{p})$, $\tilde{x}(t^1) = \omega^1(\tilde{p})$).

The value of the time interval considered (the difference $t^1 - t^0$) is also determined for the situation when the final instant is considered free. This will be the case in all the applications treated.

12.3. MAXIMUM PRINCIPLE

12.3.1. The Pontryagin Theorem

For the problems treated in Sections 12.4–12.7 and partly for those of the subsequent sections, recourse will be had to the mathematical apparatus offered by the method of Pontryagin's school (the maximum principle, [6]). As in the classical calculus of variations, we will start by seeking the necessary conditions of optimality, that is, the properties of optimal controls permitting us to choose, from the set of all admissible controls transferring the system from the state $x \in S^0$ to the state $x \in S^1$, the controls that can be optimal.

The maximum principle expresses such a property. It will be stated hereunder for autonomous systems without its proof. (For the demonstration, see [6, Chapter 2].)

Consider the control system (12.2), (12.3) and the functional (12.1). Let $\tilde{u}(t)$ be an optimal control, $\tilde{x}(t)$ the corresponding optimal solution, and $\tilde{p} \in P$ such that $\tilde{x}(t^0) = \omega^0(\tilde{p})$ and $\tilde{x}(t^1) = \omega^1(\tilde{p})$. Let, further,

$$\frac{dy}{dt} = \frac{\partial f}{\partial x}(\tilde{x}(t), \tilde{u}(t))y \tag{12.4}$$

be the corresponding variation system.

Consider also the adjoined system

$$\frac{d\xi}{dt} = -\xi \frac{\partial f}{\partial x}(\tilde{x}(t), \tilde{u}(t)) - \xi_0 \frac{\partial f_0}{\partial x}(\tilde{x}(t), \tilde{u}(t)) \tag{12.5}$$

the row vector ξ having n components (viz., $\xi_1, \xi_2, \ldots, \xi_n$). The terminal instant t^1 is assumed free.

Then the maximum principle can be stated thus:

THEOREM. *If $\tilde{u}$ is an optimal control, then there exists a solution $\tilde{\xi}(t)$ of the system* (12.5) *and a constant $\xi_0 \leqslant 0$ such that*

(a) $\tilde{\xi}(t)f(\tilde{x}(t), \tilde{u}(t)) + \xi_0 f_0(\tilde{x}(t), \tilde{u}(t)) = \max_{u \in U} [\tilde{\xi}(t)f(\tilde{x}(t), u) + \xi_0 f_0(\tilde{x}(t), u)]$

in any point t in which the function $u(t)$ is continuous;

(b) $\max_{u \in U} [\tilde{\xi}(t)f(\tilde{x}(t), u) + \xi_0 f_0(\tilde{x}(t), u)] = 0$

for $t^0 \leqslant t \leqslant t^1$;

(c) *the transversality conditions expressed by the relations*

$$\sum_{i=1}^{n} \xi_i(t^1) \frac{\partial \omega_i^1}{\partial p_\sigma}(\tilde{p}) - \sum_{i=1}^{n} \xi_i(t^0) \frac{\partial \omega_i^0}{\partial p_\sigma}(\tilde{p}) = 0$$

hold for $\sigma = 1, 2, \ldots, s$.

For brevity, we will use the notation

$$H(x, \xi, u) = \xi f(x, u) + \xi_0 f_0(x, u).$$

REMARK. As mentioned already, in what follows there will be situations in which a number of initial conditions and a number of terminal conditions will be set for the state variables. In these situations, from the transversality conditions written above it follows that the values of the components of the vectorial function ξ corresponding to the components of x for which no initial conditions are fixed are zero; so are those corresponding to the components of the vector x that are free at the terminal instant.

12.3.2. Comments

Certain aspects will be outlined here, from which the practical utility of the method becomes apparent. Obviously, the maximum principle, supplying a necessary condition of optimality, does not generally permit us to determine uniquely the optimal solution sought. But, as will be seen from the following remarks, it generally allows us to contain the problem up to the point where it is possible to find the optimal solution by comparing a restricted number of possible solutions.

Finding the optimal solution amounts to determining $2n + r$ functions ($\tilde{u}_i(t)$, $i = 1, 2, \ldots, r$; $\tilde{x}_j(t)$, $j = 1, 2, \ldots, n$; $\xi_k(t)$, $k = 1, 2, \ldots, n$) and a numerical parameter t^1.

With $\tilde{x}(t)$ and $\xi(t)$ having been found, $\tilde{u}(t)$ is determined from the condition (a). Indeed, this condition reduces the problem to finding the maximum of a function of r variables: $u_1, u_2, \ldots, u_r$, with t appearing as a parameter. It is assumed that a maximum can be found for any t. The condition (a) is thus equivalent to r relations. (For instance, if $u(t)$ is in the interior of the region U, finding the maximum amounts to solving a set of r equation in u_i obtained by setting to zero the partial derivatives $\partial H/\partial u$.)

The $2n$ functions x_j, ξ_k are determined by the system of differential equations (12.2), (12.3). For these, $2n$ initial or terminal conditions are required. A part of these are given by setting a number of initial and terminal conditions, say, $2n - s$, for the variables x, while the remainder up to $2n$ is determined from the s transversality conditions (c), corresponding to the components of x that are free at the initial instant and to those that are free at the terminal instant (amounting to s in the aggregate). For determining the numerical parameter t^1 use will be made of the additional relation supplied by the condition (b).

It follows that for determining the $2n + r + 1$ unknowns we dispose of the same number of $2n + r + 1$ relations. Consequently, the problem will generally admit of a finite number of solutions, of which it will be possible to determine the optimal solution by trials (numerical computations or even flight tests).

The comparison is compulsory, so that it is necessary to find effectively all extremal solutions. If there exists an optimal solution, it will be found among the latter, which does not mean that the other extremal solutions obtained by means of the maximum principle come near the optimal one. In this respect the discussion in [7] carried out for a simple case of controlled motion is of interest. The maximum principle is applied to a minimum time problem for the motion of a material point having a constant velocity, by using as a control variable the normal acceleration or its time rate of change ($\dot{\gamma}$ or $\ddot{\gamma}$, γ being the flight path inclination). It is shown that certain extremal solutions found on the basis of the maximum principle are far from optimal. On the contrary, it is shown that certain types of extremal solutions in the sense of the maximum principle cannot be optimal whatever might be the numerical data of the problem. If such extremals can be found by a qualitative study, as is the case of the simple example considered, they can be excluded a priori, the comparison being carried out for the numerical data of the problem only between those extremal solutions in the sense of the maximum principle that can be optimal.

There are simple cases in which the optimal solution appears uniquely determined. This implies proof of the existence of an optimal control for the problem considered, and the unique determination of the control from the maximum principle. This, for instance, is the case of the minimum time problem of linear systems (see [6, Chapter 3]) or of the optimal stabilization problem, dealt with in Sections 12.8–12.11.

12.3.3. Switching Conditions

It can be seen that for obtaining the effective solution of the optimum problem, the proposed method conduces to a two-point boundary value problem, which is generally nonlinear. The problem is further complicated since the system of equations (12.2), (12.3) for which the solution is sought (through numerical integration), depends on the optimal control $u(t)$ we are looking for. Since the function $u(t)$ can, and generally does, present finite discontinuities, in the points of discontinuity of the control the system changes discontinuously, too. For carrying out the numerical integration (i.e., for writing the programming algorithm of the computer) it is necessary to determine the switching points (in which the system jumps discontinuously). In other words, it is necessary to deduce some switching functions that will be taken up in the algorithm (or, when the optimization process is automated, in the program of the computer block of the automatic pilot). These functions are obtained from the condition (a) of the maximum principle.

Consequently, the numerical problem is solved as follows. We start with a value of the control resulting as a rule from the set initial conditions or from physical considerations. The system is then integrated numerically over an initial interval, and the switching function is computed in parallel. The proceeding is continued until the switching function reaches a critical value. Further, over the next interval beginning from this moment the proceeding is repeated, but starting from the new value of the control, until the switching function again reaches a critical value, and so on.

In simple cases, from an examination of the switching function it is possible to determine also the number of switchings of the optimal control over the interval $[t^0, t^1]$. In what follows, the switching conditions will be given in the form of tables for each case studied.

12.4. OPTIMAL PROGRAMMING OF THE BASIC CONTROLS IN CLIMB AND ACCELERATED FLIGHT

12.4.1. Introduction

In the previous chapters the object for study consisted, to use an accepted expression, of the flying qualities. The problem was stated as follows. If the basic motion, including the program of basic controls required for its implementation under standard conditions, be set, what inherent properties—of stability—must the aircraft have, and what correction controls must be had recourse to in order that this motion also be ensured under conditions presenting deviations from the standard. Sections 12.5, 12.6, and 12.7 will be devoted to the question of choosing the basic motion. It will be stated as an optimum problem. Namely, according to the general statements in the previous section, we will seek the law of basic controls, which induces an evolution both possible and technically admissible that should best achieve a given flight objective (a given mission) from a certain defined viewpoint.

The optimal trajectory, if it exists,† is achieved by controls, either automated or carried out by the pilot. In the first case, programming of the automatic pilot's computing block and its task in general would be greatly facilitated if the result in the optimum problem were obtained in the form of a synthesis, that is, by finding the functional dependence of the control (hence of the control variables) on the state of the system (on the state variables). In the present state of the general theory, however, the problem of the analytical construction of these functions has not yet been solved for most cases of practical interest, which are usually too complicated. For the time being we should be content with a solution of the problem for open systems. Automatic control requires a complete solution that implies the use of numerical methods.

† An optimal control policy does not always exist, or if it exists, cannot always be achieved in flight practice. In either case it appears useful to seek what is referred to as a suboptimal policy, representing an existing and possible practical optimum. This aspect will not be considered in the present book.

For manual controls certain general qualitative information that does not involve a complete numerical solution of the problem may become useful. Frequently, such information can appear to be of a greater practical utility than prescribing exactly the motion of the controls to be carried out as a time function, when considering open systems, or even of the state of the system (if the synthesis of optimal controls has been obtained). Indeed, to comply exactly with a rigorous prescription is in most cases too difficult a task for the pilot, while general indications as a rule can easily be associated with the ordinary piloting reflexes, reactions, and habits. Indications may thus be given concerning the operating conditions of the engine (such as avoidance of intermediate régimes and approximate states of aircraft motion at which a discontinuous jumping in the operating conditions of the engine has to be ordered).

Consequently, the way in which the result is presented depends, on the one hand, on the theoretical possibilities of approaching the question and, on the other hand, on the practical aim pursued. The most precise or most detailed presentation of the solution is not always the best.

In this section an attempt will be made to obtain certain qualitative information concerning the controls that will provide the optimal régime aimed at. Then we will seek to obtain the data required for programming a computer (system of equations, end conditions, conditions of discontinuous variation in the value of certain parameters, etc.). To obtain these results we will have recourse to the mathematical apparatus supplied by the maximum principle stated in the previous section.

12.4.2. Technical Problem Formulation; Assumptions

The problem of flight dynamics considered initially is that of the optimization of climbing and accelerated flight so that the difference of altitude be covered and the intended velocity be reached either within a minimum interval of time or by consuming a minimum amount of fuel.

The optimum will be sought throughout the whole atmospheric space (characterized by the inequality $h \geqslant 0$) without any restriction being imposed on the horizontal displacements. This ideal case corresponds in practice to the takeoff in a zone in which the terrain features permit any evolution (clear terrain). In addition to facilitating the mathematical treatment, such a scheme affords another advantage. By selecting the initial point corresponding to the height above which the constraint of horizontal movement ceases, we achieve a situation where the aircraft conjoins the flight capabilities of both conventional and VTOL types. This permits us to compare the performances of these two solutions from the point of view of the optimum problems considered (climbing time and fuel consumption during climb).

Vertical takeoff and landing (VTOL) aircraft will be considered in addition to the conventional types. As a matter of fact, under the assumption of unrestricted horizontal displacement, there is no difference, at least not theoretically, in the approach to the problem for VTOL aircraft with the main thrust axis fixed in the aircraft body from that for ordinary aircraft. Certain differences, reflected in the mathematical model used, do appear, however, in VTOL aircraft with rotating thrust.

Considering the illustrating purpose of this chapter, we will not make an exhaustive study of the climbing optimization of VTOL aircraft. Among VTOL aircraft, we will confine ourselves to an outline of the problem for machines whose main thrust provided by the engines ensures both the propulsion and (in part or completely) the lift in evolutions specific to VTOL aircraft. The main thrust is not implicated in the attitude control of the machine, assuming that the resultant of thrust passes constantly through the center of mass of the aircraft. The angular attitude against the ground is controlled either by conventional controls (elevator, rudder, ailerons), when the aerodynamical velocity is high enough to ensure their efficiency, or by additional gas jets placed adequately, or by a combination of these.

Also for simplicity, with all the machines considered we assume that:

—mass of the machine and air density are constant;

—aerodynamic forces and moments, for all the remaining parameters being set, are proportional to the square of the velocity at all the régimes at which they are taken into account;

—flight velocity is small enough to allow of neglecting the variation of aerodynamic coefficients with the Mach number;

—a linear variation is admitted for the coefficient of lift, and a parabolic one for the coefficient of drag with the incidence, except for machines with rotating thrust line. For the latter, when the incidence value exceeds a certain limit ($\alpha \leqslant -\alpha^{(1)}$ or $\alpha \geqslant \alpha^{(2)}$), the aerodynamic resultant is admitted parallel to the flight direction and independent of incidence. The variation admitted for the pitching moment coefficient is piecewise linear with respect to α (piecewise constant static stability) and to the rate of pitch (the damping coefficient also piecewise constant), and with the derivative $d\alpha/dt$, the switching taking place for all these parameters according to the variation of the incidence in the points $\alpha = -\alpha^{(1)}$ and $\alpha = \alpha^{(2)}$. In particular, it can be admitted that for $\alpha < -\alpha^{(1)}$ and $\alpha > \alpha^{(2)}$, the aerodynamic pitching moment is zero, or depends—linearly—only on the pitching velocity q;

—thrust is independent of the flight régime (particularly of the velocity), being considered dependent only on the gas supply controlled by the throttle valve;

—the controls are free of inertia, so that the passage from one value of the control parameter to any other is supposed to take place instantaneously.

NOTE. As will be shown, any motion parameter that is disposed of arbitrarily will be considered as a control. The last assumption referring to the lack of inertia will be extended to include all those quantities (e.g., total thrust, incidence) when these quantities are considered "controls," hence when it is considered that their value can be imposed upon the system arbitrarily.

12.5. VTOL AIRCRAFT WITH ADJUSTABLE ORIENTATION OF THRUST VECTOR

12.5.1. Equations of Motion

The case of vertical takeoff and landing aircraft with rotating thrust line will first be presented, since for such machines it is possible to use a simplified model, like that resorted to in the next subsection. The technique adopted can thus be presented on the basis of a simple example.

Under the assumptions of previous section, the longitudinal controlled motion of the aircraft will be described by a system of nonlinear differential equations of the fifth order (see system (2.1) to which is added the kinematic relationship connecting the flight altitude with the rate of climb):

$$\frac{dh}{dt} = V \sin\gamma,$$

$$\frac{dV}{dt} = -\frac{\mathbf{D}}{m} - g\sin\gamma + \frac{T}{m}\cos(\alpha + \varphi),$$

$$\frac{d\gamma}{dt} = \frac{\mathbf{L}}{mV} - \frac{g\cos\gamma}{V} + \frac{T\sin(\alpha + \varphi)}{mV},$$

$$\frac{d\alpha}{dt} = -\frac{d\gamma}{dt} + q,$$

$$\frac{dq}{dt} = \frac{M}{B} + \frac{M_c}{B}, \tag{12.6}$$

or, by setting off the state and control variables of the system, and by noting $T = mg\tau$ and $M_c/B = \kappa$:

$$\frac{dh}{dt} = V \sin\gamma \equiv f_1(V, \gamma),$$

$$\frac{dV}{dt} = -(1-\eta)k_1 V^2 - \eta(k_{\text{II}} + k_{\text{III}}\alpha^2)V^2 - g\sin\gamma$$
$$+ g\tau\cos(\alpha+\varphi) \equiv f_2(V, \gamma, \alpha; \tau, \varphi),$$

$$\frac{d\gamma}{dt} = (1-\eta)k_{\text{IV}}V + \eta k_{\text{V}}V\alpha - \frac{1}{V}g\cos\gamma + \frac{1}{V}g\tau\sin(\alpha+\varphi)$$
$$\equiv f_3(V, \gamma, \alpha; \tau, \varphi),$$

$$\frac{d\alpha}{dt} = -\frac{d\gamma}{dt} + q \equiv -f_3 + q \equiv f_4(V, \gamma, \alpha, q; \tau, \varphi),$$

$$\frac{dq}{dt} = (1-\eta)V^2(k_{\text{VI}}q + k_{\text{VII}}\dot{\alpha}) + \eta V^2(k_{\text{VIII}}\alpha + k_{\text{IX}}q + k_{\text{X}}\dot{\alpha})$$
$$+ \kappa(V, \alpha; \delta_e, \delta_r) = (1-\eta)V^2(k_{\text{VI}}q + k_{\text{VII}}f_4) + \eta V^2(k_{\text{VIII}}\alpha$$
$$+ k_{\text{IX}}q + k_{\text{X}}f_4) + \kappa(V, \alpha; \delta_e, \delta_r) \equiv f_5(V, \gamma, \alpha, q; \tau, \varphi, \delta_e, \delta_r), \tag{12.7}$$

where

$$\eta(\alpha) = \begin{cases} 1 & \text{for } -\alpha^{(1)} \leqslant \alpha \leqslant \alpha^{(2)}, \\ 0 & \text{for the remainder.} \end{cases}$$

The constants $k_{\text{I}}, k_{\text{II}}, \ldots, k_{\text{X}}$ take on the following approximate expressions.

$$k_{\text{I}} = \frac{\rho S C_{D_1}}{2m}, \quad k_{\text{II}} = \frac{\rho S C_{D_0}}{2m}, \quad k_{\text{III}} = \frac{\rho S C_{L_\alpha}^2}{2\pi\lambda m}, \quad k_{\text{IV}} = \frac{\rho S C_{L_1}}{2m},$$

$$k_{\text{V}} = \frac{\rho S C_{L_\alpha}}{2m}, \quad k_{\text{VI}} = \frac{\rho S \bar{c}^2 C_{m_{q_1}}}{8BV_{\text{crit}}}, \quad k_{\text{VII}} = \frac{\rho S \bar{c}^2 C_{m_{\dot{\alpha}_1}}}{8BV_{\text{crit}}}, \quad k_{\text{VIII}} = \frac{\rho S \bar{c} C_{m_\alpha}}{8B},$$

$$k_{\text{IX}} = \frac{\rho S \bar{c}^2 C_{m_q}}{8BV_{\text{crit}}}, \quad k_{\text{X}} = \frac{\rho S \bar{c}^2 C_{m_{\dot{\alpha}}}}{8BV_{\text{crit}}}.$$

The sum of the first two terms of the right-hand side of the last equation is proportional to the aerodynamic pitching moment and κ to the longitudinal control moment, the proportionality factor being $1/B$.

12.5.2. A Simplified Case: the Constant Attitude

Since in VTOL aircraft the resulting thrust line can be rotated about a point, assumed here to be the mass center of the machine, the flight attitude may remain all the time normal. For simplicity, the attitude will be assumed constant ($\theta = \theta_c =$ const.) [8]. Then, instead of the fifth-order system (12.7) it will be possible to describe the longitudinal motion by means of the following third-order system of differential equations.

$$\frac{dh}{dt} = V(k_{XI} \cos \alpha - k_{XII} \sin \alpha),$$

$$\frac{dV}{dt} = -(1-\eta)k_I V^2 - \eta(k_{II} + k_{III}\alpha^2)V^2 - k_{XIII} \cos \alpha$$
$$+ k_{XIV} \sin \alpha + g\tau \cos(\alpha + \varphi),$$

$$\frac{d\alpha}{dt} = -(1-\eta)k_{IV}V - \eta k_V V\alpha + \frac{k_{XIV}}{V} \cos \alpha$$
$$+ \frac{k_{XIII}}{V} \sin \alpha - \frac{g\tau}{V} \sin(\alpha + \varphi). \tag{12.8}$$

The magnitude of the thrust force or, since the weight of the machine is assumed constant, the ratio $\tau = T/mg$, and the angle φ defining the orientation of the thrust are considered to be the control variables. Both τ—under the assumptions admitted—and φ are direct controls; hence they can be used instead of the "natural" ones (position of the gas lever and that of the thrust line inclination lever) without any additional assumptions. Since the lack of inertia of the controls (response time zero) is admitted by hypothesis, the functions $\tau(t)$ and $\varphi(t)$ may be considered piecewise continuous. As the thrust varies between a minimum value T_{min}, which can in particular be zero, and a maximum value T_{max}, and the rotation of the thrust axis is also limited, the closed region U of the admissible controls will appear in the form of the following set of inequalities, expressing the respective constraints.

$$\tau_{min} \leqslant \tau \leqslant \tau_{max}, \qquad -\varphi^{(1)} \leqslant \varphi \leqslant \varphi^{(2)}$$

or, for $\tau_{min} = 0$, $\varphi^{(1)} = 0$, $\varphi^{(2)} = \pi/2 + \varepsilon$ $(\varepsilon < \pi/4)$,

$$0 \leqslant \tau \leqslant \tau_{\max}, \qquad 0 \leqslant \varphi \leqslant \frac{\pi}{2} + \varepsilon \tag{12.9}$$

where $\tau_{\min} = T_{\min}/mg$, $\tau_{\max} = T_{\max}/mg$.

The flight program consists in a climb and an acceleration from the altitude h^0, at which the machine flies with the velocity V^0, to the altitude h^1, at which the flight continues level with the velocity V^1. The initial and terminal conditions will therefore be those shown in Table 12.1.

Table 12.1. End Conditions for System (12.7)

t	h	V	α
t^0	h^0	V^0	p (free)
t^1 (free)	h^1	V^1	$\alpha^1 = \theta_c$

By using the parametrical representation from Section 12.2, the s-space of the parameters reduces to a straight line ($s = 1$) and the manifolds S^0 and S^1 in the three-dimensional space (h, V, α) degenerate S^0 into the straight line of intersection of the planes $h = h^0$ and $V = V^0$ ($m^0 = 2$) and S^1 in the point (h^1, V^1, θ_c), ($m^1 = 3$).

In this case, the control system will therefore consist of the set of equations (12.7) and of the boundary conditions from Table 12.1.

The initial and terminal conditions from Table 12.1 have been completed by means of a transversality condition that makes up for the lack of an initial value set for one of the state variables, namely, for α. The only parameter being $p = \alpha^0$, we have:

$$\omega_1^0(p) = h^0 = \text{const.}, \qquad \omega_2^0(p) = V^0 = \text{const.}, \qquad \omega_3^0(p) = p = \alpha^0,$$

$$\omega_1^1(p) = h^1 = \text{const.}, \qquad \omega_2^1(p) = V^1 = \text{const.}, \qquad \omega_3^1(p) = \alpha^1 = \theta_c = \text{const.}$$

Hence

$$\frac{\partial \omega_1^0}{\partial p} = \frac{\partial \omega_2^0}{\partial p} = \frac{\partial \omega_1^1}{\partial p} = \frac{\partial \omega_2^1}{\partial p} = \frac{\partial \omega_3^1}{\partial p} = 0, \qquad \frac{\partial \omega_3^0}{\partial p} = 1$$

and the transversality condition (see point (c) of the theorem in Section 12.3.1) reduces to

$$\xi_3(t^0) = 0. \tag{12.10}$$

Finally, by writing the condition (b) for $t = t^1$ one more relation is obtained, namely,

$$\max_{\substack{0 \leqslant \tau \leqslant \tau_{max} \\ 0 \leqslant \varphi \leqslant \pi/2+\varepsilon}} H(h^1, V^1, \alpha^1, \xi_1^1, \xi_2^1, \xi_3^1; \tau, \varphi) = 0. \tag{12.11}$$

The values of the auxiliary variables at the terminal instant t^1 were denoted by ξ_1^1, ξ_2^1, and ξ_3^1. The last relation is necessary because the terminal instant t^1 is free.

It is required to determine the law of variation of the controls τ and φ so that the flight program defined above should be carried out either (i) in a minimum interval of time $[t^0, t^1]$, or (ii) with a minimum of fuel. In the first case, therefore, the performance index will be the difference between the terminal (undetermined) instant and the initial set instant

$$I = t^1 - t^0 \tag{12.12}$$

while in the second case, this index will be represented by the integral functional

$$J = \int_{t^0}^{t^1} f_0(\tau)\, dt. \tag{12.13}$$

As mentioned in Section 12.2, the function $f_0(\tau)$ represents, by the approximation of a constant proportionality factor, the fuel consumption per unit of time under the assumption that this latter only depends on the magnitude of thrust. The following additional assumptions are admitted in connection with this function.

(a) For small values of the argument the function $f(\tau)$ is supposed to be small (possibly $f_0(0) = 0$), and the first derivative very large (in particular, $\lim_{\tau \to 0} f_0'(\tau) = \infty$) (prime is taken to mean derivative with respect to the argument in parentheses).

(b) Further, that

$$\lim_{\tau \to \infty} \frac{f_0(\tau)}{\tau} = 0$$

and that

(c) the second derivative $f_0''(\tau)$ does not change its sign.

For example, the function f_0 can be of the form $f_0(\tau) = \nu\tau^\beta$ with $\nu > 0$ and $0 < \beta < 1$.

The mathematical problem will be stated as follows. The admissible controls $\tau(t)$ and $\varphi(t)$ for which the solution of the system (12.8) will ensure the transfer from the initial to the terminal state indicated in Table 12.1 are to be determined so that (i) the time $I = t^1 - t^0$ is minimal, or (ii) the integral functional (12.13) is minimal. We will start with the application of the maximum principle.

Consider the adjoined system

$$\frac{d\xi_1}{dt} = 0,$$

$$\frac{d\xi_2}{dt} = \xi_1(-k_{\mathrm{XI}}\cos\alpha + k_{\mathrm{XII}}\sin\alpha) + \xi_2 V[(1-\eta)k_{\mathrm{I}} + \eta(k_{\mathrm{II}} + k_{\mathrm{III}}\alpha^2)]$$
$$+ \xi_3\left[(1+\eta)k_{\mathrm{IV}} + \eta k_{\mathrm{V}}\alpha + \frac{1}{V^2}(k_{\mathrm{XIV}}\cos\alpha + k_{\mathrm{XIII}}\sin\alpha) + \frac{g\tau}{V^2}\sin(\alpha+\varphi)\right],$$

$$\frac{d\xi_3}{dt} = \xi_1 V(k_{\mathrm{XI}}\sin\alpha + k_{\mathrm{XII}}\cos\alpha) + \xi_2(2\eta k_{\mathrm{III}}\alpha V^2 - k_{\mathrm{XIII}}\sin\alpha - k_{\mathrm{XIV}}\cos\alpha$$
$$+ g\tau\sin(\alpha+\varphi)) + \xi_3\left[\eta k_{\mathrm{V}} V + \frac{1}{V}(k_{\mathrm{XIV}}\sin\alpha - k_{\mathrm{III}}\cos\alpha) + \frac{g\tau}{V}\cos(\alpha+\varphi)\right].$$
$$(12.14)$$

The function H will be expressed by

$$H(h, V, \alpha; \xi_1, \xi_2, \xi_3; \tau, \varphi) = \xi_0\bar{f}_0 + \xi_1 V(k_{\mathrm{XI}}\cos\alpha - k_{\mathrm{XII}}\sin\alpha)$$
$$+ \xi_2[-(1-\eta)k_{\mathrm{I}}V^2 - \eta(k_{\mathrm{II}} + k_{\mathrm{III}}\alpha^2)V^2$$
$$- k_{\mathrm{XIII}}\cos\alpha + k_{\mathrm{XIV}}\sin\alpha + g\tau\cos(\alpha+\varphi)]$$
$$+ \xi_3\left[-(1-\eta)k_{\mathrm{IV}}V - \eta k_{\mathrm{V}}V\alpha + \frac{k_{\mathrm{XIV}}}{V}\cos\alpha\right.$$
$$\left. + \frac{k_{\mathrm{XIII}}}{V}\sin\alpha - \frac{g\tau}{V}\sin(\alpha+\varphi)\right]$$

with $\bar{f}_0 = 1$ for minimum time and $\bar{f}_0 = f_0(\tau)$ for minimum fuel.

According to condition (a) of the maximum principle (see theorem in Section 12.3.1), the optimal policy for τ and φ, if it exists, ensures at any instant that the function H, with h, V, and α fixed and ξ_0 a negative or null constant, be maximum. Since the function H will have a maximum value with respect to τ and φ at the same time as the function

$$\mathscr{H}_t = \xi_2 g\tau \cos(\alpha + \varphi) - \frac{\xi_3 g\tau \sin(\alpha + \varphi)}{V}$$

for problems of minimum time, or with the function

$$\mathscr{H}_c = \xi_0 f_0(\tau) + \xi_2 g\tau \cos(\alpha + \varphi) - \frac{\xi_3 g\tau \sin(\alpha + \varphi)}{V}$$

for those of minimum fuel consumption, $\mathscr{H}_t$ and $\mathscr{H}_c$ being obtained from H by eliminating the terms that do not contain the control variables, the optimal controls $\tilde{\tau}(t)$ and $\tilde{\varphi}(t)$ are to be sought among those that achieve at any instant the maximum of the function $\mathscr{H}_t$ or of the function $\mathscr{H}_c$, respectively.

Since the auxiliary variables are determined with the approximation of a constant factor, without restraint of generality, we can set $\xi_0 = -1$, and therefore for the maximum fuel problem $\mathscr{H}_c$ becomes

$$\mathscr{H}_c = -f_0(\tau) + g\tau\left(\xi_2 \cos(\alpha + \varphi) - \frac{\xi_3 \sin(\alpha + \varphi)}{V}\right).$$

By the change of variables

$$\xi_2 = \frac{1}{g}\sigma \sin \xi, \qquad \xi_3 = \frac{V}{g}\sigma \cos \xi, \tag{12.15}$$

the following expressions are obtained for $\mathscr{H}_t$ and $\mathscr{H}_c$.

$$\mathscr{H}_t = \tau\sigma \sin(\xi - \alpha - \varphi) \tag{12.16}$$

and

$$\mathscr{H}_c = -f_0(\tau) + \tau\sigma \sin(\xi - \alpha - \varphi). \tag{12.17}$$

We begin by seeking the optimal variation law of the control function $\varphi(t)$, obviously the same for both problems. It can be seen that $\tilde{\varphi}(t) = \varphi$ optimum having to achieve at every instant the maximum of the function $\sin(\xi - \alpha - \varphi)$ (τ and σ being positive) will be determined by the variation of the function $\xi - \alpha$. Since the value of φ is subject to the constraint (12.9), i.e., $0 \leqslant \varphi \leqslant \pi/2 + \varepsilon$, $\tilde{\varphi}(t)$ will vary as shown in Table 12.2, in which k is any whole number inclusive zero. The difference $\xi - \alpha$ plays the part of a switching function, and the ends of the intervals from each row represent the related critical values of this function.

Table 12.2. Optimum Orientation of the Thrust (Being the Same for Minimum Time and for Minimum Fuel)

Switching conditions	$\tilde{\varphi}(\xi - \alpha)$
$2k\pi \leqslant \xi - \alpha \leqslant \frac{\pi}{2} + 2k\pi$	$\varphi_{\min} = 0$
$\frac{\pi}{2} + 2k\pi \leqslant \xi - \alpha \leqslant \pi + \varepsilon + 2k\pi$	$\xi - \alpha - \frac{\pi}{2} + 2k\pi$
$\pi + \varepsilon + 2k\pi \leqslant \xi - \alpha \leqslant \frac{3\pi}{2} + \varepsilon + 2k\pi$	$\varphi_{\max} = \frac{\pi}{2} + \varepsilon$
$\frac{3\pi}{2} + \varepsilon + 2k\pi \leqslant \xi - \alpha \leqslant 4k\pi$	$\varphi_{\min} = 0$

It is to be noted that, if within the interval $[t^0, t^1]$ there exists a finite subinterval on which the function $\xi - \alpha$ results by integration identically equal to one of the critical values (which, e.g., cannot take place for $\xi(t)$ and $\alpha(t)$ analytical functions, the critical values being constant), the optimal control policy $\tilde{\varphi}$ cannot be determined by means of the maximum principle on that subinterval. The observation holds for any switching function.

The optimal law of thrust magnitude $\tilde{\tau}(t)$ is obtained now depending on $\tilde{\varphi}(t)$. The result, as will be shown, is a *bang-bang* type variation (either maximum or minimum, intermediate values being excluded).

Indeed, for the minimum time problem τ will be $\tau_{\max}$ or zero, according to whether the function $\sigma \sin(\xi - \alpha - \tilde{\varphi})$ or, σ being positive, the function $\sin(\xi - \alpha - \tilde{\varphi})$ is positive or negative. (If on a certain finite interval $\sigma \equiv 0$ or $\sin(\psi - \alpha - \tilde{\varphi}) \equiv 0$, $\tilde{\tau}$ cannot be determined in this way.) It can readily be seen that for values of the function $\xi - \alpha$, corresponding to the first three rows of Table 12.2, $\sin(\xi - \alpha - \tilde{\varphi}) > 0$, hence $\tilde{\tau} = \tau_{\max}$, and for those in the interval indicated in the last row $\sin(\xi - \alpha - \tilde{\varphi}) < 0$, hence $\tilde{\tau} = 0$.

For climbing and accelerated flight ($h^1 > h^0$, $V^1 > V^0$) the situation $\tilde{\tau} = 0$ is unlikely, if not—by reasoning intuitively—downright absurd. However it may be, the actual conclusion results from computation. For the case in which $\tilde{\tau} = \tau_{\max}$ all the time, the situation $3\pi/2 + \varepsilon + 2k\pi \leqslant \xi - \alpha \leqslant 4k\pi$ excludes itself automatically.

For the minimum total fuel problem the switching function is the same as for minimum time, that is, $\Sigma = \sigma \sin(\xi - \alpha - \varphi)$, but it has a different critical value. Let us suppose that $f_0(t)$ satisfies the assumptions (a), (b), and (c) stated above. Then the critical value of the switching function Σ will be $\Sigma_{\text{crit}} = f_0(\tau_{\max})/\tau_{\max}$.

For enhanced intuitiveness we will have recourse to a graphical representation. The function

$$\mathscr{H}_c(\tau) = -f_0(\tau) + \Sigma\tau$$

where $f_0(\tau)$ satisfies the foregoing assumptions may be represented in the plane $(\tau, \mathscr{H}_c)$ by a set of curves depending on the parameter Σ (Figure 12.1). In compliance with the constraint (12.9), referring to τ (i.e., $0 \leqslant \tau \leqslant \tau_{\max}$),

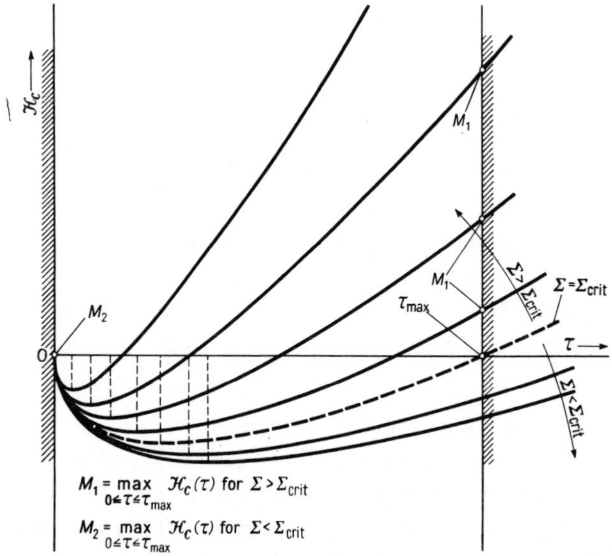

Figure 12.1.

we will consider only that part of the plane $(\tau, \mathscr{H}_c)$ that is included between the straight lines $\tau = 0$ and $\tau = \tau_{\max}$. In this region, the curves corresponding to the values of the parameter Σ higher than the critical value intersect the axis of the abscissas in two points each, $\max_{0 \leqslant \tau \leqslant \tau_{\max}} \mathscr{H}_c(\tau)$ being the ordinates of the points of intersection with the straight line $\tau = \tau_{\max}$, where $\mathscr{H}_c > 0$. On the contrary, the curves corresponding to the values $\Sigma < \Sigma_{\text{crit}}$ meet the axis $O\tau$ within the region of admissible controls only in the origin, being situated in the zone $\mathscr{H}_c < 0$. Hence $\max_{0 \leqslant \tau \leqslant \tau_{\max}} \mathscr{H}_c(\tau) = 0$. The conclusions above are epitomized in Table 12.3.

Table 12.3.
Optimum Thrust Value

Switching conditions	$\tilde{\tau}(\Sigma)$
$\Sigma \geqslant \Sigma_{\text{crit}}$	$\tau_{\max}$
$\Sigma \leqslant \Sigma_{\text{crit}}$	0

For minimum time $\Sigma_{\text{crit}} = 0$; for minimum fuel $\Sigma_{\text{crit}} = f_0(\tau_{\max})/\tau_{\max}$.

The optimal control policy and the corresponding motion are determined effectively by numerical integration of the system consisting of (12.8) and (12.14) and depending on the parameters τ and φ, in their turn functions of solution through the agency of the switching relations given by Tables 12.2 and 12.3. Therefore, as we have already stated, the algorithm must set up so that we will compute for every instant t the switching function $\xi - \alpha$, from which ensues $\varphi(t)$ and, by means of it, $\Sigma = \sigma \sin(\xi - \alpha - \varphi)$, which determines $\tau(t)$.

Integration is carried out for the initial conditions set out in Table 12.1, completed with (12.10), which after the substitution (12.15) becomes

$$\xi(t^0) = \cos^{-1} 0 \tag{12.18}$$

and for $\alpha(t^0)$, $\xi_1(t^0)$, and $\sigma(t^0)$ arbitrary. The initial values corresponding to the control variables $\varphi(t^0)$ and $\tau(t^0)$ result from Tables 12.2 and 12.3.

NOTE. Sometimes certain physical considerations regarding the values at the initial instant of the control variables restrain the arbitrariness in selecting the free initial conditions. For instance, on the ground that at the initial instant the value of the thrust, and sometimes its inclination, cannot be zero, there follow the additional conditions

$$\Sigma(t^0) \equiv \sigma(t^0) \sin(\xi^0 - \alpha^0 - \varphi^0) \geqslant \Sigma_{\text{crit}}, \tag{12.19}$$

$$\frac{\pi}{2} + 2k\pi \leqslant \xi^0 - \alpha^0 \leqslant \frac{3\pi}{2} + \varepsilon + 2k\pi, \tag{12.20}$$

and if we start with a vertical thrust (tail-sitters) instead of the inequalities (12.20), we will have the equalities

$$\varphi(t^0) = \frac{\pi}{2} - \theta_c, \quad \xi(t^0) - \alpha(t^0) = -(2k-1)\pi - \theta_c, \qquad (12.21)$$

the latter on the basis of Table 12.2. Obviously, the additional relations (12.19) and (12.20) or (12.21) are only likely and do nothing more than facilitate the selection of free initial conditions. As a result of solving to the end the numerical problem they may prove to be wrong.

For solving the two-point boundary value problem stated, the only procedure available is that of *trial and error*. We start with the partially set and partially arbitrary initial conditions and the system (12.8), (12.14) is then integrated numerically by taking into account the switching conditions, until one of the terminal conditions given in Table 12.1 or the condition (12.11) is satisfied. However, the other conditions set will not generally be satisfied. By a process of iteration, for instance, such as that presented in Appendix I, the procedure is repeated with the free initial conditions modified until all the final conditions given and the relation (12.11) are verified. A finite number of extremal solutions is thus obtained of which the optimal one results by comparison. The initial value of the optimal incidence, which may or may not satisfy the relations from (12.19) to (12.21) (also resulting obviously are $\xi(t^0)$ and $\sigma(t^0)$), as well as the final instant t^1 (the time interval $t^1 - t^0$) will also result implicitly.

12.5.3. Variable Longitudinal Attitude

The assumption $\theta =$ const. was agreed on in order to illustrate the procedure by the example of a simpler system. Giving it up, however, does not involve an excessive complication of the calculations. (For the main results presented here and in Section 12.6, see [9].)

Consider the system (12.7). Consequently, the adjoined system will also be of the fifth order:

$$\begin{aligned} &\frac{d\xi_1}{dt} = -\frac{\partial H}{\partial h}, \quad \frac{d\xi_2}{dt} = -\frac{\partial H}{\partial V}, \quad \frac{d\xi_3}{dt} = -\frac{\partial H}{\partial \gamma}, \\ &\frac{d\xi_4}{dt} = -\frac{\partial H}{\partial \alpha}, \quad \frac{d\xi_5}{dt} = -\frac{\partial H}{\partial q}, \end{aligned} \qquad (12.22)$$

where the function H is now

$$H = -\bar{f}_0 + \xi_1 f_1 + \xi_2 f_2 + \xi_3 f_3 + \xi_4 f_4 + \xi_5 f_5, \tag{12.23}$$

$\bar{f}_0$ being, as above, either identically equal to 1 for the minimum time problem, or equal to $f_0(\tau)$ for the minimum fuel problem (f_0 representing, as above, the fuel consumption in unit time). If we consider the quantities τ, φ, and κ to be control variables, condition (a) of the maximum principle amounts to maximizing the function

$$\mathscr{H} = -\bar{f}_0(\tau) + \xi_2 g\tau \cos(\alpha + \varphi) + (\xi_3 - \xi_4)\frac{g \sin(\alpha + \varphi)}{V}$$

$$+ \xi_5(k_{\mathrm{XV}} V\tau \sin(\alpha + \varphi) + \kappa) \tag{12.24}$$

where $k_{\mathrm{XV}} = -g[(1 - \eta)k_{\mathrm{VII}} + \eta k_{\mathrm{X}}]$, obtained by retaining from (12.23) only the terms that contain these variables.

To the constraints expressed by the inequalities (12.9) will be added those referring to the control moment. Generally, the control moment depends on the state of the system through the flight velocity and the incidence, and is consequently an indirect control variable. In order that it might be considered an admissible control, hence for $\kappa(t)$ to be a piecewise continuous function, it is necessary that the following additional assumption be agreed on: whatever might be the value of arguments V and α, by appropriately selecting the parameters δ_e and δ_r, we can have κ take arbitrary values between certain limits. For small velocities, or for values of the incidence that are external to the region $[-\alpha^{(1)}, \alpha^{(2)}]$ (supercritical incidences), the assumption is implemented by means of control jets represented by the parameter δ_r. Obviously, the assumption must be completed with a supposed quasi-instantaneous response of the control moment to the actuation of the elevator or the control jets.

It is to be noted that the limits between which κ may vary at a certain instant also depend on the value of the variables V and α at that instant

$$\kappa^{(1)}(V, \alpha) \leqslant \kappa \leqslant \kappa^{(2)}(V, \alpha). \tag{12.25}$$

It is necessary to admit the existence of an upper bound for the function $\kappa^{(1)}$ and of a lower bound for $\kappa^{(2)}$.

By substituting

$$g\xi_2 = \sigma \sin \xi, \qquad \frac{g(\xi_3 - \xi_4)}{V} + k_{\mathrm{XV}}\xi_5 V = \sigma \cos \xi, \tag{12.26}$$

we get, for (12.24),

$$\mathscr{H} = -\bar{f}_0(\tau) + \sigma\tau \sin(\xi - \varphi - \alpha) + \xi_5 \kappa. \tag{12.27}$$

It can be seen that in order to ensure at every instant a maximum with respect to τ and φ of the function $\mathscr{H}$, the laws expressed in Tables 12.2 and 12.3 hold, while for achieving a maximum with respect to the control variable κ, the law of variation will be given in Table 12.4, the auxiliary variable ξ_5 appearing as a switching function.

Table 12.4. Law of Optimal Control of the Longitudinal Control Moment

Switching condition	$\tilde{\kappa}(\xi_5)$
$\xi_5 \geqslant 0$	$\kappa^{(2)}(V, \alpha)$
$\xi_5 \leqslant 0$	$\kappa^{(1)}(V, \alpha)$

Integration is carried out for the initial and terminal conditions given in Table 12.5. If the flight is required to be uniform, horizontal, and straight, $\alpha(t^1) = \alpha^1$ will correspond to the equilibrium condition for the constant horizontal velocity V^1.

Table 12.5. End Conditions for System (12.7)

t	h	V	γ	α	q
t^0	h^0	V^0	p_1 (free)	p_2 (free)	p_3 (free)
t^1	h^1	V^1	$\gamma^1 = 0$ or free	α^1 or free	$q^1 = 0$ or free

If it is required to reach only the altitude h^1 with the velocity V^1, without other additional conditions imposed on the flight régime at the terminal instant, only $h(t^1)$ and $V(t^1)$ will appear set, while γ, α, and q remain free for $t = t^1$.

The transversality conditions (c) (see Section 12.3) for $\gamma(t^0)$, $\alpha(t^0)$, and $q(t^0)$ free and $\gamma(t^1)$, $\alpha(t^1)$, and $q(t^1)$ set reduce to

$$\xi_3(t^0) = 0, \quad \xi_4(t^0) = 0, \quad \xi_5(t^0) = 0. \tag{12.28}$$

For $\gamma(t^1)$, $\alpha(t^1)$, and $q(t^1)$ free, in addition to (12.28) we will have

$$\xi_3(t^1) = 0, \quad \xi_4(t^1) = 0, \quad \xi_5(t^1) = 0. \tag{12.29}$$

The final instant t^1 being free, one more relation similar to relation (12.11) obtained from condition (b) of the maximum principle will be required. The algorithm will be constructed in a manner similar to the one above, and the same technique will be used in proceeding further.

12.6. ORIENTATION OF THE THRUST FORCE FIXED AGAINST THE MACHINE

12.6.1. Equations of Longitudinal Motion

The treatment of the problem is in principle the same for both VTOL aircraft and conventional machines. The differences that might occur are only quantitative, being reflected more particularly in the value of the thrust force: in many conventional aircraft τ is smaller than the unit, while in VTOL aircraft it is necessarily larger than the unit. This can give rise to other differences, reflected, for instance, in the initial value of the path inclination. All these quantitative differences, however, are not of a kind to impede a unitary theoretical treatment of the flight optimization problem, particularly of the climb and acceleration. For this reason, by aircraft we will mean, in what follows, both the conventional airplane and the vertical takeoff and landing machine.

On the basis of the agreed-on assumptions (see Section 12.4) and supposing in addition that the direction of the resulting thrust is all the time parallel to the zero-lift reference axis of the aircraft (with undeflected control surfaces), the equations of the controlled motion in the plane of symmetry of the machine, with τ and κ considered to be control variables, are obtained from (12.7) by setting $\eta \equiv 1$ and $\varphi = 0$. We obtain

$$\frac{dh}{dt} = V \sin \gamma \equiv f_1(V, \gamma),$$

$$\frac{dV}{dt} = -(k_{\rm II} + k_{\rm III}\alpha^2)V^2 - g \sin \gamma + g\tau \cos \alpha \equiv f_2(V, \gamma, \alpha, \tau),$$

$$\frac{d\gamma}{dt} = k_{\rm V} V\alpha - \frac{g \cos \gamma}{V} + \frac{g\tau \sin \alpha}{V} \equiv f_3(V, \gamma, \alpha, \tau),$$

$$\frac{d\alpha}{dt} = q - k_{\rm V} V\alpha + \frac{g \cos \gamma}{V} - \frac{g\tau \sin \alpha}{V} \equiv f_4(V, \gamma, \alpha, q, \tau),$$

$$\frac{dq}{dt} = V^2(k_{\text{VIII}}\alpha + k_{\text{IX}}V + k_{\text{X}}f_4) + \kappa(V, \alpha, \delta_e, \delta_r) \equiv f_5(V, \gamma, \alpha, q, \tau, \delta_e, \delta_r). \tag{12.30}$$

Constants k with subscripts from II to X are those from the previous section. For facility of writing, the notation f was maintained in the right-hand sides.

Subsequently, τ and κ will appear as control variables, by considering therefore the system above. Whenever other quantities, such as incidence or path inclination, are used as control variables, the system will be altered accordingly.

12.6.2. Optimal Climb and Acceleration by Thrust and Longitudinal Moment Control

The problem considered here is a particular case of the problem treated in the previous section corresponding to the constant null value of the angle φ and to the value one of the parameter η. Consequently, the results found will hold also for this case. Thus, for both minimum time and minimum fuel (having the performance indices (12.12) and (12.13), respectively), the optimal law of the control moment κ is given by Table 12.4, the switching function being the variable ξ_5 of the adjoined system corresponding to the state variable q. Likewise, for determining the optimal thrust we may use Table 12.3, where the switching function Σ takes the expression $\Sigma = \sigma \sin(\xi - \alpha)$, having, for the minimum time and minimum fuel problem, respectively, the same critical values as in the case of φ variable, namely, $\Sigma_{\text{crit}} = 0$ for minimum time and $\Sigma_{\text{crit}} = f_0(\tau_{\max})/\tau_{\max}$ for minimum fuel.

The initial conditions are altered as follows. For conventional aircraft starting from the ground, by considering an instant prior to takeoff as initial,† $\gamma(t^0) = 0$. Correspondingly, the equality $\xi_3(t^0) = 0$ will be missing from the additional relations (12.28) obtained from the transversality conditions. Conversely, for VTOL aircraft the initial inclination of the path is arbitrary, no matter what moment is considered initial; it can, in particular, also have the value $\pi/2$.

† For formal reasons connected with the mathematical model used, $V = 0$ cannot be taken as the initial velocity, the functions f_3, f_4, and f_5 being defined for $V \geqslant \varepsilon > 0$. Hence the motion is not considered from the starting moment, but from a subsequent moment, which otherwise may be, however, proximate to that of the start of the machine.

The final instant t^1 being free, the additional condition of the type (12.11) will hold good as well; that is,

$$\max_{\substack{0 \leqslant \tau \leqslant \tau_{\max} \\ \kappa^{(1)}(V,\alpha) \leqslant \kappa \leqslant \kappa^{(2)}(V,\alpha)}} H(t^1) = 0$$

where the function $H = -f_0 + \xi_1 f_1 + \xi_2 f_2 + \xi_3 f_3 + \xi_4 f_4 + \xi_5 f_5$, with f_i as in (12.30), $f_0 = 1$ for minimum time, and $f_0 = f_0(\tau)$ for minimum fuel, is taken for the final instant t^1.

In what follows, however, new aspects will result from utilizing, in addition to the direct control variables or in their stead, the indirect control variables, namely, the incidence of the aircraft and the flight path angle.

12.6.3. Minimum Time to Climb and Accelerate by Control of the Incidence

Consider the incidence of the aircraft as the only control variable. Thrust is taken as constant and equal to its maximum value. Incidence is obviously an indirect control variable, which replaces the control variable κ (which is also indirect). For α to be considered a control variable, the additional assumption relating to κ (from Section 12.5) must be strengthened as follows. Whatever the variables V and α might be, by choosing conveniently the values of the parameters δ_e and δ_r, we can increase the control moment sufficiently to ensure any incidence in the interval

$$-\alpha^{(1)} \leqslant \alpha \leqslant \alpha^{(2)} \tag{12.31}$$

within a negligible time interval. The inequalities (12.31) represent the admissible range of the control, $\alpha^{(1)}$ being the maximum negative incidence and $\alpha^{(2)}$ the maximum positive incidence of the aircraft.

With α a control variable ($\alpha(t)$ considered an external constraint imposed upon the system) the fourth equation is removed from system (12.30) and the system consisting of the first three equations that describe the motion of the mass center of the aircraft is separated from the last equation, which represents the longitudinal motion about the center of mass. The system of equations modeling the motion of the center of mass will be considered as a matter of course. Angles $\alpha^{(1)}$ and $\alpha^{(2)}$ are admitted sufficiently small for neglecting the terms in α^ν with $\nu \geqslant 3$. Then the system considered will be

$$\frac{dh}{dt} = V \sin \gamma,$$

$$\frac{dV}{dt} = -(k_{\text{II}} + k_{\text{III}}\alpha^2)V^2 - g\sin\gamma + k_{\text{XVI}}\left(1 - \frac{\alpha^2}{2}\right),$$

$$\frac{d\gamma}{dt} = k_{\text{V}}\alpha V - \frac{g\cos\gamma}{V} + k_{\text{XVI}}\frac{\alpha}{V}, \tag{12.32}$$

where $k_{\text{XVI}} = g\tau_{\max}$.

Table 12.6.
End Conditions for System (12.32)

t	h	V	γ
t^0	h^0	V^0	γ^0 set or free
t^1 free	h^1	V^1	$\gamma^1 = 0$ or free

Table 12.6 gives the boundary conditions. The value of the path angle at the moment t^0 can be set (to 0, $\pi/2$, or some other value). For $\gamma(t^0)$ free we obtain from a transversality condition

$$\xi_3(t^0) = 0 \tag{12.33}$$

and for $\gamma(t^1)$ free

$$\xi_3(t^1) = 0, \tag{12.34}$$

ξ_3 being the variable in the adjoined system

$$\frac{d\xi_1}{dt} = -\frac{\partial H}{\partial h}, \quad \frac{d\xi_2}{dt} = -\frac{\partial H}{\partial V}, \quad \frac{d\xi_3}{dt} = -\frac{\partial H}{\partial \gamma}, \tag{12.35}$$

corresponding to the state variable γ. Function H for system (12.32) will be $H = -\bar{f}_0 + \xi_1 V\sin\gamma + \xi_2[-(k_{\text{II}} + k_{\text{III}}\alpha^2)V^2 - g\sin\gamma + k_{\text{XVI}}(1 - \alpha^2/2)] + \xi_3(k_{\text{V}}\alpha V - g\cos\gamma/V + k_{\text{XVI}}\alpha/V)$.

According to the maximum principle, α optimal, if it exists, achieves at any moment $t \in [t^0, t^1]$, the maximum of function H with h, V, and γ fixed. Therefore the optimal law of variation of the incidence, if it exists, is found among those that maximize the following function of α.

$$\mathscr{H} = -\xi_2\alpha^2\left(k_{\text{III}}V^2 + \frac{k_{\text{XVI}}}{2}\right) + \xi_3\alpha\left(k_{\text{V}}V + \frac{k_{\text{XVI}}}{V}\right), \tag{12.36}$$

representing the terms of the function H that contain the variable α.

The problem of establishing the optimal law of variation of the incidence and of the related switching conditions through the condition (a) of the maximum principle therefore amounts to seeking the maximum of a function of the form

$$\mu(\lambda) = a\lambda^2 + b\lambda \tag{12.37}$$

in the region

$$-\lambda^{(1)} \leqslant \lambda \leqslant \lambda^{(2)} \tag{12.38}$$

when a and b vary. For definiteness, let $\lambda^{(2)} \geqslant \lambda^{(1)} > 0$. (This restriction corresponds to the cases considered here. If $\lambda^{(1)}$ and $\lambda^{(2)}$ were indifferent, an additional number of situations against those considered below would have to be examined, the latter generally not having a real correspondent.)

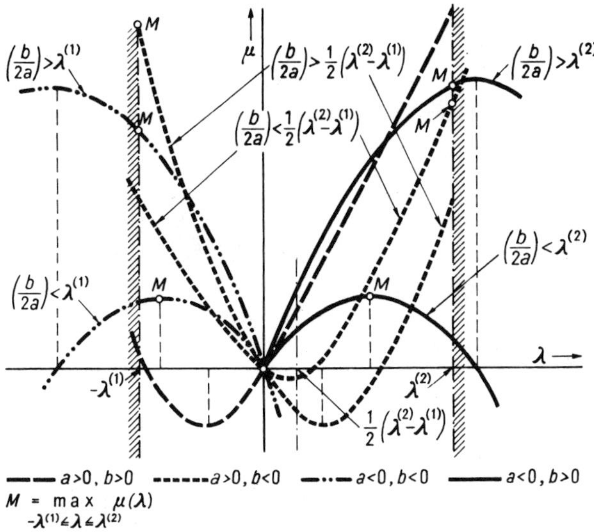

Figure 12.2.

Equation (12.37) defines a moving parabola in the plane (λ, μ). For more intuitiveness recourse will be had to a geometrical representation. Let $\tilde{\lambda}$ be the value of the variable λ for which the function μ reaches its highest value in the interval (12.38): $\max_{-\lambda^{(1)} \leqslant \lambda \leqslant \lambda^{(2)}} \mu(\lambda)$. The absolute maximum or minimum all along the axis of abscissas of the function $\mu(\lambda)$ occurring for $\lambda = c = -b/2a$ when a and b vary, $\tilde{\lambda}$ is obtained from Table 12.7. The results may be followed readily by examining the graphs in Figure 12.2.

Table 12.7

Switching conditions			
a	b	$\|c\|$	$\tilde{\lambda}(a, b, c)$
> 0	> 0	Indifferent	
			$\lambda^{(2)} = \lambda_{\max}$
		$< (\lambda^{(2)} - \lambda^{(1)})/2$	
> 0	< 0		
		$> (\lambda^{(2)} - \lambda^{(1)})/2$	
			$-\lambda^{(1)} = \lambda_{\min}$
		$> \lambda^{(1)}$	
< 0	< 0		
		$< \lambda^{(1)}$	$c =$ variable
		$< \lambda^{(2)}$	$c =$ variable
< 0	> 0		
		$> \lambda^{(2)}$	$\lambda^{(2)} = \lambda_{\max}$

Let $\lambda = \alpha$, $\mu = \mathscr{H}$, $a = -\ \xi_2(k_{\text{III}}V^2 + k_{\text{XIV}}/2)$, $b = \xi_3(k_{\text{V}}V^2 + k_{\text{XVI}}/V)$. For the problem considered it follows, then, that the optimal path will consist of arcs along which the incidence may be constant and equal to its maximum value, of arcs with a constant incidence equal to the minimum one (maximum negative incidence), and of arcs with a variable incidence having the expression $\xi_3(k_{\text{V}}V^2 + k_{\text{XVI}})/[\xi_2 V(2k_{\text{III}}V^2 + k_{\text{XVI}})]$. The various arcs are joined, depending on the solution of system (12.32), (12.35) corresponding to the end conditions from Table 12.6 and to the conditions (12.33) and (12.34) (the terminal instant being free, the relation $\max_{-\alpha^{(1)} \leqslant \alpha \leqslant \alpha^{(2)}} H(t^1) = 0$ is also added), in agreement with the conditions from Table 12.7.

12.6.4. Climb and Acceleration in Minimum Time by Path Inclination Control

The optimal law of variation of the state of the system is found, as shown already, at the same time as that of the controls by numerically solving the related two-point boundary value problem. Certain qualitative information concerning the control may be obtained also without integration. The possible nature of the various fragments that go to make up the optimal control, their joining conditions, and so on can thus be established merely on the basis of condition (a) of the maximum principle. If, while taking the necessary precautions, we consider certain parameters of motion to be the control variables, this information will concern the mechanical motion of the machine itself. Thus by choosing above the incidence of the aircraft as the control variable, hence as a quantity that, under the agreed-on assumptions, can be disposed of arbitrarily, the possible forms of the function $\tilde{\alpha}(t)$ and the way in which they are joined may be established. To obtain certain similar qualitative information concerning the physical path of the mass center itself, an essential parameter of this, its inclination measured by the angle γ, will be chosen as the control variable in what follows. Obviously, in order to "dispose arbitrarily" of the path slope, new constraints have to be imposed upon the system for removing the possibility of certain incompatibilities.

Such an additional constraint—sufficient in relation to the mathematical model used—results obviously from an examination of the equation of motion (12.30). By substituting the indirect control variable γ for the direct control variables δ_e and δ_r, those equations are canceled from the system of equations that determine the behavior of the former if the law of variation of the direct controls be given. Thus, in addition to the third and fifth equations (relating to γ and q, respectively), the fourth equation relating to α, which contains the variable q, will be canceled. But, since the variable α occurs in the first equation too, canceling of the three equations mentioned above must be set off by imposition of an assumed relation between the variables α, V, and γ. Thus, in the first equation a given function of V and γ will occur instead of the variable α. The result of the optimum problem will depend obviously on the choice of this function.

For instance, under the theoretical assumption $\alpha = \text{const.}$, there follows for $\tilde{\gamma}$ a bang-bang variation: the optimal path inclination of VTOL aircraft will either reduce to a straight vertical climb, or consist of a sequence of straight vertical climbs and dives. Indeed, with $\alpha = \text{const.}$, γ optimal achieves the maximum of the function $\mathscr{H} = (\xi_1 V - g\xi_2)\sin\gamma$ and as a result $\tilde{\gamma}$ will always be on the boundary of the admissible region (i.e., equal to $\pi/2$ if $\xi_1 V - g\xi_2 \geqslant 0$, or to $-\pi/2$ if $\xi_1 V - g\xi_2 \leqslant 0$).

An assumption closer to reality, which it can satisfactorily approximate in certain cases, would be to consider the centripetal acceleration along the optimal path to be negligible. Then, if the incidence is sufficiently small to allow the replacement of $\sin\alpha$ by its argument, we may write

$$\alpha = \frac{g\cos\gamma}{k_{\mathrm{V}}V^2 + k_{\mathrm{XVI}}} \tag{12.39}$$

and the control system reduces to

$$\frac{dh}{dt} = V \sin \gamma,$$

$$\frac{dV}{dt} = -\left[k_{\rm II} + \frac{k_{\rm III} g^2 \cos^2 \gamma}{(k_{\rm V} V^2 + k_{\rm XVI})^2}\right] V^2 - g \sin \gamma + k_{\rm XVI}\left[1 - \frac{g^2 \cos^2 \gamma}{2(k_{\rm V} V^2 + k_{\rm XVI})^2}\right], \tag{12.40}$$

with the end conditions $h(t^0) = h^0$, $V(t^0) = V^0$, $h(t^1) = h^1$, $V(t^1) = V^1$.

The optimal law of variation of the angle γ, under the foregoing assumptions, will therefore have to be sought among those that achieve the maximum of the function

$$\mathscr{H} = \frac{g^2(2k_{\rm III} V^2 + k_{\rm XVI})\xi_2}{2(k_{\rm V} V^2 + k_{\rm XVI})^2} \sin^2 \gamma + (\xi_1 V - g\xi_2) \sin \gamma. \tag{12.41}$$

It can be seen that, if

$$\frac{-\pi}{2} \leqslant -\gamma^{(1)} \leqslant \gamma^{(2)} \leqslant \frac{\pi}{2}, \tag{12.42}$$

$\sin \tilde{\gamma}$, hence also $\tilde{\gamma}$, are determined by means of Table 12.7, where this time $\lambda = \sin \gamma$, $\mu = \mathscr{H}$, $a = g^2(2k_{\rm III} V^2 + k_{\rm XVI})\xi_2/[2(k_{\rm V} V^2 + k_{\rm XVI})^2]$, $b = \xi_1 V - g\xi_2$, $\lambda^{(1)} = \sin^{-1}\gamma^1$, $\lambda^{(2)} = \sin^{-1}\gamma^{(2)}$.

For VTOL aircraft† $\gamma^{(1)} = \gamma^{(2)} = \pi/2$, hence it follows that the optimal trajectory will consist of straight vertical climbs and dives, as well as of arcs with a variable inclination, equal at every instant to $(\xi_1 V - g\xi_2) \cdot (k_{\rm V} V^2 + k_{\rm XVI})^2/[g^2(2k_{\rm III} V^2 + k_{\rm XVI})\xi_2]$, the switching conditions being those indicated in Table 12.7.

12.7. STATE VARIABLE CONSTRAINT

In the questions so far treated restrictions were only set on the control variables. As to the state variables, it was assumed that the values resulting in compliance with the motion equations are admissible; in other words, that the accessibility region is coincident with that of admissible values. The assumption is not obvious and, rigorously speaking, not always justified. Thus, for instance, the flight altitude results unlimited from the motion

† For conventional aircraft $\gamma^{(1)}$ and $\gamma^{(2)}$ depend at every instant on the flight régime: $\gamma^{(1)} = \gamma^{(1)}(V, \alpha, \ldots)$, $\gamma^{(2)} = \gamma^{(2)}(V, \alpha, \ldots)$ and the analysis would be complicated by the requirement to examine a larger number of situations that would bring no new element from the point of view of principle.

equation in which the variation of density was neglected (by supposing that the aerodynamic forces depend on the flight velocity and aerodynamic characteristics only, and the thrust on the velocity alone); or, the maximum velocity resulting from the equations on the basis of the available thrust, the aerodynamic characteristics, and the weight of the aircraft may exceed the velocity admitted on considerations of strength of the structure or on physiological grounds (in the case of a curved trajectory), and so on. Hence, it appears necessary to check the applicability of this assumption a posteriori. But there are situations in which this assumption is a priori invalid with respect to the motion equations used. For example, the equations written for the longitudinal motion of VTOL aircrafts in the preceding sections permit any horizontal movement independent of the altitude h (any point in level flight is accessible). The operating conditions of these machines may, however, impose a restriction on horizontal movements. This case will be outlined in the present section. In situations of this kind the imposition of constraints on the state variables becomes unavoidable.

Although there is a mathematical theory relating to the necessary extremum conditions for problems with state variable constraint (see [6, Chapter 6], and [5] as representing a more recent contribution), it is apparently very difficult to use the necessary conditions for obtaining information about optimum conditions. Hence, considering the degree of approximation of the model in general, the replacement of state variable constraints with some penalty imposed for infringement of these constraints introduced into the performance index itself appears to be preferable. (With regard to the mathematical theory of the penalty method in optimization problems, see [10, 11].) In this way, the problem reduces to a similar one treated previously in this chapter.

For definiteness, consider the minimum time problem under incidence control. The problem was treated in the previous section without taking into account the horizontal position coordinate of the center of mass. This coordinate will now be introduced as an additional state variable. We will denote it by x. One more equation will thereby be added to (12.32) and the system to be hereunder considered will be

$$\frac{dx}{dt} = V\cos\gamma, \qquad \frac{dh}{dt} = V\sin\gamma,$$

$$\frac{dV}{dt} = -(k_{\mathrm{II}} + k_{\mathrm{III}}\alpha^2)V^2 - g\sin\gamma + k_{\mathrm{XVI}}\left(1 - \frac{\alpha^2}{2}\right),$$

$$\frac{d\gamma}{dt} = k_{V\alpha}V - \frac{g\cos\gamma}{V} + \frac{k_{XVI}\alpha}{V}. \tag{12.43}$$

Specifically, VTOL aircraft can take off from any point, whatever the configuration of the ground (the local orographic conditions, the distance and height of the surrounding buildings, etc.). Hence, in stating the problem it might become necessary to include certain constraints on the horizontal movement (i.e., on the variable x), depending on the configuration of the ground. Such restrictions obviously cannot be implied by system (12.43); they have therefore to be modeled separately, for example, in the form of an inequality such as $r(x, h) \leqslant 0$. For simplicity the function r is considered linear, namely, $r = x - x_0 - kh$, with $k = \text{const.}$ (Suppose that the aircraft has to take off from the bottom of a valley. Then the equation $r(x, h) = 0$ will represent the generatrix corresponding to the intersection of the plane in which the evolution takes place with a cone inscribed in the surface of the valley.) Hence the additional constraint will be

$$r \equiv x - x_0 - kh \leqslant 0, \qquad k = \text{const.} \tag{12.44}$$

Now we will add to the performance index, which for the minimum time problem reduces to $t^1 - t^0$, the functional $\sigma \int_{t^0}^{t^1} p(r)\, dt$ with $\sigma = \text{const.}$ The penalty function $p(r)$ will be chosen of the form

$$p(r) = \begin{cases} 0 & \text{for} \quad r \leqslant -\varepsilon \\ (r + \varepsilon)^2 & \text{for} \quad r \geqslant -\varepsilon \end{cases}$$

with ε constant. Since the performance index is

$$\hat{I} = \int_{t^0}^{t^1} (1 + \sigma p(r))\, dt \tag{12.45}$$

for a sufficiently high value of the constant σ, however small might be the time interval $t^1 - t^0$, $\hat{I}$ cannot be a minimum if the aircraft approaches the boundary imposed, $r = 0$, at a distance on the level smaller than ε, ε playing here the part of a safety margin.

Hereafter the problem will be treated as one without state variable constraint. Changing the control system, including the performance index with respect to the case treated in Section 12.6, will entail a modification of the optimal trajectory. Indeed, according to the maximum principle, the optimal incidence, if it exists, will ensure at every instant the maximum of the function

$$\hat{H} = - \sigma p(r) + \hat{\xi}_0 V \cos \gamma + H$$

where H corresponds to system (12.32) considered in the previous section and $\hat{\xi}_0$ is the auxiliary variable corresponding to the additional state variable x. That part of H which contains the control variable α is formally identical with $\mathscr{H}$ in (12.36), differing from the latter, however, in that the auxiliary variables corresponding to V and γ (viz., $\hat{\xi}_2$ and $\hat{\xi}_3$) are no longer the same as a result of the modification of the adjoint system. Indeed, this latter will be (the prime denotes derivative)

$$\frac{d\hat{\xi}_0}{dt} = - \frac{\partial \hat{H}}{\partial x} \equiv \sigma p'(r),$$

$$\frac{d\hat{\xi}_1}{dt} = - \frac{\partial \hat{H}}{\partial h} \equiv - \sigma k p'(r),$$

$$\frac{d\hat{\xi}_2}{dt} = - \frac{\partial \hat{H}}{\partial V} \equiv - \hat{\xi}_0 V \cos \gamma - \frac{\partial H}{\partial V} (h, V, \gamma, \hat{\xi}_1, \hat{\xi}_2, \hat{\xi}_3),$$

$$\frac{d\hat{\xi}_3}{dt} = - \frac{\partial \hat{H}}{\partial \gamma} \equiv \hat{\xi}_0 V \sin \gamma - \frac{\partial H}{\partial \gamma} (h, V, \gamma, \hat{\xi}_1, \hat{\xi}_2, \hat{\xi}_3).$$

Consequently, the optimal trajectory in the conditions of a penalty being imposed upon approaching the safety boundary, as in the case of the system without constraint, will consist of arcs along which the incidence is constant, equaling its maximum value; of arcs along which α is constantly equal to its minimum level; and finally of arcs with a variable incidence having the expression $\hat{\xi}_3(k_V V^2 + k_{XVI})/[\hat{\xi}_2 V(2k_{III} V^2 + k_{XVI})]$. The switching conditions are those indicated in Table 12.7, where $\tilde{\lambda}$ denotes the optimal incidence, $a = - \hat{\xi}_2(k_{III} V^2 + k_{XVI}/2)$, $b = \hat{\xi}_3(k_V V^2 + k_{XVI}/V)$, and $c = - b/2a$.

12.8. OPTIMAL STABILIZATION: GENERAL

12.8.I. Introduction

Sections 12.4–12.7 considered the question of optimal programming of the basic motion by establishing an optimal policy for the basic controls. The subsequent sections of this chapter are concerned with the determination of the optimal synthesis of correction controls and the respective optimal system in certain basic motions. That is, the problem will be investigated for the flight situations discussed under different aspects in the previous chapters: the angular stabilization of VTOL aircraft in hovering flight and the artificial improvement of longitudinal stability characteristics of conventional aircraft in the phugoid motion.

The conditions for certain basic motions that are unstable or have unsatisfactory response characteristics in free flight to be artificially stabilized by means of controls were analyzed in Sections 4.3, 5.5, 5.6, 5.7, and Chapters 6 and 7. The conditions stated in Sections 4.3 and 5.7 are assumed to be satisfied.

Simplified models will be used, enabling us to solve the problem effectively by obtaining the synthesized optimal controls. Thus, for describing the rotation induced about the center of mass of a hovering VTOL aircraft (see Section 7.3), we have recourse to the simplest model, namely, the free rigid body in equilibrium, whereby the resultants of the external forces and moments are assumed to be vanishingly small in both the basic and the disturbed motion. (The simplified model corresponding to the assumption of neglecting the gyroscopic moments was termed standard scheme in Section 7.3.) The artificial stabilization of the motion about its center of mass will be investigated, under the assumption that both the restoring and the damping moment are supplied completely by the controls. To illustrate a case of quasi-neutral stability (slightly damped, or gently divergent disturbed motion) the phugoid motion of a conventional aircraft will be modeled in a manner consistent with the classical hypothesis that the incidence angle is canceled throughout the motion.

As the value of the control variable will appear—in the form of cost—in the performance index, it will be possible to forego the assumption that the domain of the controls is closed: the limitation of the control variables will implicitly appear in the results. Linear systems will be used for rotation damping in hovering in the ideal case, which will here be called the standard case, as well as for the phugoid induced motion. Thus, it will be possible to have recourse to the mathematical model furnished by the problem stated by Letov [12] and Kalman [13, 14] considered over an infinite interval of time. A procedure due to Popov [15, Section 30], permitting us to work out to the end the problems proposed, will subsequently be applied.

In the problem of artificially damping the free rotation—in the absence of some restoring or damping moment—we will use the classical nonlinear model of the rotation of a rigid body about its center of mass, which supplies a more truthful description of the rotation of hovering VTOL aircraft or of a flying machine in extra-atmospheric space. The case will be treated by Bellman's method of dynamic programming [16].

12.8.2. Statement of the Kalman–Letov Problem

This problem will here be stated in a form that takes into account the models that are to be used. Consider the control system

$$\frac{dx}{dt} = Ax + b\bar{u}, \qquad x(0) = x_0 \tag{12.46}$$

where the column vector x represents the state variables and $\bar{u}$ the control variables, while A is a constant matrix and b a constant row vector. Consider also the functional

$$J = \int_0^\infty (x_j^2 + \kappa u^2)\, dt \tag{12.47}$$

where x_j and u are given scalar components of the vectors x and $\bar{u}$, respectively, while κ is a strictly positive scalar constant. In particular, the vector $\bar{u}$ has all its components canceled out except the component u, whereas the only component of the vector b that is nonzero and equal to 1 will be that of the same order.

We seek the control $u(t)$, belonging to the class of piecewise continuous functions, for which

$$\lim_{t \to \infty} x(t) = 0$$

while the integral J converges and is minimal.

12.9. HOVERING FLIGHT OF VTOL AIRCRAFT

12.9.1. Linear Models

In writing the equations, the method of partially controlled motions is once more applied. The linear auxiliary system corresponding to the constraint (or "partial control") of the motion of the mass center under the mentioned assumption will be

$$\frac{dp}{dt} = -c_1 q + u_1, \qquad \frac{dq}{dt} = c_2 p - c_3 r + u_2,$$

$$\frac{dr}{dt} = c_4 q + u_3, \qquad \frac{d\phi}{dt} = p - r \tan\theta_0,$$

$$\frac{d\theta}{dt} = q, \qquad \frac{d\psi}{dt} = r \sec \theta_0; \tag{12.48}$$

u_1, u_2, and u_3 are the components of the control moment about the three body axes;† the constants c_1, c_2, c_3, and c_4 depend on the moments and products of inertia of the machine about the aircraft body axes with respect to which the equations were written. They will not be involved in the considerations of the present chapter.

System (12.48) simplifies in a manner consistent with the following assumptions.

(i) The undisturbed angle of pitch is assumed to be negligible ($\theta_0 = 0$), by considering VTOL aircraft flying at normal attitude.

(ii) Lift is achieved by an even number of turbojet engines, and the corresponding rotors on opposite sides about the medium plane of symmetry have the same rotation speed and are opposite handed; then $c_1 = c_2 = c_3 = c_4 = 0$.

Therefore, system (12.48) decomposes into three independent systems, corresponding to the three axes of the reference frame adopted: one for the Ox axis, namely,

$$\frac{dp}{dt} = u, \qquad \frac{d\phi}{dt} = p, \tag{12.49}$$

and two other similar systems corresponding to the Oy and Oz axes.

In order to solve the problem completely, the simplest model will be considered first.

$$\frac{dp}{dt} = u, \qquad p(t^0) = p^0 \tag{12.50}$$

and a functional of the type (12.47), namely,

$$J = \int_0^\infty (p^2 + \kappa u^2)\, dt$$

where the constant positive coefficient κ measures the cost, will be taken as the performance index. According to the maximum principle, if there

† In Section 7.6 the case of the simultaneous utilization of rotation dampers and static stabilizers is considered, hence taking $u_1 = -c_p p - c_\phi \phi$, $u_2 = -c_q q - c_\theta \theta$, $u_3 = -c_r r - c_\psi \psi$. The translation of the mass center was assumed canceled by manual control.

exists a control u that is optimal in the sense shown, it must be sought among those securing at every instant the maximum value of the function

$$H = -p^2 - \kappa u^2 + \xi u$$

where the additional variable ξ satisfies the adjoint equation

$$\frac{d\xi}{dt} = -\frac{\partial H}{\partial p} \equiv 2p.$$

The function H reaches its maximum at the points at which $\partial H/\partial u = 0$ (the region of controls being open), hence for $u = \xi/2\kappa$.

A synthesis will be obtained by integration of the system

$$\frac{dp}{dt} = \frac{1}{2\kappa}\xi, \qquad \frac{d\xi}{dt} = 2p. \tag{12.51}$$

Since the characteristic roots of the system, namely $\lambda_{1,2} = \pm 1/\sqrt{\kappa}$ are not situated on the imaginary axis, a solution will be framed for the negative characteristic root

$$p(t, p^0, \xi^0) = p^0 e^{-t/\sqrt{\kappa}}, \qquad \xi(t, p^0, \xi^0) = \xi^0 e^{-t/\sqrt{\kappa}} \tag{12.52}$$

whence $\xi/p = \xi^0/p^0$. By substituting the solution in any of the equations (12.51), we have $\xi^0/p^0 = -2\sqrt{\kappa}$ and therefore the synthesis of optimal control will be

$$u = \frac{-p}{\sqrt{\kappa}}. \tag{12.53}$$

Consequently, if the model (12.50) can be considered adequate, the equation of rotation damper minimizing the functional is, with the admitted approximations, (12.53). Equation (12.50) becoming $dp/dt = -p/\sqrt{\kappa}$, the angular velocity of roll—if the model (12.50) is acceptable—will decay exponentially according to the first equation in (12.52).

Let us now resume the model (12.49). As a performance index it is possible to use either the functional $\int_0^\infty (p^2 + \kappa u^2)\, dt$ for the optimization of damping, or

$$J = \int_0^\infty (\phi^2 + \kappa u^2)\, dt \tag{12.54}$$

for the optimization of the static stabilization of the lateral attitude, or $\int_0^\infty (p^2 + \kappa_1\phi^2 + \kappa_2 u^2)\, dt$ for the optimization of both damping in roll and static stabilization about the longitudinal axis simultaneously applied to the aircraft motion. By using the same procedure as above, it can readily be seen that whatever might be the performance index chosen, the optimum system has characteristic roots on the imaginary axis, and hence the procedure outlined for the model (12.49) cannot be used. The problem can be solved by applying a method due to Popov. An account of this method is beyond the scope of this book. The reader is advised to consult Chapter 5, Section 30, of the monograph [15]. It will suffice here to recall the conditions in which Popov's procedure is applicable. To this end, the following two functions of two complex variables λ and σ will be constructed for (12.46) and (12.47).

$$\chi(\lambda, \sigma) = \kappa + b^*(\lambda I - A^*)^{-1} M(\sigma I - A)^{-1} b \tag{12.55}$$

and

$$\pi(\lambda, \sigma) = \frac{1}{\tilde{\kappa}} \det(\lambda I - A^*) \det(\sigma I - A). \tag{12.56}$$

An asterisk denotes the transpose. In χ, M denoted a constant matrix depending on the coefficients of the state variables occurring in the functional J. In the studied case, for the performance index (12.54), M is expressed by

$$M = \begin{pmatrix} 0 & 0 \\ 0 & 1 \end{pmatrix}.$$

I is the unity (identity) matrix, $\tilde{\kappa}$ is a positive norming factor that, however, will not occur in the arguments.

The condition for the problem to be solved is

$$\pi(-i\omega, i\omega) \neq 0 \tag{12.57}$$

for any real ω.

For precision, the performance index (12.54) will be used for the system (12.49). Then

$$\chi(\lambda, \sigma) = \kappa + \frac{1}{\sigma^2\lambda^2}, \quad \pi(\lambda, \sigma) = \frac{1}{\tilde{\kappa}}(1 + \kappa\sigma^2\lambda^2).$$

We have

$$\pi(-i\omega, i\omega) = \frac{1}{\tilde{\kappa}}(1 + \kappa\omega^4)$$

and therefore condition (12.57) is satisfied.

Application of this procedure yields a synthesis of optimal control under the form (for certain details of deduction, see Appendix II)

$$u = -\frac{2}{(4\kappa)^{1/4}}p - \frac{1}{\kappa^{1/2}}\phi. \tag{12.58}$$

The optimal system will be

$$\frac{dp}{dt} = -\left(\frac{4}{\kappa}\right)^{1/4} p - \frac{1}{\kappa^{1/2}}\phi, \quad \frac{d\phi}{dt} = p$$

and the optimal value of the performance index (12.54)

$$J_{\mathrm{opt}} = \kappa \left[\left(\frac{4}{\kappa}\right)^{1/4} p^{02} - \frac{2}{\kappa^{1/2}} p^0\phi^0 + \left(\frac{4}{\kappa^3}\right)^{1/4} \phi^{02}\right].$$

The optimal control (12.58) implies the simultaneous use of static stabilizers and rotation dampers in a certain proportion (given by the coefficients of the right-hand side), although the performance index refers only to the correction of the deviation of the angle of bank.

12.9.2. Dynamic Programming Approach: Partially Controlled Hovering VTOL Aircraft Modeled as a Rigid Body with Fixed Point: A Nonlinear Model

Consider the rotation of a rigid body about the center of mass assumed to be fixed. This simple physical model acceptably describes the motion induced by a disturbance of a VTOL aircraft of which the basic motion is hovering under the constraint of the translation of mass center. In this case the known equations of the rigid body with fixed point appear as an auxiliary system for the partially controlled motion. Such a model seems to be closer to the actual situation than the linear one used in the preceding section.

The problem of optimum damping will be stated as follows. Let M_1, M_2, M_3 be the control moments about the three axes of the reference frame fixed in the aircraft (body axes). As in the preceding section, the control variables u_1, u_2, u_3 will be used in their stead: $u_1 = M_1/A$, $u_2 = M_2/B$, $u_3 = M_3/C$. We note $h_1 = Ap$, $h_2 = Bq$, $h_3 = Cr$, and $K_1 = 1/A$, $K_2 = 1/B$, $K_3 = 1/C$. (Here A, B, and C stand, as in some of the preceding chapters, for the aircraft moments of inertia about the body axes.)

Consider the control system

$$\frac{dh_1}{dt} = (K_3 - K_2)h_2h_3 + u_1 \equiv f_1,$$

$$\frac{dh_2}{dt} = (K_1 - K_3)h_3h_1 + u_2 \equiv f_2,$$

$$\frac{dh_3}{dt} = (K_2 - K_1)h_1h_2 + u_3 \equiv f_3, \tag{12.59}$$

with the initial conditions $h_1(0) = a_1$, $h_2(0) = a_2$, $h_3(0) = a_3$. The control variables will be determined to minimize the integral functional $\int_0^\infty [\lambda(p^2 + q^2 + r^2) + \mu(M_1^2 + M_2^2 + M_3^2)]\, dt$, or, with the above notation

$$J(u_1, u_2, u_3) = \int_0^\infty \sum_i (\beta_i h_i^2 + \gamma_i u_i^2)\, dt$$

where $\beta_1 = \lambda/A^2$, $\beta_2 = \lambda/B^2$, $\beta_3 = \lambda/C^2$, $\gamma_1 = \mu A^2$, $\gamma_2 = \mu B^2$, $\gamma_3 = \mu C^2$. Symbols λ, μ, β_i, and γ_j denote positive constants indicating the desired weight ratio of components of the damping rotation and of the cost of control.

The existence of an optimal control is assumed, no matter what the initial conditions. No constraint is imposed a priori upon the control. The problem is solved via a dynamic programming method [16], by following the reasoning outlined hereafter.

Consider the system

$$\frac{dx}{dt} = f(x, u), \qquad x(t^0) = x^0$$

where the vector x represents the state variables and the vector u the control variables. It is required to determine the control u so that the functional

$$J(u) = \int_0^\infty f_0(x(t), u(t))\, dt$$

be minimized.

Let $\varphi(x)$ be a differentiable function such that $\varphi(0) = 0$ and $\min_u\{\sum_i \partial\varphi/\partial x_i (x) f_i(x, u) + f_0(x, u)\} = 0$, where x_i and f_i are the components of vectors x and f, respectively. Let $x(t)$ be a solution of the system corresponding to a certain control u. We have

$$J(u) = \int_0^\infty f_0(x(t), u(t))\, dt \geqslant -\int_0^\infty \sum_i \frac{\partial\varphi}{\partial x_i}(x(t)) f_i(x(t), u(t))\, dt$$

$$= -\int_0^\infty \frac{d\varphi}{dt}(x(t))\,dt = \varphi(x(0)) - \varphi(x(\infty)).$$

If $\lim_{t\to\infty} x(t) = 0$, then $J(u) \geqslant \varphi(x^0)$.

On the other hand, if $\tilde{u}(x)$ is the function for which $\sum_i \partial\varphi/\partial x_i\,(x) f_i(x, \tilde{u}(x)) + f_0(x, \tilde{u}(x)) \equiv 0$, that is, the function for which the minimum is reached, it follows by virtue of the above inequalities that $J(\tilde{u}) = \varphi(x^0) \leqslant J(u)$, whatever u may be; hence the control $\tilde{u}$ is optimal.

It follows that if a function φ, with $\varphi(0) = 0$, and a control $\tilde{u}(x)$ are found such that $\min_u\{\sum_i \partial\varphi/\partial x_i\, f_i(x, u) + f_0(x, u)\} = 0$ and $\sum_i \partial\varphi/\partial x_i\, f_i(x, \tilde{u}(x)) + f_0(x, \tilde{u}(x)) \equiv 0$ and zero solution of the system $dx/dt = f(x, \tilde{u}(x))$ is asymptotically stable with x^0 in the attraction region, then the control $\tilde{u}(x)$ is optimal.

Resuming the considered problem, let $\varphi(h_1, h_2, h_3) = (\lambda\mu)^{1/2} \sum_i h_i^2$. We obviously have $\lambda\mu = \beta_i\gamma_i$, hence $\sum_i (\partial\varphi/\partial h_i\, f_i + \beta_i h_i^2 + \gamma_i u_i^2) = \sum_i(\sqrt{\beta_i}h_i + \sqrt{\gamma_i}u_i)^2$.

The minimum of this sum, equal to zero, will be reached for $\tilde{u}_i = -\,h_i(\beta_i/\gamma_i)^{1/2}$, or by resuming the initial notations, for

$$\tilde{M}_1 = -\left(\frac{\lambda}{\mu}\right)^{1/2} Ap, \quad \tilde{M}_2 = -\left(\frac{\lambda}{\mu}\right)^{1/2} Bq, \quad \tilde{M}_3 = -\left(\frac{\lambda}{\mu}\right)^{1/2} Cr. \tag{12.60}$$

It can readily be seen that the corresponding system

$$\frac{dh_1}{dt} = -\left(\frac{\beta_1}{\gamma_1}\right)^{1/2} h_1 + (K_3 - K_2)h_2h_3,$$

$$\frac{dh_2}{dt} = -\left(\frac{\beta_2}{\gamma_2}\right)^{1/2} h_2 + (K_1 - K_3)h_3h_1,$$

$$\frac{dh_3}{dt} = -\left(\frac{\beta_3}{\gamma_3}\right)^{1/2} h_3 + (K_2 - K_1)h_1h_2,$$

is asymptotically stable in the large (globally)† since the sum $\sum_i h_i^2$ is a Liapunov function for this system having the derivative with respect to t along the integral curves equal to $-\,2\sum_i (\beta_i/\gamma_i)^{1/2}h_i^2$. Consequently, as the conditions stated above are fulfilled, the control (12.60) is optimal for the system (12.59), while the performance index J represents a desired compromise

† The local asymptotic stability results, according to the theory of stability by the first approximation, from the linearized system, resolved into three independent exponentially stable systems.

between the effect of stabilization and its cost. Although the system (12.59) is nonlinear, the ensuing optimal control is linear as a combination of linear rotation dampers about the three body axes of the reference frame fixed in the aircraft proportional each to the corresponding moment of inertia.

The damping velocity can be estimated by means of the Liapunov function $\mathscr{V} = \sum_i h_i^2$ using the known procedure. The estimation will obviously be given by the most slightly damped mode. Let $\nu = \min_i 2(\beta_i/\gamma_i)^{1/2}$. Then damping will take place at least with the same velocity as that of the exponential $e^{-\nu t}$. For instance, if the greatest inertia moment is the one about the Oz axis, then $\nu = (2/C)(\lambda/\mu)^{1/2}$.

12.10. OPTIMAL STABILIZATION OF LONGITUDINAL ATTITUDE IN THE PHUGOID PHASE

The phugoid motion generally constitutes a quasi-neutral behavior with respect to the deviation induced by the disturbance of flight velocity and of the angle of pitch with an almost constant incidence. Rigorously, it is a question in most cases of a slow stable or divergent variation of these variables, usually countered without particular effort by the controls. The effort can become appreciable in the case of a heavier divergence. In the case of an automatic stabilization the attempt at an optimal programming of the stabilizer appears to be justified even for slight instabilities. A suitable stabilization of the velocity and of the attitude implies the stability of the longitudinal motion on the whole, according to the theory of the stability of partially controlled motions.

The phugoid motion will be modeled by supposing the incidence strictly constant. It is assumed that the automatic stabilizer operates on the angular displacement about the lateral axis. The following system will therefore be considered.

$$\frac{d\Delta V}{dt} = p_{11}\Delta V + p_{12}\Delta\theta, \qquad \frac{d\Delta\theta}{dt} = p_{21}\Delta V + u, \tag{12.61}$$

with u a (scalar) control variable. It is required to secure the minimum of a performance index representing a suitable compromise between the stabilization of the longitudinal attitude and the cost of the respective control:

$$J(u) = \int_0^\infty (\Delta\theta^2 + \kappa u^2)\, dt$$

where κ is a positive constant. The optimal programming will be sought by a procedure based on the above-mentioned method of Popov [15]. In this case, according to the definitions (12.55) and (12.56),

$$\chi(\lambda, \sigma) = \kappa + (0 \quad 1)\begin{pmatrix} \lambda - p_{11} & -p_{21} \\ -p_{12} & \lambda \end{pmatrix}\begin{pmatrix} 0 & 0 \\ 0 & \mu \end{pmatrix}\begin{pmatrix} \sigma - p_{11} & -p_{12} \\ -p_{21} & \sigma \end{pmatrix}^{-1}\begin{pmatrix} 0 \\ 1 \end{pmatrix}$$

$$= \kappa + \frac{(\lambda - p_{11})(\sigma - p_{11})}{(\lambda^2 - p_{11}\lambda - p_{12}p_{21})(\sigma^2 - p_{11}\sigma - p_{12}p_{21})},$$

$$\pi(\lambda, \sigma) = \frac{1}{\tilde{\kappa}}\,[\kappa(\lambda^2 - p_{11}\lambda - p_{12}p_{21})(\sigma^2 - p_{11}\sigma - p_{12}p_{21}) + (\lambda - p_{11})(\sigma - p_{11})].$$

Then

$$\pi(-i\omega, i\omega) = \frac{1}{\tilde{\kappa}}\,\{\kappa[(\omega^2 + p_{12}p_{21})^2 + p_{11}^2\omega^2] + p_{11}^2 + \omega^2\} \neq 0$$

for any real ω, and hence the procedure is applicable. The synthesis of optimal control $\tilde{u}(\Delta V, \Delta\theta)$ is obtained in the form (see Appendix II)

$$\tilde{u} = \frac{\sigma_1\sigma_2 - p_{11}(\sigma_1 + \sigma_2)}{p_{12}}\,\Delta V - (\sigma_1 + \sigma_2)\,\Delta\theta \tag{12.62}$$

where σ_1 and σ_2 are the roots with a negative real part of the equation

$$\kappa\sigma^4 - (2\kappa p_{12}p_{21} + p_{11}^2 + 1)\sigma^2 + \kappa p_{12}^2 p_{21}^2 + p_{11}^2 = 0. \tag{12.63}$$

There results accordingly an optimal control linear in ΔV and $\Delta\theta$.

12.11. SOME CONCLUSIONS

In modeling the disturbed motion of VTOL aircraft in the case of undisturbed hovering flight, we assumed that the aerodynamic terms are negligible according to the theory of stability by the first approximation.

For attitude stabilization and rotation damping about each of the body axes, the optimal control, inferred on the basis of a linearized set of equations, is for each axis a combination of a static stabilizer and a rotation damper in a proportion resulting from (12.58). For the artificial damping of the rotation of this machine, considered as a rigid body, about its center of mass, the optimal control inferred on the basis of the classical nonlinear system is a set of three linear rotation dampers, one for each axis of the reference frame. The corresponding optimal system results asymptotically stable. The velocity of damping is estimated depending on the performance index.

The method used for solving the optimum problem was that of the dynamic programming [16] for the nonlinear model (12.59) and a technique due to Popov [15] for the linear systems of the type (12.49).

The latter method was also used in the problem of optimal programming of the controls for the stabilization of longitudinal attitude of conventional aircraft in the phugoid phase of the induced motion. On the basis of the known model corresponding to the variation of incidence being neglected in this phase, there results an optimal control linear with respect to the dynamic variables ΔV and $\Delta \theta$ of the system.

APPENDIX I

COMPUTATIONAL ASPECTS

For definiteness we will consider the problem of minimum fuel in climbing and accelerated flight for fixed-thrust VTOL aircraft. The flight program is assumed to be achieved by thrust and longitudinal moment control. This case of optimal control was examined in Section 12.6.

The control system considered will therefore consist of the set of equations (12.30) together with the end conditions given in Table A.1.

Table A.I. End Conditions for System (12.30)

t	h	V	γ	α	q
t^0	h^0	V^0	free	free	free
t^1 free	h^1	V^1	$\gamma^1 = 0$	α^1	$q^1 = 0$

The corresponding adjoint system will be

$$\frac{d\xi_1}{dt} = -\frac{\partial H}{\partial h} = 0, \quad \frac{d\xi_2}{dt} = -\frac{\partial H}{\partial V}, \quad \frac{d\xi_3}{dt} = -\frac{\partial H}{\partial \gamma}, \quad \frac{d\xi_4}{dt} = -\frac{\partial H}{\partial \alpha}, \quad \frac{d\xi_5}{dt} = -\frac{\partial H}{\partial q} \tag{1}$$

where

$$H = -f_0(\tau) + \xi_1 f_1 + \xi_2 f_2 + \xi_3 f_3 + \xi_4 f_4 + \xi_5 f_5. \tag{2}$$

The end conditions given in Table A.1 are further completed by three initial conditions for the auxiliary variables, that is, by

$$\xi_3(t^0) = 0, \quad \xi_4(t^0) = 0, \quad \xi_5(t^0) = 0, \tag{3}$$

which are deduced from the transversality conditions corresponding to the free initial values of the state variables γ, α, and q. Consider also the relation

$$\max_{\substack{0\leqslant\tau\leqslant\tau_{\max} \\ -\kappa^{(1)}(V,\alpha)\leqslant\kappa\leqslant\kappa^{(2)}(V,\alpha)}} H(t^1) = 0 \tag{4}$$

expressing condition (b) of the maximum principle (see Section 12.3).

To solve the optimum problem stated, the system (12.30) and (1) above will be integrated for the two set initial conditions (Table A.1), for the three deduced concerning the auxiliary variables (3), and for the remaining three state variables and two auxiliary variables taken arbitrarily. (Since the system is autonomous, integration could also be carried out backward from the instant t^1 considered as initial, hence set, to the instant t^0 taken as terminal and free. In such a case, the values of all of the state variables set for the instant t^1 (i.e., h^1, V^1, γ^1, and q^1) would be considered as known "initial" conditions, while the values at instant t^1 of all the auxiliary variables ξ_i would have to be chosen arbitrarily. This is, however, an obvious drawback, since auxiliary variables are usually exempt from any definite physical meaning, hence guessing their right initial values or some values close to these is practically impossible from the start. As regards the state variables, on the contrary, the initial guess may be greatly expedited by physical considerations. For instance, for fixed-thrust VTOL aircraft we can try to start with initial values such as $\gamma(t^0) = \pi/2$, $\alpha(t^0) = 0$, $q(t^0) = 0$.)

We lead off with values of the control variables that ensue, for these values of the state and auxiliary variables, from Tables **12.3** and **12.4**. Obviously, the algorithm also contains (i) the switching relations from Tables **12.3** and **12.4**, according to which the system (12.30), (1) is modified according to the solution, (ii) the set terminal conditions for the state variables (Table A.1), and (iii) the relation (4).

As has been recalled, the technique for solving a two-point boundary value problem similar to that stated is by trial and error. Integration is continued until part of the set terminal conditions (including relation (4)) is satisfied. The operation may be repeated with the arbitrary initial conditions altered until all the final relations supplied are verified to the required approximation. Because of the implied numerical differentiation, it is inconvenient to compute in this way the effect of the variation of the (free) initial conditions on the values corresponding to the instant t^1. For this reason, various authors have been seeking a possibility of accelerating the convergence of successive approximations by using procedures in which computation of the sensitivity of terminal values to changes of the initial conditions should imply integration instead of differentiation. This is the method that will be applied here, on the basis of a technique suggested by Jazwinski in [17].

As already shown, the solution of the problem implies a determination by trial and error of eleven free parameters: the values of the state variables γ, α, and q, and of the auxiliary variables ξ_1 and ξ_2 at the instant t^0; the values for $t = t^1$ of the five auxiliary variables as well as the difference $t^1 - t^0$. Therefore, the algorithm will contain eleven relations: five referring to the moment t^0:

$$\begin{aligned} &F_1(y(t^0)) \equiv y_1(t^0) - h^0 = 0, \quad F_2(y(t^0)) \equiv y_2(t^0) - V^0 = 0, \\ &F_3(y(t^0)) \equiv y_8(t^0) = 0, \qquad F_4(y(t^0)) \equiv y_9(t^0) = 0, \\ &F_5(y(t^0)) \equiv y_{10}(t^0) = 0, \end{aligned} \tag{5}$$

and six corresponding to the free final instant t^1:

$$\begin{aligned} &G_1(y(t^1)) \equiv y_1(t^1) - h^1 = 0, \quad G_2(y(t^1)) \equiv y_2(t^1) - V^1 = 0, \\ &G_3(y(t^1)) \equiv y_3(t^1) = 0, \qquad G_4(y(t^1)) \equiv y_4(t^1) = 0, \\ &G_5(y(t^1)) \equiv y_5(t^1) = 0, \qquad G_6(y(t^1)) \equiv \max_{\substack{0 \leqslant \tau \leqslant \tau_{\max} \\ -\kappa^{(1)} \leqslant \kappa \leqslant \kappa^{(2)}}} H(y(t^1)) = 0 \end{aligned} \tag{6}$$

where the column vector y stands for $y = (h, V, \gamma, \alpha, q, \xi_1, \xi_2, \xi_3, \xi_4, \xi_5)'$ (the prime denotes the transpose), hence $y_1 = h$, $y_2 = V, \ldots, y_{i+5} = \xi_i$. The expression of the function H is provided by (2). The symbol Y is also used to denote the column vector representing the right-hand sides of the system (12.30), (1); that is,

$$Y = \left(f_1, f_2, f_3, f_4, f_5, -\frac{\partial H}{\partial h}, -\frac{\partial H}{\partial V}, -\frac{\partial H}{\partial \gamma}, -\frac{\partial H}{\partial \alpha}, -\frac{\partial H}{\partial q}\right)'.$$

$F(y(t^0))$ will be a column vector whose components are the left-hand sides of system (5), that is, F_1, F_2, F_3, F_4, and F_5, and $G(y(t^1))$ the 6-column vector $(G_1, G_2, G_3, G_4, G_5, G_6)$, with G_i in (6).

Hence the system considered will be

$$\frac{dy}{dt} = Y(y) \tag{7}$$

while the end conditions (conditions (5) and (6)) will be written

$$F(y(t^0)) = 0 \tag{8}$$

and

$$G(y(t^1)) = 0, \tag{9}$$

respectively.

Let $y^{(1)}(t)$ be the solution corresponding to a set of initial values partly determined by conditions (5) and partly selected on the basis of a physical criterion concerning the state variables γ, α, and q, or taken arbitrarily. With these initial values, the numerical integration of system (7)—depending on the solution in agreement with the switching conditions—is carried out up to an instant $t^{1(1)} > t^0$ in which part of the equalities (6) are satisfied. For instance, we integrate up to the moment when the value h^1 (or the value V^1, etc.) is reached for the first time by the respective state variable. Generally, not all conditions (6) are verified by the solution $y^{(1)}(t)$, thus determined.

Further, the system in variation with respect to the solution $y^{(1)}(t)$ is framed, namely,

$$\frac{dz}{dt} = \frac{\partial Y}{\partial y}(t, y^{(1)})z \tag{10}$$

as well as the related adjoint system

$$\frac{d\lambda}{dt} = -\lambda \frac{\partial Y}{\partial y}(t, y^{(1)}) \tag{11}$$

with the row vector $\lambda = (\lambda_1, \lambda_2, \ldots, \lambda_{10})$. The general properties of adjoint systems imply

$$\lambda(t^{1(1)})z(t^{1(1)}) - \lambda(t^0)z(t^0) = 0. \tag{12}$$

Now, let $y^{(2)}(t)$ be a solution of system (7) corresponding to another set of initial values (for t^0) of the variables y_i and defined up to another terminal instant $t^{1(2)}$. We will have $y^{(2)}(t) - y^{(1)}(t) = z(t; t^0, y^{(2)}(t^0)) + o(y^{(2)}(t^0) - y^{(1)}(t^0))$. As $y^{(2)}(t^{1(2)}) - y^{(1)}(t^{1(1)}) = y^{(2)}(t^{1(1)}) - y^{(1)}(t^{1(1)}) + y^{(2)}(t^{1(2)}) - y^{(2)}(t^{1(1)}) = z^{(2)}(t^{1(1)}; t^0, y^{(2)}(t^0) - y^{(1)}(t^0)) + \dot{y}^{(2)}(t^{1(1)})(t^{1(2)} - t^{1(1)}) + o(y^{(2)}(t^0) - y^{(1)}(t^0)) + o(t^{1(2)} - t^{1(1)})$ (overdot denotes time derivative), and hence $\lambda(t^{1(1)})z^{(2)}(t^{1(1)}) = \lambda(t^{1(1)})[y^{(2)}(t^{1(2)}) - y^{(1)}(t^{1(1)}) - \dot{y}^{(2)}(t^{1(1)})(t^{1(2)} - t^{1(1)}) + \cdots]$ and instead of relation (12) we will have

$$\lambda(t^{1(1)})[y^{(2)}(t^{1(2)}) - y^{(1)}(t^{1(1)}) - \dot{y}^{(2)}(t^{1(1)})(t^{1(2)} - t^{1(1)}) + \cdots] - \lambda(t^0)z(t^0) = 0. \tag{13}$$

The adjoint system (11) is integrated backward, by considering $t^{1(1)}$ as the initial instant and t^0 as the terminal instant, for the following initial conditions.†

$$\lambda_k(t^{1(1)}) = \frac{\partial G_k}{\partial y}(t^{1(1)}), \qquad k = 1, 2, 3, 4, 5, 6. \tag{14}$$

Denoting the components of the row vector λ_k by $\lambda_{k,j}$, $j = 1, 2, \ldots, 10$, the initial conditions (14) amount, in scalar transcription, to

$$\lambda_{k,j}(t^{1(1)}) = \begin{cases} 1 & \text{for } j = k \\ 0 & \text{for } j \neq k \end{cases} \qquad k = 1, 2, 3, 4, 5. \tag{15}$$

Instead of (13) we will have

$$\frac{\partial G_k}{\partial y}(t^{1(1)})\,[y^{(2)}(t^{1(2)}) - y^{(1)}(t^{1(1)}) - \dot{y}^{(2)}(t^{1(1)})(t^{1(2)} - t^{1(1)}) + \cdots] - \lambda_k(t^0)z(t^0) = 0. \tag{16}$$

On the other hand, $G_k(y^{(2)}(t^{1(2)})) - G_k(y^{(1)}(t^{1(1)})) = \partial G_k/\partial y\,(y^{(1)}(t^{1(1)}))\cdot(y^{(2)}(t^{1(2)}) - y^{(1)}(t^{1(1)})) + \cdots$; hence (16) is written

$$G_k(y^{(2)}(t^{1^{(2)}})) - G_k(y^{(1)}(t^{1^{(1)}})) - \frac{\partial G_k}{\partial y}(y^{(1)}(t^{(1)^{(1)}}))\dot{y}^{(2)}(t^{1^{(1)}})(t^{1^{(2)}} - t^{1^{(1)}}) + \cdots - \lambda_k(t^0)z(t^0) = 0$$

or, since $\dot{y}^{(2)}(t^{1(1)}) = Y(y^{(2)}(t^{1(1)})) = Y(y^{(1)}(t^{1(1)})) + \cdots = \dot{y}^{(1)}(t^{1(1)})$, the relation becomes

$$G_k(y^{(2)}(t^{1(2)})) - G_k(y^{(1)}(t^{1(1)})) - \dot{G}_k(t^{1(1)})(t^{1(2)} - t^{1(1)}) + \cdots - \lambda_k(t^0)z(t^0) = 0. \tag{17}$$

We note that the difference $z(t^0) = y^{(2)}(t^0) - y^{(1)}(t^0)$ is not arbitrary; it satisfies the relation

$$\frac{\partial F}{\partial y}(y^{(1)}(t^0))(y^{(2)}(t^0) - y^{(1)}(t^0)) = 0. \tag{18}$$

Thus, if the difference $G_k(y^{(2)}(t^{1(2)})) - G_k(y^{(1)}(t^{1(1)})$ be specified, the variation of the initial conditions is determined from relations (17) and (18),

† If n is the number of state variables (in the specified case $n = 5$) and p the dimension of vector F (here $p = n = 5$), then system (11) is integrated backward for $2n - p + 1$ initial conditions (here $2n - p + 1 = 6$), the number $2n - p + 1$ being the dimension of vector G.

representing $p + (2n - p + 1)$ equations with $2n + 1$ unknown quantities: $2n$ (in the given case, equal to 10) unknown quantities $z_i(t^0) = y_i^{(2)}(t^0) - y_i^{(1)}(t^0)$, to which a further unknown quantity $t^{1(2)} - t^{1(1)}$ is added.

The problem is considered concluded when the initial conditions have been found satisfying (8) for which the solution of system (7) verifies to the set approximation the conditions (9). In order to approach this situation gradually, we choose in (17)

$$G_k(y^{(2)}(t^{1(2)})) - G_k(y^{(1)}(t^{1(1)})) = - c_k G_k(y^{(1)}(t^{1(1)}))$$

with $0 < c_k \leqslant 1$. Then the system of equations in $z_i(t^0)$ and $t^{1(2)} - t^{1(1)}$ will be written, by neglecting the additional nonlinear terms, as follows.

$$\begin{aligned} &\sum_{j=1}^{10} \frac{\partial F_i}{\partial y_j}(y^{(1)}(t^0))z_j(t^0) = 0, \qquad i = 1, 2, 3, 4, 5, \\ &\sum_{j=1}^{10} \lambda_{k,j}(t^0)z_j(t^0) + \dot{G}_k(t^{1(1)})(t^{1(2)} - t^{1(1)}) = - c_k G_k(y^{(1)}(t^{1(1)})) \qquad k = 1, 2, \ldots, 6. \end{aligned} \tag{19}$$

The initial conditions thus determined will have a corresponding solution that more nearly approximates the terminal conditions (9) because for $0 < c_k \leqslant 1$ we have obviously $G_k(y^{(2)}(t^{1(2)})) = (1 - c_k)G_k(y^{(1)}(t^{1(1)})) < G_k(y^{(1)}(t^{1(1)}))$.

It is admissible to confine oneself to the linear part of the variation provided that the numbers c_k are sufficiently small. But the nonlinear terms were disregarded above without a rigorous substantiation, and without an estimation of the admissible magnitude of the number c_k. Hence, their selection will have to be justified a posteriori, by computation. Namely, a trial will be made by groping with different values for c_k, and the greatest will be selected for which the solution corresponding to the new initial conditions $y^{(2)}(t^0)$ will verify to the required approximation the relations

$$G_k(y^{(2)}(t^1)) = (1 - c_k)G_k(y^{(1)}(t^{1(1)})).$$

If the numerical factors c_k so chosen were all equal to unity, the two-point problem would be solved after this first step. However, such situations obviously do not generally appear. For this reason the operations described above are repeated until a set of initial values $y^{(m)}(t^0)$ is found for which the respective solution verifies the terminal conditions $G(y^{(m)}(t^{1(m)}))$ acceptably (with a set precision).

Particularizing the problem for the system considered, the operations described are simplified. Thus, in the first five equations (19) the coefficients $\partial F_i/\partial y_j\,(y^{(1)}(t^0))$ are defined by the relations

$$\frac{\partial F_i}{\partial y_j}(y^{(1)}(t^0)) = \frac{\partial F_i}{\partial y_j}(y^{(s)}(t^0)) = \begin{cases} 1 & \text{for } i = j, \\ 0 & \text{for } i \neq j, \end{cases}$$

and these equations become

$$z_i(t^0) = 0, \qquad i = 1, 2, \ldots, s. \tag{20}$$

For the last six equations of the system, let us note that, by taking into account relations (14) and (15) and the obvious equality

$$\dot{G}_k(y^{(1)}(t^{1(1)})) = \sum_{j=1}^{10} \frac{\partial G_k}{\partial y_j}(y^{(1)}(t^{1(1)}))\, Y_j(y^{(1)}(t^{1(1)})),$$

we have

$$\dot{G}_k(y^{(1)}(t^{1(1)})) = Y_k(y^{(1)}(t^{1(1)})), \qquad k = 1, \ldots, 5,$$

$$\dot{G}_6(y^{(1)}(t^{1(1)})) = \sum_{j=1}^{10} \frac{\partial G_6}{\partial y_j}(y^{(1)}(t^{1(1)}))\, Y(y^{(1)}(t^{1(1)})).$$

Then, with (20) these equations become

$$\sum_{j=6}^{10} \lambda_{k,j}(t^0) z_j(t^0) + Y_k(y^{(1)}(t^{1(1)}))(t^{1(2)} - t^{1(1)}) = -\, c_k G_k(y^{(1)}(t^{1(1)})), \quad k = 1, \ldots, 5$$

$$\sum_{j=6}^{10} \lambda_{6,j}(t^0) z_j(t^0) + \left[\sum_{i=1}^{10} \frac{\partial G_6}{\partial y_i}(y^{(1)}(t^{1(1)}))\, Y_i(y^{(1)}(t^{1(1)}))\right](t^{1(2)} - t^{1(1)}) = -\, c_6 G_6(y^{(1)}(t^{1(1)})).$$

APPENDIX II

DEDUCTION OF CERTAIN FORMULAS

Details will be given concerning the deduction of the equations (12.58) and (12.62), supplying the optimal synthesis of correction controls for the relating problems.

Consider the control system (12.46) and the performance index (12.47). Construct the functions (12.55) and (12.56) of two complex variables λ and σ, and assume the condition (12.57) to be fulfilled. Then, there exists (see [15]) at least one polynomial $\psi(\sigma)$ having only roots with negative real parts, such that

$$\pi(-\sigma, \sigma) = \overline{\psi(-\bar{\sigma})}\psi(\sigma) \tag{1}$$

(bars denote complex conjugates); there exists also† w such that

$$w^*(\sigma I - A)^{-1}b = \frac{\tilde{\psi}(\sigma)}{\det(\sigma I - A)} \tag{2}$$

where $\tilde{\psi}(\sigma) = \sqrt{\kappa}\psi(\sigma) - \gamma \det(\sigma I - A)$, the number γ being defined as follows.

$$\gamma = \lim_{\sigma\to\infty} \frac{\sqrt{\kappa}\psi(\sigma)}{\det(\sigma I - A)}. \tag{3}$$

The optimal control is obtained from the relation

$$\gamma u + w^* x = 0. \tag{4}$$

System (12.49), (12.54). We have

$$x = \begin{pmatrix} p \\ \phi \end{pmatrix}, \quad A = \begin{pmatrix} 0 & 0 \\ 1 & 0 \end{pmatrix}, \quad b = \begin{pmatrix} 0 \\ 1 \end{pmatrix},$$

and in (12.47) $j = 2$ (viz., $x_j = \phi$). Functions χ in (12.55) and π in (12.56) are to be deduced.

The form chosen for $\psi(\sigma)$ is

$$\psi(\sigma) = \gamma_0 \left(\sigma^2 + \frac{2}{(4\kappa)^{1/4}}\sigma + \frac{1}{\kappa^{1/2}} \right)$$

where γ_0 results from (1) above, equal to $-(\kappa/\tilde{\kappa})^{1/2}$. The number γ in (3) and the function $\tilde{\psi}(\sigma)$ are obtained by means of the function $\psi(\sigma)$, and these are in turn used to determine the vector w^* in (2). Then, the optimal synthesis (12.58) results directly from (4).

Deduction of formula (12.63). In the case of the system (12.61) we have

$$x = \begin{pmatrix} \Delta V \\ \Delta\theta \end{pmatrix}, \quad A = \begin{pmatrix} p_{11} & p_{12} \\ p_{21} & 0 \end{pmatrix}, \quad b = \begin{pmatrix} 0 \\ 1 \end{pmatrix}, \quad j = 2$$

(viz., in (12.47), x_j represents $\Delta\theta$); $\tilde{\kappa}\pi(-\sigma, \sigma)$ is the left-hand side of the equation (12.63). If σ_1 and σ_2 are the roots with negative real parts of this equation, $\psi(\sigma)$ will be taken equal to $\gamma_0(\sigma - \sigma_1)(\sigma - \sigma_2)$, where γ_0, obtained

† Under the assumption of complete controllability, this latter being implied by the condition $\det(b, Ab, - A^{n-1}b) \neq 0$. This assumption is inherently satisfied if the system is equivalent to a higher-order equation.

from (1), is, as above, $-(\kappa/\tilde{\kappa})^{1/2}$. Further γ and $\tilde{\psi}$ result equal to $-\sqrt{\kappa}$, and $\sqrt{\kappa}[-(\sigma_1+\sigma_2)\sigma+\sigma_1\sigma_2]$, respectively.

The components of the vector w will be $w_1 = \sqrt{\nu}[\sigma_1\sigma_2 - p_{11}(\sigma_1+\sigma_2)]/p_{12}$ and $w_2 = -\sqrt{\nu}(\sigma_1+\sigma_2)$. In the problem considered, $p_{12} \neq 0$.

The optimal synthesis (12.62) follows readily on the basis of relation (4) above.

REFERENCES

1. A. Miele, "Some recent advances in the mechanics of terrestrial flight," *ARS Jet Prop.* **28**(1958), 581–587.
2. G. Leitman (Ed.), *Optimization Techniques with Applications to Aerospace Systems.* Academic Press, New York, 1962.
3. R. Bellman (Ed.), *Mathematical Optimization Techniques.* Univ. of California Press, Berkeley, 1963.
4. C. T. Leondes (Ed.), *Advances in Control Systems: Theory and Applications*, Vols. 1, 2, 3. Academic Press, New York. 1964–1966.
5. L. W. Neustadt, "An abstract variational theory with applications to a broad class of optimization problems, Part 2: Applications," *J. SIAM* (Ser. on Control) **5**(1967), 1.
6. L. S. Pontryagin, V. G. Boltyanskii, R. V. Gamkrelidze, and E. F. Mishchenko, *The Mathematical Theory of Optimal Processes.* Wiley, New York, 1962.
7. J. W. Burrows and R. W. Rishel, "The strange extremals of a normal acceleration control problem," *IEEE Trans. Automatic Control* **AC-10**(1965), 344–346.
8. T. Hacker, "Optimal control in climb of rotating-thrust VTOL aircrafts" (in Romanian), *Stud. Cercetări Mec. Apl. Acad. R. P. R.* **15**(1964), 433–440.
9. T. Hacker, "Problèmes de dynamique et de commande des avions VTOL; Guidage optimal et stabilité," *Proc. ICAS 4th Congr. (Paris), 1964* **1965**, 495–517.
10. K. Okamura, "Some mathematical theory of the penalty method for solving optimal control problems," *J. SIAM* (Ser. on Control) **2**(1965), 317–331.
11. D. L. Russel, "Penalty functions and bounded phase coordinate control," *J. SIAM* (Ser. on Control) **2**(1965), 409–422.
12. A. M. Letov, "Analytical design of regulators" (in Russian), *Avtomat. Telemekh.* **21**(1960).
13. R. E. Kalman, "Contribution to the theory of optimal control," *Bol. Soc. Mat. Mexicana* **1960**, 102–119.
14. R. E. Kalman, "When is a linear control system optimal?" *J. Basic Eng.* (*Trans. ASME*) **D86**(1964), 51–60.
15. V. M. Popov, *Hyperstability of Automatic Systems* (in Romanian). Editura Academiei R. S. Romania, 1966. English version in press, Springer.
16. R. Bellman, *Dynamic Programming.* Princeton Univ. Press, Princeton, New Jersey, 1957.
17. A. H. Jazwinski, "Optimal trajectories and linear control of nonlinear systems," *AIAA J.* **2**(1964), 1371–1379.

NOTATION

Chapter 1

t	time; the independent variable in equations of motion
x, y, z	state-variable vectors
u	control-variable, vector or scalar

Chapter 2

V	velocity of aircraft, e.g., in undisturbed flight relative to the atmosphere at rest or true airspeed
u, v, w	components of V with respect to the body axes of aircraft (see Appendix to Chapter 2)
a	local speed of sound in the atmosphere
$\mathbf{M} = V/a$	Mach number
α	angle of incidence of the zero lift line of the aircraft with undeflected elevator, in undisturbed flight or angle of attack
$\beta = \sin^{-1}(v/V)$	angle of sideslip (see Figure 7.1)
γ	upward inclination of flight path to the horizontal in undisturbed flight (angle of the center of gravity velocity vector V with the horizontal)
θ	angle of pitch (longitudinal attitude; see Figure 2.1)
ϕ	angle of bank (lateral attitude; see Figure 2.1)
ψ	angle of yaw (azimuth of longitudinal axis of aircraft; see Figure 2.1)
$\boldsymbol{\omega}$	vector of aircraft angular velocity about the center of gravity
p, q, r	scalar components about the aircraft body axes of $\boldsymbol{\omega}$ (rates of roll, pitch, and yaw, respectively; see Appendix to Chapter 2)
m	total mass of aircraft
A, B, C	moments of inertia about body axes
D, E, F	products of inertia about body axes
k_x, k_y, k_z	radii of gyration about body axes
S	total lifting surface area
b	length of wingspan (double of reference length for lateral equations)
$\bar{c}$	length of mean wing chord (double of reference length for longitudinal equations)
z_T	thrust vector arm (distance between center of gravity and thrust vector situated above the center of gravity)

φ — thrust line setting angle (angle between thrust vector line and the zero-lift line of the aircraft; $\varphi > 0$ corresponds to the component of thrust vector with respect to Oz body axis having the same sense as the respective component of positive lift)

$\mathbf{h}$ — total angular momentum vector of the aircraft

$\mathbf{h}_r$ — angular momentum vector of spinning rotors

$\mathbf{h}_a = \mathbf{h} - \mathbf{h}_r$ — angular momentum vector of the aircraft supposed rigid body, without its spinning rotors

$h_{r_x}, h_{r_y}, h_{r_z}$ — scalar components of $\mathbf{h}_r$ referred to body axes

δ_Y — control parameter for longitudinal moment (e.g., the elevator setting δ_e)

δ_T — control parameter for thrust (e.g., the position of the throttle valve lever)

τ — time lag in the control system

$\mathbf{F}$ — resultant external force vector acting on the aircraft (see Appendix to Chapter 2)

F_x, F_y, F_z — scalar components of $\mathbf{F}$ about the body axes

$-X, Y, Z$ — components with respect to body axes of resultant external aerodynamic force

D — aerodynamic drag

L — aerodynamic lift

$\mathcal{M}$ — resultant external moment vector about the center of mass (see Appendix to Chapter 2)

M_x, M_y, M_z — scalar components of $\mathcal{M}$ about body axes

L, M, N — external aerodynamic rolling, pitching, and yawing moments

T — thrust

M_c — longitudinal control moment

g — gravity acceleration

ρ — local air density

NF — Froude number

μ — relative density of aircraft

i_A, i_B, i_C — nondimensional moments of inertia about body axes

C_L — lift coefficient

C_D — drag coefficient

C_y — lateral force coefficient

C_l, C_m, C_n — nondimensional coefficients of rolling, pitching, and yawing moments

C_T — thrust coefficient

$C_{T_V} = (\partial T/\partial V)(\rho S V_0)^{-1}$ — nondimensional form of derivative $\partial T/\partial V$

subsubscripts **M**, α, β, $\dot{\alpha}$, p, q, r, etc. denote partial derivatives with respect to the respective nondimensional quantity (e.g., $C_{D_{\mathbf{M}}} = \partial C_D/\partial \mathbf{M}$, $C_{m_q} = (\partial C_m/\partial q)(\bar{c}/2V_0)$)

C_{m_α} longitudinal static stability coefficient
C_{l_β} lateral static stability coefficient or dihedral effect
C_{n_β} directional or weathercock static stability coefficient
C_{l_p}, C_{m_q}, C_{n_r} coefficients of damping in roll, pitch, and yaw, respectively

Capital Greek delta Δ denotes disturbance-induced deviation; square brackets [ ], dimensional unit; French circumflex ^, nondimensional quantity; a dot over a symbol stands for the derivative with respect to the independent variable (t or $\hat{t}$).

Chapter 3 (Section 3.6)

$\mathbf{V}_g$ local gust velocity vector
V_{g_x}, V_{g_y}, V_{g_z} its scalar components referred to wind axes (tangent, normal, and binormal to the path at the aircraft center of gravity)

Subscript g stands for gust, ad for admissible, ef for effective.

Chapter 6 (Section 6.4)

h flight altitude
$\zeta = \Delta h/(\bar{c}/2)$ nondimensional height disturbance
$\sigma = \rho/\rho_0$ nondimensional air density, where ρ_0 stands for a reference magnitude (e.g., air density at sea level or at flying altitude)

$\sigma_\zeta = \dfrac{\bar{c}}{2\rho_0}\dfrac{d\rho}{dh}$, $a_\zeta = \dfrac{\bar{c}}{2V_0}\dfrac{da}{dh}$, $C_{T_\sigma} = \dfrac{1}{V_0^2 S}\dfrac{\partial T}{\partial \rho}$ dimensionless parameters of system (6.21)

p air pressure
T absolute temperature
κ adiabatic constant (= 1.405)

Subscript T means absolute temperature; str, stratosphere.

Chapter 7

S_F rudder-fin area
$\hat{S}_F = S_F/S$ its nondimensional form
l_F arm of vector of aerodynamic force acting on vertical tail, referred to the aircraft center of gravity

γ wing dihedral
σ setting angle of axes of the mean spinning rotors
$c_p, c_q, c_r, c_\phi, c_\theta, c_\psi$ coefficients of linear correction terms (gearing ratios for proportional control)

Chapter **11**

τ time lag (response time of the control system)
$\kappa = c_\phi / c_p, \ \hat{\kappa} = (k_x g)^{1/2} \kappa$

Chapter **12**

δ_e elevator setting
φ setting angle of thrust vector line
$\tau = T/mg$ thrust-to-weight ratio
$\kappa = M_c/B$ longitudinal moment parameter (B = const.)
x horizontal space movement of the aircraft center of gravity referred to its position at initial instant

Subscript crit means critical.

AUTHOR INDEX

Numbers in parentheses indicate the numbers of the references when these are cited in the text without the names of the authors.

Numbers set in *italics* designate the page numbers on which the complete literature citation is given.

Babister, A. W., 88(8), 89(8), 102(8), *122*, 141(2), *154*, 173(5), *182*
Barbălat, I., 201(4), *218*
Bautin, N. N., 98, *123*
Bellman, R., xxi, xxii, 46, 47(10), *67*, 92(14), *122*, 232(5), 247, *253*, *254*, 300 (16), 304(16), *312*
Birkhoff, G., 20(5), *30*
Boltyanskiy, V. G., 263(6), 266(6), 291(6), *312*
Bourbaki, N., 46(12), *67*
Brown, B. P., 69(4), *86*
Bryan, G. H., xvii, xviii, 87, 92, *122*
Buckingham, E., 20, *30*
Burrows, J. W., 265(7), *312*

Chebotarev, N. G., 247(12), *254*
Chernobrovkin, L. S., 90(11), 102(11), *122*, 184(1), *191*
Collar, A. R., 209(6), *218*
Cooke, K. L., 232(5), *253*
Cremer, L., 95, *123*

Draper, C. S., xix
Duncan, W. J., 65(25), *67*, 88(6), 89(6), 102(6), 107(6), *122*, 129(8), *154*, 162(3), *182*

El'sgol'ts, L. E., 232(1), 235(1), 247(1), *253*
Etkin, B., xix, 10, 13(1), *30*, 33(6), 37, 51, *67*, 88(7), 89(7), 95(7), 112(7), *122*, 125(1), *154*

Flatto, L., 187(2), *191*
Fourier, J., 20(3), *30*

Gamkrelidze, R. V., 263(6), 266(6), 291 (6), *312*
Gates, S. B., 95, *122*
Glauert, H., xvii, xviii, 19, *30*, 253
Gronwall, T. H., 46, *67*, 196, 197

Hacker, T., 44(9), 51(13, 14), 59(14), 65(22), *67*, 68(1), *86*, 130(9), *154*, 200(1, 2), 201(3), 213(2), *218*, 237(8), 249(8), 251(13), *254*, 272(8), 280(9), *312*
Hage, R. E., 89(10), *122*, 127(5), 151(5), *154*
Halanay, A., 7(2), *9*, 31(5), 39(5), 41(5), 43(5), *67*, 68(1), *86*, 232(6), 235, *253*
Hale, J. K., 187(3, 4), 188, *191*, 200
Hurwitz, A., 118, 119, *123*

Jazwinski, A. H., 306, *312*
Jones, B. M., 63, 65, *67*, 126, 132, *154*

Kalatchev, G. S., xx, 127, *154*, 205(5), *218*
Kalman, R. E., 294, 295, *312*
Kamenkov, G. V., 65(19), *67*, 102(23, 24), *123*
Korenev, G. V., 4(1), *9*
Krasovskiy, N. N., 232(4), 247, 250(4), *253*

Lanchester, F. W., xvii, 65, *67*, 126, *154*
La Salle, J., 31(4), 39(4), *67*
Lebedev, A. A., 65(20, 21), *67*, 90(11), 102(11), *122*, 184(1), *191*
Lefschetz, S., 31(4), 39(4), *67*
Leonhard, A., 95, *123*
Letov, A. M., 294, 295, *312*
Levinson, N., 187(2), *191*
Liapunov, A. M., xx, xxi, 31, 37, 39, 58, 60, *66*, 87, 92, 99, 102(13), *122*, 134, *154*, 203, 243, *254*

Li Yueh-sheng, 187(5), 189(5), *191*

Malkin, I. G., 31(2), 39(2), 41, 43(2), 49, *67*, 100, 101
Markus, L., 189(5), *191*
Massera, J. L., 31(3), *67*
Matthews, J. T., 69(4), *86*
Meyman, N. N., 247(12), *254*
Miele, A., 255(1), *312*
Mikhaylov, A. V., 95, *123*
Mises, R. von, 65(24), *67*, 151(14), *154*
Mishchenko, E. F., 263(6), 266(6), 291(6), *312*
Myshkis, A. D., 232(2), *253*

Neumark, S., 68, 79, 83(2), *86*, 92, 141, 146(11), 147(11), 148(11), *154*
Neustadt, L. W., 260(5), 291(5), *312*
Nyquist, H., 88, *123*

Okamura, K., 291(10), *312*
Ostoslavskiy, I. V., xx, 89(9), *122*, 127(7), *154*, 205(5), *218*

Perkins, C. D., 69(3), *86*, 89(10), *122*, 127(5), 151(5), *218*
Phillips, W. H., 69(4), *86*
Pinney, E., 232(3), *253*
Pinus, N. Z., 150(13), *154*
Pontryagin, L. S., xxii, 263, 266(6), 291(6), *312*
Popov, V. M., 43(8), *67*, 294, 298, 303, 304, 310(15), *312*
Pyshnov, V. S., xx

Renaudie, J., 91(12), *122*
Repin, Yu. M., 235(7), *254*
Reshetov, V. D., 150(13), *154*
Rishel, R. W., 265(7), *312*
Routh, E. J., xvii, xx, 87, 90, 118, 119, *122*
Rumyantsev, V. V., 61(17), *67*
Russel, D. L., 291(11), *312*

Scheubel, F. N., 141, *154*
Sedov, L. I., 20(4), *30*
Shmeter, S. M., 150(13), *154*
Stokes, A. P., 187(3), *191*

Vedrov, V. S., xviii, xx, 88(5), 89(5), 90(5), 102(5), *122*
Vetter, H. C., 173(6), *182*

Westerwick, R., 173(4), *182*
White, R. J., 161(2), *181*
Williams, W. E., xviii, 87, 92, *122*
Wright, E. M., 247, *254*

Yamabe, H., 189(6), *191*

Zimmerman, C. H., 95, *122*, *123*

SUBJECT INDEX

Admissible control, 260
Aerodone, Lanchester's, 126
Aerodynamic damping, 57, 128, 242
Aerodynamic derivatives, 98
Aerodynamic forces, 10, 21
Aerodynamic moments, 10
  pitching (longitudinal), 12, 21
Aerodynamic time, unit of, 24, 221
Aircraft dynamic stability, conventional theory, 91
Aircraft response characteristics, 69
Airscrew, angular momentum, 180
Altitude variation
  effect for standard troposphere and stratosphere, 153
  influence on phugoid motion, 226–229
Angle of
  attack (incidence), 11
  bank, pitch and azimuth (yaw), 27–28
  bank constraint, 159
  elevator, 259
  pitch as a control variable, 260
  sideslip, 26, 166
Angular momentum
  of spinning elements (rotors, airscrews), 14, 180
  total, 14
Angular position of aircraft, 27–28
Artificial stability of VTOL aircraft rotation, 176
Artificial stabilization of aircraft rotation, 294
Asymmetrical variables, 17
Asymptotic behavior, 90
Asymptotic stability, 32, 42, 89, 92
  definition, 32
  global (in the large), 301
  and instability, boundary between, 92
  in systems with time lag, 246–247
  theorem, 40
Asymptotic ε_0-stability, 53
Atmosphere, standard, 150–152
Atmospheric turbulence
  intensity, 52
  as persistent disturbance, 51
Attitude of aircraft, 27–28
Automatically controlled aircraft, 68
Automatic pilot characteristics, 68
Automatic stabilization, 302
Automatic stabilizer characteristics, 68
Autonomous equations systems, 183
Auxiliary system, 6, 71–73, 81
Azimuth angle, *see* Yaw angle

Bang-bang type variation, 277
Bank angle, 27
Basic controls, 3
  program, 34
Basic motion, 1, 8
Body axes, 13, 27
Boundary of the asymptotic stability region, 127
Boundary value problem, two-point, 280, 305
Buckingham's pi theorem, 20

Center of mass (gravity) location, 97, 183
Characteristic equations, transcendental, of systems with time lag, 246
Characteristic roots, 93
Closed-loop system, 5, 7, 230
Constant attitude VTOL aircraft, 272
Constrained stability, Neumark's theory, 79
Constrained variables, 70, 74
  according to Neumark, 80
Constraint of
  angular attitude and angular velocity, 110
  flight speed, 133–134
  pitch angle and angular velocity, 131–132
Constraints, 260–261
  on control variables, 260
  external, 257
  on state variables, 290–293
  system of, 257
Continuity of solution with respect to

initial conditions, 59, 103
Control, 3, 270
admissible, 260
coordinate, function, or variable, 4, 5
efficiency, 77, 106
force, 10
moment, 10, 28
moment, longitudinal, 12
moment, longitudinal, optimal, 282 (Table 12.4)
optimal, 263
policy, 257
policy, optimal, 267
strength, 115
system, 4, 68, 262
variable, 4, 5, 257, 258–260
variable, direct and indirect, 259
variable constraint, 260
Controlled motion, 2
Controlled (constrained) variable, 70, 74, 80
Controls
effect of, 68, 81
secondary effect of, 115–118
Conventional theory of aircraft dynamic stability, 91
Correction controls, 3
Cost of controls, 294, 296
Coupled motions, 172–173
Coupling terms, 17
Criteria of similitude, 21, 24
Criteria of stability for linear systems with constant coefficients
Nyquist type, 88, 247
Routh-Hurwitz type, 88
Criterion of Hurwitz, 119
Criterion of optimality, 256, 257, 263
Criterion of Routh, 118
Critical test functions, 95

Damping and frequency, suitable relation between, 112–113
Damping velocity (decrement) of induced deviations, 44, 78
Dangerous and safe stability boundary, 100
Danger point of the stability boundary, 100
Differential inequalities with time lag, Halanay's lemma concerning, 235
Dihedral
coefficient, 97
convention of signs, 167
effect, 97, 161, 167–169, 178
of wing, 97, 166–168
Dimensional independence, 20
Dimensional unit, 20, 221
Dimension of a physical quantity, 20
Direct control variable, 259
Directional static stability, 178
Disturbance, 2, 33
as a command signal, 77
persistent, 2
short duration, 2, 33
Disturbed motion, 2, 8
equations, 8, 12
with time lag, 246
Dutch roll, 161
Dynamic programming, 294, 300
Dynamic stability margin, 43, 56, 76, 195, 242, 243
estimate, 57
a measure of, 44, 195

Effect of controls, 68, 81
Efficiency of controls, 77, 106
Elevator
angle, 116, 259
deflection, 116
setting, convention of signs, 116
Equations
of asymmetrically disturbed motion, 17
of the disturbed motion, 8, 12
in variations, 87
Estimates
of additional terms, 45–51
of damping velocity, 44
numerical, 41
of varying speed effect, 210–218
Euler's equations, 29

Feedback system, 5, 230
Fixed thrust axis VTOL aircraft, 269, 283
Flight altitude, *see* Height
Flight mission, purpose of, 261
Flight path inclination, 11
optimal control of, 288–290
Flight program, 1, 2
Flight régime, 34
Flight speed constraint
direct under throttle control, 130
partial control through, 133–134
Free flight, 3, 7
with locked controls, 7
Free variables, 70, 74, 80
"Freezing of coefficients" process, 187–189, 198

Frequency of oscillations of induced deviations, 78
Froude's number, 22
Fundamental (dimensional) units, 20, 22

Glauert's transformation (dimensionless variables and system parameters), 19, 253
Global stability, 37, 301
Gravity center location, 97, 183
Gronwall's lemma (inequality), 46, 196
Gust intensity, allowable, 51–59
Gyroscopic
  action (effect), 10
  coupling terms, 173
  moments, 181
  terms, 176

Halanay's lemma, 235
Hale's lemma, 188, 200
Handling characteristics (qualities), 35, 127
Heading
  control, 157
  in free flight, 157
Height (flight altitude)
  disturbance in horizontal basic motion, 141–154
  as a system parameter, 219
  see also Altitude variation
Hovering flight of VTOL aircraft, 173–181, 295
  controlled with time lag, 251–253
Human pilot, 75
  intervention efficiency and response characteristics, 69
Hurwitz's
  criterion, 119
  theorem, 120

Ideal control, 231
Imaginary characteristic roots, 90
Incidence, 11
  constraint, direct, 129
  as a control variable, 260, 285
Inclination of the flight path, 11
Incomplete stability, 60–66, 73–74
Independent dimensional units, system of, 20
Indirect control variable, 259
Induced motion, 2, 31
Inertia moment, 13, 183
Inherent stability, 4, 5, 60, 68
Initial conditions, 232, 259, 261, 264
Initial set and initial function, 232
Instability, 40, 90, 93
  by the first approximation, theorem, 93
  theorem, 40
Instantaneous disturbance, 2, 31
International Standard Atmosphere, 150–152
Intrinsic reference system, 11

Jones, B. Melvill, simplifying assumptions for pure short-period oscillations, 63, 126

Kalman-Letov problem, 295

Lanchester's
  phugoid, 126
  simplified theory, 65
Large and small characteristic roots of longitudinal disturbed motion system, 126
Lateral attitude, 27
  constraint, 113, 115
  and rolling velocity constraint, partial control through, 170–172
Lateral damping coefficient, 169
Lateral disturbed motion equations, 13
Lateral oscillation, 122, 161
Lateral stability in partial control conditions, 158–172
Lateral static stability coefficient, 160, 167
Lateral static stabilizer, 251
Letov's problem, 295
Liapunov
  asymptotic stability, 89
  function, 40, 44, 49, 190
  functional, 250
  function method, 38–40, 49, 190, 200–205, 243
  stability, 31–32, 37, 59–60, 99, 101
Linear approximation, 92
Linear differential equations system with constant coefficients, 37, 87, 91
Linear first approximation system, 38
Linear model, 87, 91
Linear theory of control systems, 88
Local atmospheric velocity, 52
Local stability, 38
Locked controls
  free flight with, 7
  lateral stability, 155–157
  longitudinal stability, 124–130
Longitudinal aerodynamic moment, 21

Longitudinal attitude (pitch angle), 11, 27, 124
constraint, partial control through, 130–131
and flight speed constraint, partial control through, 132–133
Longitudinal control moment, 12, 116
Longitudinal damping coefficient, 97
Longitudinal and lateral stability equations, their separation into two independent sets, 17
under persistent disturbances, 51
for unsteady basic motions, 192
Longitudinal motion, 10
Longitudinal response to elevator control, 35
Longitudinal stability
diagram, 128
with locked control, 124–130
in partial control conditions, 130–134
Longitudinal static stability, 35, 128–129, 179

Main effect of controls, 81, 83
Margin of dynamic stability, *see* Dynamic stability margin
Mass center location, 97, 183
Maximum flight speed, 206
Maximum principle, 263–264
Maximum relative variation of the system parameters, 183
Mean aerodynamic chord, 21
Mechanical behaviour of the airframe, 69
Mechanical motion of a rigid body, 10
Method of the Liapunov function, *see* Liapunov function method
Method of variation of the parameters, 46, 238
Mission (of flight), 256
Model of partially controlled motion, validity of, 105
Moment of inertia, 13, 183

Negative definite and semidefinite functions, 39
Neumark's theory of constrained stability, 79
Neutral behavior, 91, 93
Neutral dynamic stability, 101–105
Neutral phugoid stability, 140–141
Neutral stability, 90, 94, 101–105
theoretical, 135
Nonautonomous systems, 38
with time lag, 244

Nondimensional aerodynamic rotary derivatives, 22
Nondimensional equations, 19
Nondimensional form, 19–21, 221, 253
Nondimensional parameter, 21
Nondimensional system, 19, 221
for hovering flight, 253
Normal-attitude VTOL aircraft, 179
Numerical estimates, 41
Nyquist criteria, 88, 247

Open-loop system, 6
Optimal control, 263
Optimal programming, 256
of the stabilizer, 302
Optimal solution, 265–266
Optimal synthezis of correction controls, 293
Optimal trajectory, 267
Optimum
damping, 299
longitudinal control moment, 282 (Table 12.4)
orientation of VTOL aircraft thrust, 277 (Table 12.2)
problem, 267
thrust value for rotating thrust VTOL aircraft, 279 (Table 12.3)
Ordinary differential equations with time lag (retarded argument), 232
Orientation of aircraft, 27–28
Oscillatory instability
lateral, 161
longitudinal, 60
Oscillatory mode (of lateral disturbed motion), 163–164
Oscillatory stability (lateral), 162, 168

Parasite effects of controls, 115
Partial control, 74–75
efficiency, 74, 79, 81
Partially controlled motion, 6, 69
equations system of, 70
Partially controlled systems with time lag, 233
Partial stability of flight, 66
Penalty function, 292
method, 291–293
Performance index (quality), 256, 262, 294
Period of oscillation of induced deviations, 108
Persistent disturbance, 2, 33, 69
Perturbation, *see* Disturbance
Perturbation forces, 33

Perturbed system, 33
Phugoid
divergence, 129, 236–237
instability, 127, 129
Lanchester's, 126
motion, 215, 294, 302
motion, influence of height variation on, 226–229
oscillation, 154
oscillation frequency, influence of speed variation on, 214–218
paths, 65
roots, 126–127
stability, neutral, 140–141
Piecewise continuous time function, 258
Pilot-aircraft system, 69
Pilot, human, *see* Human pilot
Pitch angle
and angular velocity constraint, partial control through, 131–132
as a control variable, 260
see also Longitudinal attitude
Pitch damper, 176–177
Pi theorem, 20, 23
Planform of the aircraft, 162
Pontryagin's theorem, 264
Positive definite function, 39, 61
Positive semidefinite function, 39
Partical damping (ε-damping, κ-damping), 110, 113
Practical instability, 63, 103
Practical optimum, 267
Practical stability, 43, 99, 103
Product of inertia, 13
Program of
basic controls, 3
flight, 1–3
Programmed motion (flight), 2–3
Programming, optimal, 256
Purpose of the flight mission, 261

Quality of performance, 256
Quantitative characteristics of free disturbed motion, 78
Quasi-autonomous systems, 100
Quasi-linear systems, 92
Quasi-neutral behavior, 302
Quasi-neutral stability, 294
Quasi-zero speed, 177, 179

Rapid aperiodic mode (of lateral disturbed motion), 163
Rapid convergence in roll, 158, 165, 169–170
Rapid incidence adjustment, 35, 63, 66, 126, 147
in unsteady flight, 192 and following, 223, 225
Rapid mode of the longitudinal disturbed motion, 126
Reaction time, 107, 109
Reference speed, 25
Régime of flight, 34
Region of uniform asymptotic stability in systems with time lag, 247–248
Relative density of aircraft, 24, 162, 221
Repeated vertical gusts, flight under, 52
Response
calculation, 104
characteristics of aircraft, and pilot, 69
to the controls, transitory period of the, 82
longitudinal, 35
of the pilot (human or automatic), 77
time, 107
Retarded argument, ordinary differential equations with, 232
Rigid-body-with-fixed-point model, 299
Roll damper, 176, 177, 251
Rolling
mode, 158
motion, 163
pure, 158, 169–170
Rotary derivatives, 22, 98
Rotating thrust VTOL aircraft, 269, 270
Rotation dampers, 299, 302
their effect on aircraft stability, 175–181
Routh-Hurwitz criteria, 88, 95, 190
Routh's
critérion, 118
theorem, 119
Rudder-fin area, 160

Safe and dangereous stability boundary, 98–101, 104, 120–122
Safety limit of vertical gust velocity, 57
Safe point of the stability boundary, 100
Secondary effects of controls, 76, 79, 105, 115–118
Secondary variation of lift and drag induced by elevator deflection, 116–117
Short-duration disturbance, 2, 33
Short-period oscillation
mode (of incidence and rate of pitch), 126
pure, 132
Sideslip angle, 26, 166
convention of signs, 166–167

Significant modes of induced deviations, 109
Similitude criteria, 21, 24
Slow aperiodic mode (of lateral disturbed motion), 163
Slow variation of the system parameters, 184, 187–189
Small disturbance method, 87
Small variation of the system parameters, 186
Sound speed, 152
Spinning rotors
  angular momentum, 180
  gyroscopic moments induced by, 173
Spiral divergence, 113, 159, 163
Spiral instability, 161
Spiral mode, 113, 159
Spiral stability, 162, 168
  boundary, 160
Stability, 3, 4, 32, 42, 90
  and asymptotic behavior, 90
  axes, 11, 192
  boundary in the parameter space, 96, 98
  and control, relation between, 34
  of the control system, 4
  in conventional theory of flying qualities, 90
  criteria, 88
  definition, 32
  diagram, 97, 159
  within a finite time interval, 65
  global (in the large), 37
  of hovering flight controlled with time lag, 251–253
  incomplete, 60–66
  inherent, 4, 5, 60
  model, 7
  neutral, 90, 94, 101–105, 135
  of nonautonomous systems with time lag, 250
  with respect to a part of the variables, 61–62
  under persistent disturbances, 5, 33, 41
  region in the parameter space, 93–98
  of a state of motion, 37
  in systems with time lag, 233–234
  theorem of Liapunov, 40
  of zero solution of systems with time lag, 233
ε_0-Stability, 43
Stability by the first approximation, 8, 38, 93
  in systems with time lag, 247
  theorem, 38, 93
  theory, 95
Stability of partially controlled motions, 6, 66, 91, 105
  lateral, 158–172
  longitudinal, 130–134
  mathematical formulation, 69–73
Stability of partially controlled systems
  investigation of, 106
  with time lag, 233–234
Stabilization through partial control, 78
  its efficiency, 76
Stabilizing controls, 3
Stabilizing intervention of controls, 66
Standard aircraft, 21
Standard atmosphere, 150–152
Standard basic motion, 21
State variables (coordinates), 4, 5, 257–259
  constraint, 290–293
Static and damping coefficients, 180–181
Static stability
  coefficient, 51, 57, 97
  directional (weathercock), 160
  lateral, 160, 161, 167
  longitudinal, 35, 128–129, 178
Static stabilizers, 299
  their effect on VTOL aircraft stability, 177–181
Stationary basic motion, 91
Steady basic motion, 185
Strength of control, 115
Suboptimal policy, 267
Switching conditions, 266–267
  for optimum longitudinal control moment, 282 (Table 12.4)
  for optimum orientation of thrust, 277 (Table 12.2)
  for optimum thrust value, 279 (Table 12.3)
Switching function and switching point, 266–267
Synthezis of optimal control, 6–7, 267–268, 293
System of constraints, 257
System of control, 262
System parameters, 94
System with time lag, 232

Tangential velocity of the mass center as a state variable, 11
Terminal conditions, 261, 264
Test functions, 95
Throttle control, flight speed constraint by, 130
Thrust control, its main and secondary effect, 117

Thrust as a direct control variable, 259
Time lag, 77, 234
  allowable, 108
  of control system, 11, 71, 79, 81–82
  ordinary differential equations with, 232
  in throttle control, 133
Time of reaction, 107, 109
Time of response, 105, 107
Total time derivative of a function along the integral curve, 39
Trajectory, optimal, 267
Transcendental characteristic equations of systems with time lag, 246
Transmission of the control, 78
Transversality conditions, 264
Trial-and-error technique, 280, 305

Undisturbed (basic) motion, 1, 8
Uniform accelerated motion model, 207
Uniform asymptotic stability, 32, 44
  definition, 32
  of a partially controlled motion, definition, 71
  of a partially controlled motion, proposition, 72
  in partially controlled systems with time lag, 234
  under persistent disturbances, 41
  region, 247–248
  in systems with time lag, 233
  theorem, 40
Uniform decelerated motion model, 207
Uniform stability
  definition, 32
  of a partially controlled motion, definition, 71
  of a partially controlled motion, proposition, 72
  under persistent disturbances, 34
  in systems with time lag, 233–234
  theorem, 40
Unit of aerodynamic time, 24, 221
Units, fundamental (dimensional), 20
Unsteady basic motion, 185
  (nonautonomous system) with time lag, 244
Variable-attitude VTOL aircraft, 280
Variable dimensional unit, 221
Variables, 257
  of control, 4, 5, 257, 258–260
  controlled (constrained), 70
  independent, 257
  of state, 4, 5, 257–259
  uncontrolled (free), 70
Variation (increase or damping) speed of induced deviations, 78
Variation of the parameters method, 46
Vertical gust velocity, safety limit of, 58
Vertical take-off and landing (VTOL) aircraft
  constant-attitude, 272
  fixed thrust-axis, 269, 283
  hovering, 173–181, 295
  rotating thrust, 269, 270
  variable-attitude, 280

Wagner effect, 78
Weathercock (directional) stability, 97, 160, 161, 178
Weight ratio of modes, 110
Wind axes, 11
Wing dihedral, 160–162
  convention of signs, 167
Wingsweep angle, 162, 166–168

Yaw (azimuth) angle, 27–28
  and bank angle constraint, partial control through, 162–164
  constraint, partial control through, 158–162
  and rate of yaw constraint, partial control through, 166–169
  and sideslip angle constraint, partial control through, 165–166
Yaw damper, 113, 161, 169, 171, 176, 177

Zero and imaginary characteristic roots, 90